INTERNATIONAL CENTRE FOR MECHANICAL SCIENCES

COURSES AND LECTURES - No. 423

CRASHWORTHINESS

ENERGY MANAGEMENT
AND OCCUPANT PROTECTION

EDITED BY

JORGE A.C. AMBROSIO
INSTITUTO SUPERIOR TÉCNICO

 SpringerWienNewYork

This volume contains 207 illustrations

In order to make this volume available as economically and as
rapidly as possible the authors' typescripts have been
reproduced in their original forms. This method unfortunately
has its typographical limitations but it is hoped that they in no
way distract the reader.

ISBN 3-211-83334-X Springer-Verlag Wien New York

PREFACE

Crashworthiness ensures the vehicle structural integrity and its ability to absorb crash energy with minimal diminution of survivable space. Restraint systems limit occupant motion mitigating injuries that may result from contact with the vehicle interior during sudden acceleration conditions. Both structural crashworthiness and occupant protection technologies are multi-disciplinary and highly specialized, including complex technical fields spanning from the areas of mechanics to biological sciences. This book presents numerical procedures relevant to the structural and biomechanical topics pertinent to the occupant protection.

The book covers the fundamentals of the impact mechanics and biomechanics while providing an overview of modern analysis and design techniques used in the areas of impact energy management and occupant protection. The nonlinear structural response is thoroughly presented with emphasis on the use of nonlinear finite elements for crash analysis, conceptual modeling techniques, nonlinear multibody procedures and recent trends in crash simulation. The importance of the topics covered can be appraised in the framework of the occupant protection. Therefore, the issues related with impact biomechanics are presented in detail, from the fundamentals to the state-of-art, including occupant kinematics, injury mechanisms, occupant mathematical modeling using finite elements and multibody dynamics methods, human surrogates and their biofidelity. The sequence of calculations required in the prototyping of energy absorbing structures are emphasized enabling the reader to appraise the industrial applications and trends for the future in crashworthiness.

The first part of the book, authored by Prof. Norman Jones, addresses methodologies ranging from static methods in simple structures, for which a rigid perfectly plastic material behavior is assumed, to fully dynamic approaches. Special emphasis is given to the quasi-static methods, which allow for the description of many impact situations. The material strain rate sensitivity is presented in the framework of the structural impact. The use of special structural components, particularly tubes subjected to axial crushing loads, is discussed and demonstrated in general crashworthiness applications. The interested reader seeking a more complete treatment of the topics of structural impact is refered to Prof. Jones' book Structural Impact (Cambridge University Press, 1989), on which this part is largely based.

The second part of this book, authored by Dr. Wlodek Abramowicz, presents a detailed description of the macro element method for the design of crashworthiness structures. The characteristic feature that distinguishes this approach from all other classical formulations in nonlinear mechanics is that trial deformation functions are postulated on the basis of experimental observations. It is shown how experimentally determined shape functions are incorporated into a consistent, mathematically tractable engineering method suitable to the design of complex crashworthiness structures. This methodology provides a very powerfull design tool that can be used alternatively or in conjunction with the finite element method.

The nonlinear explicit finite element methods are introduced in the third part of this book by Prof. Ted Belytschko. Here the reader will find a treatment of the governing equations leading to finite element discretizations, which are solved, in the context of structural impact, using explicit integration methods. The analysis of structural impact involves the contact with the structural components, for which appropriate contact models are proposed and discussed. Readers interested in more detailed and complete treatment of nonlinear finite elements in crashworthiness analysis are referred to Prof. Belytschko's book Nonlinear Finite Elements for Continua and Structures (Wiley, Chichester, U.K., 2000).

The numerical tools for crashworthiness based on multibody methodologies are efficient in the design and analysis of both structural and biomechanical models. The fourth part of this book presents the fundamentals of rigid and flexible multibody dynamics for crash applications. In a first approach, the structural deformations are described by using the concept of plastic hinges. More complex models are obtained by means of a methodology that couples the finite element method and the multibody dynamics in the framework of flexible multibody systems. This is described and exemplified making use of relevant applications to automotive and railway vehicle impact. In the process, several contact models are presented and discussed. Multibody dynamics provides an unified methodology for the integrated simulation of biomechanical and structural systems. This topic is discussed and exemplified in this part through the use of a complete vehicle with occupants inside, which is simulated in complex crash environments, including vehicle rollover. This part deals also with the application of advanced methodologies based on optimization to the design of crashworthiness environments.

Prof. Jac Wismans, main developer of the computer code MADYMO, contributes to this book with its fifth part where the crashworthiness biomechanics is discussed. Starting with an introduction to injury biomechanics, where injury models, injury scaling and risk and the whole body tolerance to impact are discussed, this part progresses with a detailed description of the mathematical tools for human body and dummy modelling. Here, special emphasis is given to the topics of the biofidelity of the biomechanical models. Finally, the approaches proposed are used in the development of models for the human neck and applied to whiplash situations.

The final part of this book, authored by Dr. David Viano, presents the fundamentals of injury biomechanics, occupant restraints and crash compatibility. The different chapters deal with the specific injury mechanisms of the human body, starting by the head and going through the spine, with emphasis on the cervical spine, chest and abdomen. The biomechanical implications of the occupant restraints are discussed taking into account their performance and effectiveness. The last chapter of this part presents a comprehensive discussion on the topics of crash compatibility, mathematical description of vehicle crashes and real world accident data.

This book is a result of the Advanced School on Crashworthiness: Energy Management and Occupant Protection, which took place at the Centre International des Sciences Méchaniques (CISM), Udine, Italy, during September 11-15, 2000. The course, lectured by the authors of the different parts of this book, brought together a large number of participants ranging from doctoral and postgraduate students to researchers, developers and young faculty, concerned with advanced theoretical and design issues in crashworthiness of transportation systems. The lecture notes used to support the course were thoroughly revised by the authors, taking into account the discussions generated and the interests of the target readers.

I am indebted to the lecturers of the Advanced School not only for putting together excellent presentations that greatly motivated the active participation of those that attended the course but also for their contribution to the lecture notes and to this book. I am grateful to all participants in the Advanced School for their excellent contributions to the discussions that took place during and after the course. A word of acknowledgement is also due to the CISM scientific council for supporting the Advanced School and recognizing the importance of topics related with crashworthiness in the framework of the Mechanical Sciences. Finally, special thanks are due to Prof. Arantes e Oliveira not only for the initial discussions that led to the proposal and organization of the Advanced School in Crashworthiness but also for his continued support.

Jorge A.C. Ambrósio

CONTENTS

Part I

STRUCTURAL IMPACT

N. Jones
University of Liverpool, Liverpool, UK

1. RIGID, PERFECTLY PLASTIC METHODS

1.1   Introduction

In many energy absorbing systems and structural crashworthiness problems, the external impact energy is absorbed through the inelastic behaviour of a ductile material.   Moreover, in these extreme conditions of interest, the inelastic strains are large and dominate the elastic behaviour. Thus, the idealisation of a rigid, perfectly plastic material may be used for theoretical and numerical methods of analysis with little sacrifice in accuracy.

The rigid, perfectly plastic method of analysis is introduced in this chapter to study the dynamic plastic response of beams. However, the general procedure is similar to that used for the analysis of more complex structural members such as frames, plates, shells and other components of energy absorbing systems subjected to dynamic loads.

Many books have been written on the theoretical behaviour of structural members which are modelled using a rigid, perfectly plastic material and subjected to static loads [1.1-1.4]. In particular, the limit theorems of plasticity have been used to obtain bounds on the exact static collapse load.   This procedure is illustrated in the next section for the static behaviour of a simply supported beam. It is shown in the subsequent sections how these ideas can be extended to examine the dynamic plastic response of beams. Further information on the accuracy and limitations of this method of analysis can be found in Reference [1.5].

1.2   Static plastic collapse of a simply supported beam

The limit theorems of plasticity are now used to obtain the limit load of the simply supported beam in Figure 1.1(a) which is made from a rigid, perfectly plastic material.

If a plastic hinge forms at the beam centre owing to the action of a uniformly distributed pressure p^u as shown in Figure 1.1(b), then an upper bound calculation (i.e., external work rate equals internal work rate) gives

$$2(p^u L)(L\dot\theta/2) = M_o 2\dot\theta$$

or

$$p^u = 2M_o/L^2, \tag{1.1}$$

where M_o is the plastic collapse moment for the beam cross-section.

The bending moment distribution in the region $0 \le x \le L$ of the beam in Figure 1.1(a) is

$$M = p(L^2 - x^2)/2, \tag{1.2}$$

which has the largest value

$$M = pL^2/2 \tag{1.3}$$

at the beam centre. Thus, the bending moment distribution (Eq. (1.2)) is statically admissible (i.e., $-M_o \le M \le M_o$) for a pressure

$$p^l = 2M_o/L^2, \tag{1.4}$$

which when substituted into Eq. (1.2) gives

$$M/M_o = 1 - (x/L)^2, \tag{1.5}$$

as shown in Figure 1.2.

It is evident from equations (1.1) and (1.4) that

$$p_c = 2M_o/L^2 \tag{1.6}$$

is the exact collapse pressure, since it satisfies simultaneously the requirements of the upper and lower bound theorems of plasticity [1.1-1.5]. It should be noted that equation (1.6) may be used for uniform beams with any cross-sectional shape which is symmetrical with respect to the plane of bending.

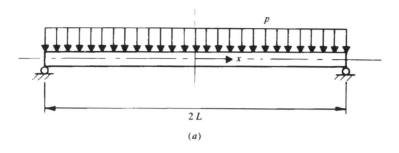

(a)

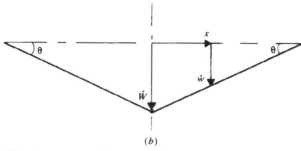

(b)

Figure 1.1 (a) Simply supported beam subjected to a uniformly distributed loading. (b) Transverse velocity profile for a simply supported beam with a plastic hinge at the mid-span.

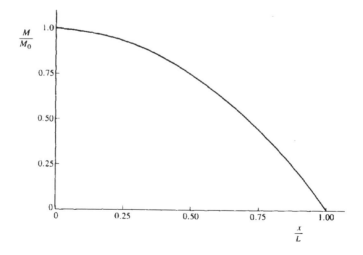

Figure 1.2 Bending moment distribution in one-half ($0 \leq x \leq L$) of the simply supported beam in Figure 1.1(a) according to equation (1.5).

1.3   Governing equations for beams

The dynamic behaviour of the beam in Figure 1.3, for infinitesimal displacements, is governed by the equations

$$Q = \partial M / \partial x \, , \tag{1.7}$$

$$\partial Q / \partial x = -p + m \partial^2 w / \partial t^2 \tag{1.8}$$

and

$$\kappa = -\partial^2 w / \partial x^2 \, , \tag{1.9}$$

which are identical to the equations for static behaviour except for the inclusion of an inertia term in the lateral equilibrium equation (1.8). (*m* is mass per unit length of beam, *t* is time.) In order to simplify the following presentation, the influence of rotatory inertia is not retained in the moment equilibrium equation (1.7), but its influence is discussed in Reference [1.5]. Moreover, it is assumed that the beam is made from a rigid, perfectly plastic material with a uniaxial yield stress σ_o, and a fully plastic bending moment M_o.

Thus, a theoretical solution for a beam must satisfy equations (1.7) and (1.8) in addition to the boundary conditions and initial conditions. The generalised stress, or bending moment M, must remain statically admissible with $-M_o \leq M \leq M_o$ and not violate the yield condition. Furthermore, the transverse velocity field must give rise to a generalised strain rate, or curvature rate $\dot{\kappa}$, which satisfies the normality requirement of plasticity. In other words, the curvature rate vector associated with an active plastic region in a beam must be normal to the yield curve at the corresponding point (i.e., $\dot{\kappa} \geq 0$ when $M = M_o$ and $\dot{\kappa} \leq 0$ when $M = -M_o$). The theoretical solution is said to be exact if the generalised stress field is statically admissible and the associated transverse velocity field is kinematically admissible.

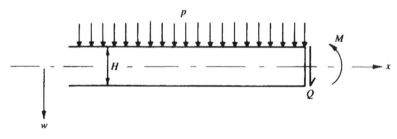

Figure 1.3   Notation for a beam.

1.4   Simply supported beam, $p_c \leq p_0 \leq 3p_c$

1.4.1 Introduction

Consider the dynamic response of the simply supported rigid, perfectly plastic beam shown in Figure 1.4(a). The entire span of this particular beam is subjected to the rectangular pressure pulse which is sketched in Figure 1.5 and which may be expressed in the form

$$p = p_o, \quad 0 \leq t \leq \tau \tag{1.10}$$

and

$$p = 0, \quad t \geq \tau \tag{1.11}$$

It was shown in §1.2 that the exact static collapse pressure of this beam is given by equation (1.6), or

$$p_c = 2M_o / L^2, \tag{1.12}$$

with the associated incipient transverse displacement field sketched in Figure 1.1(b). Clearly, a rigid, perfectly plastic beam remains rigid for pressure pulses, which satisfy the inequality $0 \leq p_0 \leq p_c$, and accelerates when it is subjected to larger pressures. However, if the pressure is released after a short time, then a beam may reach an equilibrium position (with a deformed profile) when all the external dynamic energy has been expended as plastic work. Equations (1.10) and (1.11) suggest that it is convenient to divide the subsequent analysis into the two parts $0 \leq t \leq \tau$ and $\tau \leq t \leq T$, where T is the duration of motion. Moreover, it is necessary to consider only one half of the beam $0 \leq x \leq L$ owing to its symmetry about the mid-span position $x = 0$.

1.4.2 First phase of motion, $0 \leq t \leq \tau$

A theoretical solution is sought when using the transverse or lateral velocity field which is sketched in Figure 1.4(b) and which has the same form as the displacement profile associated with the static collapse pressure p_c. Thus,

$$\dot{w} = \dot{W}(1 - x/L), \quad 0 \leq x \leq L, \tag{1.13}[1]$$

---

[1] The notation $(\dot{\ }) = \partial(\ )/\partial t$ is used throughout this chapter.

where $\dot{W}$ is the lateral velocity at the mid-span. Eqs (1.9) and (1.13) give $\dot{\kappa} = 0$ throughout the beam except at $x = 0$, where $\dot{\kappa} \to \infty$. The beam is, therefore, idealised as two rigid arms (i.e., $-M_0 < M < M_0$) which are connected by a central plastic hinge ($M = M_0$).

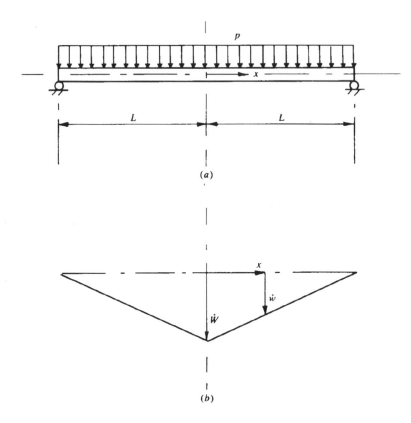

Figure 1.4   (a) Uniformly loaded beam with simple supports. (b) Transverse velocity field.

Now, substituting equation (1.7) into equation (1.8) gives

$$\partial^2 M / \partial x^2 = -p + m\, \partial^2 w / \partial t^2 ,\qquad(1.14)$$

which, when using equations (1.10) and (1.13), becomes

$$\partial^2 M / \partial x^2 = -p_0 + m(1 - x/L)\, d^2 W / dt^2 \qquad(1.15)$$

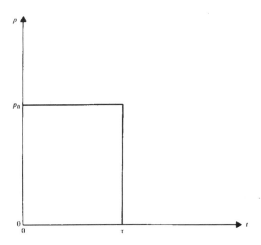

Figure 1.5   Rectangular pressure pulse.

Equation (1.15) may be integrated spatially,[2]

$$M = -p_o\, x^2/2 + m(x^2/2 - x^3/6L)\, d^2W/dt^2 + M_0, \qquad (1.16)$$

where the arbitrary constants of integration have been determined from the requirements that $M = M_o$ and $Q = \partial M/\partial x = 0$ at $x = 0$. However, it is necessary to have $M = 0$ at $x = L$ for a simply supported boundary. Thus,

$$d^2W/dt^2 = 3(\eta - 1)\, M_o/mL^2, \qquad (1.17)$$

where

$$\eta = p_o/p_c \qquad (1.18)$$

is the ratio of the magnitude of the dynamic pressure pulse to the corresponding static collapse pressure[3]. If equation (1.17) is integrated with respect to time, then

$$W = 3(\eta - 1)M_o\, t^2/2mL^2, \qquad (1.19)$$

since $W = \dot{W} = 0$ at $t = 0$. The first stage of motion is completed at $t = \tau$,

$$W = 3(\eta - 1)M_o\, \tau^2/2mL^2 \qquad (1.20)$$

and

$$\dot{W} = 3(\eta - 1)M_o\, \tau/mL^2. \qquad (1.21)$$

---

[2] Equation (1.16) is valid only for beams with a uniform cross-section, i.e., m is assumed to be independent of x.

[3] Equation (1.17) gives $\eta = 1$ (i.e., $p_o = p_c$) when $\ddot{W} = 0$, for static loads.

1.4.3 Second phase of motion, $\tau \leq t \leq T$

The external pressure is removed suddenly at $t = \tau$, so that the beam is unloaded during this stage of motion. However, the beam has a transverse velocity at $t = \tau$ according to equations (1.13) and (1.21) and, therefore, has a finite kinetic energy. Thus, the beam continues to deform for $t \geq \tau$ until the remaining kinetic energy is absorbed through plastic energy dissipation at the plastic hinges.

If the velocity field, which is indicated in Figure 1.4(b) and described by equation (1.13), is assumed to remain valid during this stage, then it is evident that equation (1.16) remains unchanged except $p_0 = 0$. Thus, equation (1.17) becomes

$$d^2W/dt^2 = -3\,M_o/mL^2, \qquad (1.22)$$

which may be integrated to give

$$\dot{W} = 3M_o\,(\eta\tau - t)/mL^2, \qquad (1.23a)$$

and

$$W = 3M_o\,(2\eta\tau t - t^2 - \eta\tau^2)/2mL^2 \qquad (1.23b)$$

when using equations (1.20) and (1.21) as initial conditions ($t = \tau$) for the displacement and velocity profiles during the second stage of motion. The beam reaches its permanent[4] position when $\dot{W} = 0$, which, according to equation (1.23a), occurs when

$$T = \eta\,\tau. \qquad (1.24)$$

Equations (1.13), (1.23b) and (1.24) predict the final deformed profile

$$w = 3\eta\,(\eta - 1)\,M_o\,\tau^2\,(1 - x/L)/2mL^2. \qquad (1.25)$$

---

[4] No elastic unloading occurs for a rigid, perfectly plastic material.

1.4.4 Static admissibility

The theoretical analysis in §§ 1.4.2 and 1.4.3 satisfies the equilibrium equations (1.7) and (1.8), initial conditions and boundary conditions for a simply supported beam subjected to a pressure pulse with a rectangular shaped pressure-time history. However, the bending moment M has only been specified at the supports and the mid-span. It is necessary, therefore, to ensure that the bending moment does not violate the yield condition anywhere in the beam for $0 \leq x \leq L$ and during both stages of motion $0 \leq t \leq \tau$ and $\tau \leq t \leq T$.

Now, equation (1.16) may be written

$$M/M_o = 1 - \eta(x/L)^2 + (\eta - 1)(3 - x/L)(x/L)^2/2 \qquad (1.26)$$

when using equation (1.17), and gives $dM/dx = 0$ at $x = 0$, as expected. Moreover,

$$(L^2/M_o)\, d^2M/dx^2 = \eta - 3 - 3(\eta - 1)x/L, \qquad (1.27)$$

which predicts $d^2M/dx^2 \leq 0$ at $x = 0$ provided $\eta \leq 3$. Thus, no yield violations occur anywhere in the beam during the first stage of motion when $\eta \leq 3$ because $d^2M/dx^2 < 0$ over the entire span, as indicated in Figure 1.6(a). The theoretical solution which is presented for the first stage of motion is, therefore, correct provided $1 \leq \eta \leq 3$. However, a yield violation occurs when $\eta > 3$ since $d^2M/dx^2 > 0$ at $x = 0$, which implies that the bending moment near the mid-span exceeds the plastic limit moment M_o.

It is evident that equation (1.26) with $\eta = 0$ gives

$$M/M_o = 1 - (3 - x/L)(x/L)^2/2 \qquad (1.28)$$

during the second stage of motion $\tau \leq t \leq T$. Equation (1.28), or equation (1.27), with $\eta = 0$, gives $d^2M/dx^2 \leq 0$, so that no yield violations occur because $M = M_o$ and $d M/d x = 0$ at $x = 0$ and $M = 0$ at $x = L$, as shown in Figure 1.6(b). Thus, the above theoretical solution, which leads to the response time predicted by equation (1.24) with the associated permanent shape given by equation (1.25), is correct provided $1 \leq \eta \leq 3$, or $p_c \leq p_o \leq 3p_c$.

1.5   Simply supported beam, $p_o \geq 3p_c$

It was observed in §1.4.4 that the theoretical solution in §1.4.2 and §1.4.3 is exact (i.e., both statically and kinematically admissible) for the beam illustrated in Figure 1.4(a), when it is subjected to the rectangular shaped pressure pulse in Figure 1.5 provided $p_o \leq 3p_c$, where p_c is the static collapse pressure given by equation (1.12). This theoretical analysis was first obtained by Symonds [1.6] who also examined a dynamic pressure loading having $p_o \geq 3p_c$. An exact theoretical rigid, perfectly plastic solution for this case was obtained using the three phases of motion illustrated in Figure 1.7, as discussed further in reference [1.5]. It turns out that

$$T = \eta\tau \tag{1.29}$$

is the duration of motion and the associated maximum permanent transverse displacement at the mid-span is

$$W_f = p_o\tau^2 \, (4\eta\text{-}3)/6m \tag{1.30}$$

1.6   Simply supported beam loaded impulsively

1.6.1 Introduction

It is evident from Figure 1.8 that the maximum permanent transverse displacement of the simply supported beam illustrated in Figure 1.4(a) is, from a practical viewpoint, insensitive to the value of the dimensionless pressure ratio when $\eta > 20$, approximately. External pressure loadings having a finite impulse with an infinitely large magnitude ($\eta \to \infty$) and an infinitesimally short duration ($\tau \to 0$) (Dirac delta function) are known as impulsive. In other words, a beam of unit breadth acquires instantaneously a uniform transverse velocity V_o, where, to conserve linear momentum[5],

$$(2mL)V_o = (p_o 2L)\tau,$$

or

$$V_o = p_o\tau/m \tag{1.31}$$

---

[5] Newton's second law requires $F = d(mv)/dt$, or $F \, dt = d(mv)$. Thus, $\int F dt$ = change in linear momentum, where $\int F dt$ is impulse.

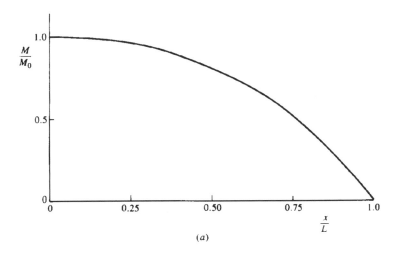

(a)

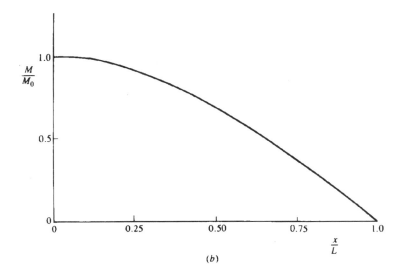

(b)

Figure 1.6   (a) Bending moment distribution ($0 \leq x/L \leq 1$) during the first phase of motion ($0 \leq t \leq \tau$) for a simply supported beam subjected to a uniformly distributed rectangular pressure pulse with $\eta = 2$. (b) Bending moment distribution ($0 \leq x/L \leq 1$) during the second phase of motion ($\tau \leq t \leq T$) for a simply supported beam subjected to a uniformly distributed rectangular pressure pulse with $1 \leq \eta \leq 3$.

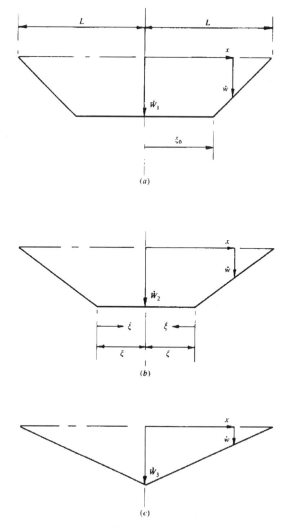

Figure 1.7 Transverse velocity profiles for a simply supported beam subjected to a
rectangular pressure pulse with $\eta \geq 3$. (a) First phase of motion, $0 \leq t \leq \tau$.
(b) Second phase of motion, $\tau \leq t \leq T_l$. (c) Third phase of motion, $T_l \leq t \leq T$.

In this circumstance, equations (1.29) and (1.30) predict

$$T = mV_o/p_c. \tag{1.32}$$

and

$$W_f = mV_o^2L^2/3M_o \tag{1.33}$$

for the duration of response and maximum permanent transverse displacement,
respectively.

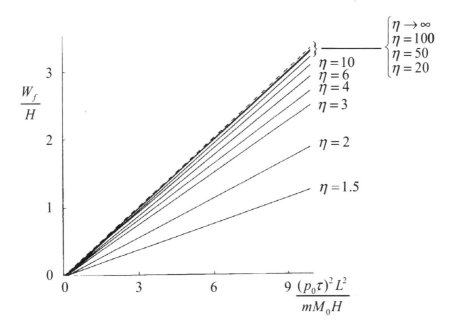

Figure 1.8   Maximum permanent transverse displacement of a simply supported beam (with unit width) subjected to a rectangular pressure pulse having various values of the dimensionless pressure ratio η.

The approximation of an impulsive loading usually simplifies a theoretical analysis and is an acceptable idealisation for many practical problems. Thus, the impulsive loading of a simply supported beam is examined in the following sections, but by using an alternative method which employs momentum and energy conservation principles.

1.6.2 First phase of motion, $0 \le t \le T_1$

Initially, a discontinuity in velocity occurs at $x = L$ since the support remains fixed spatially (i.e., $w = \dot{w} = 0$ at $x = L$ for all t). It appears reasonable to postulate that plastic hinges would develop at the supports when $t = 0$ then travel inwards with a speed $\dot{\xi}$ towards the beam centre, as indicated at a time t by the velocity profile in Figure 1.7(b). The central portion $0 \le x \le \xi$ remains rigid since it continues to travel at the initial velocity V_0 until disturbed by the inwards moving plastic hinge. This velocity field can be written in the form

$$\dot{w} = V_o, \quad 0 \le x \le \xi \tag{1.34}$$

and

$$\dot{w} = V_o (L - x)/(L - \xi), \quad \xi \le x \le L \tag{1.35}$$

Now, the conservation of angular momentum of one half of the beam about a support demands

$$\int_0^L mV_o(L - x)\, dx = \int_0^\xi mV_o(L - x)\, dx + \int_\xi^L mV_o(L - x)^2\, dx/(L - \xi) + M_o t \tag{1.36}$$

since the entire beam is rigid[6] except at the travelling plastic hinge with a limit moment M_o. Equation (1.36) gives

$$t = mV_o (L - \xi)^2 /6\, M_o , \tag{1.37}$$

which is the time required for a plastic hinge to travel inwards from a support to the location $x = \xi$.

The first stage of motion is completed at $t = T_1$ when the two travelling hinges coalesce at $x = 0$. Equation (1.37) predicts

$$T_1 = mV_o L^2/6M_o \tag{1.38}$$

when $\xi=0$. Thus, the corresponding lateral displacement at $x=0$ is $W_1=V_oT_1$, or

$$W_1 = mV_o^2 L^2 /6M_o. \tag{1.39}$$

1.6.3 Final phase of motion, $T_1 \le t \le T$

Now, at the end of the first stage of motion ($t = T_1$), the beam has a linear velocity profile with a peak value V_o and, therefore, possesses a kinetic energy $mV_o^2L/3$ which remains to be dissipated during subsequent motion. It appears reasonable to assume that this kinetic energy will be dissipated in a plastic hinge which remains stationary at $x = 0$ during the second stage of motion, as shown in Figure 1.7(c). In this circumstance, the conservation of energy requires that

$$mV_o^2 L/3 = 2M_o\theta_2 , \tag{1.40}$$

---

[6] Equations (1.9), (1.34) and (1.35) give $\dot{\kappa} = 0$.

where $2\theta_2$ is the angular change of the central plastic hinge. The displacement acquired by the beam centre during the second stage of motion is $W_2 = L\theta_2$, or

$$W_2 = \mathrm{m}V_o^2\, L^2/6M_o. \tag{1.41}$$

Finally, the total permanent lateral displacement at the beam centre is $W_f = W_1 + W_2$ which agrees with equation (1.33). This expression may be written in the dimensionless form

$$W_f/H = \lambda/3, \tag{1.42}$$

where

$$\lambda = mV_o^2\, L^2/M_o H \tag{1.43}$$

is a non-dimensionalised form of the initial kinetic energy and H is the beam depth.

It is interesting to observe that the contributions to the central lateral displacement of the beam are identical during both phases of motion (i.e., $W_1 = W_2$) even though two-thirds of the initial kinetic energy is dissipated during the first phase of motion and only one-third during the final phase. However, two travelling plastic hinges are present in the beam during the first phase of motion, as shown in Figure 1.7(b), while only one stationary plastic hinge develops at the beam centre during the second phase.

It is shown in reference [1.5] that the foregoing theoretical predictions are both statically (i.e., no yield violations) and kinematically (i.e., no geometrical violations) admissible and, therefore, exact in the context of the rigid, perfectly plastic methods of analysis.

1.7   Final comments

If the supports of the beam in Figure 1.1(a) are restrained axially and the external loads are large enough to cause finite transverse displacements, then in-plane forces are introduced into the beam. These membrane forces can cause a significant increase in the load carrying capacity for static loads and reduced permanent displacements for the dynamic case, as observed in §2.4 and §2.5 for quasi-static behaviour.

The phenomenon of material strain rate sensitivity is introduced in Chapter 3 and may also have a significant influence on the dynamic plastic response of a structure.

Many authors have written articles on the dynamic plastic response of beams, frames, plates and shells, as discussed further in References [1.5] and [1.7-1.9].

REFERENCES

[1.1]   Hodge, P. G., *Plastic Analysis of Structures*, McGraw Hill, New York, 1959.

[1.2]   Horne, M. R., *Plastic Theory of Structures*, MIT Press, Cambridge, Mass., 1971.

[1.3]   Johnson, W. and Mellor, P.B., *Engineering Plasticity*, Van Nostrand Reinhold, London, 1973.

[1.4]   Calladine, C. R., *Plasticity for Engineers*, Ellis Horwood, Chichester, and John Wiley, New York, 1985.

[1.5]   Jones, N., *Structural Impact*, Cambridge University Press, 1989; paperback edition (1997).

[1.6]   Symonds, P. S., Large plastic deformations of beams under blast type loading, *Proceedings of the Second US National Congress of Applied Mechanics*, 505-15, 1954.

[1.7]   Goldsmith, W., *Impact*, Edward Arnold, London, 1960.

[1.8]   Johnson, W., *Impact Strength of Materials*, Edward Arnold, London, and Crane Russak, New York, 1972.

[1.9]   Stronge, W. J. and Yu, T. X., *Dynamic Models for Structural Plasticity*, Springer-Verlag, London, 1993.

2. QUASI-STATIC BEHAVIOR

2.1 Introduction

A *quasi-static method* of analysis is used often in engineering practice for the design of structures subjected to dynamic loadings. It simplifies both theoretical and numerical methods of analysis, and, in the appropriate circumstances, captures the principal features of the response. The accuracy of quasi-static procedures is often acceptable, especially when acknowledging any uncertainties in the dynamic properties of the material, the structural support conditions (e.g., riveted joints) and the characteristics of the dynamic loading.

It is assumed in a quasi-static analysis that the deformation profile is time-independent for a structure subjected to dynamic loads. In fact, a quasi-static analysis uses the deformation profile for the same structural problem when subjected to the same spatial distribution of loads, but applied statically. Thus, it is assumed that any inertia forces in a structure caused by a dynamic load do not give rise to a deformation profile which changes with time[1]. Any transient behaviour caused by travelling plastic hinges, for example, as shown in Figure 2.1(b) for a mass impact on a rigid, perfectly plastic beam, is ignored.

The entire quasi-static response of this particular problem is governed by the transverse deformation profile with stationary plastic hinges in Figure 2.1(c). The quasi-static method of analysis is outlined more fully in the next section.

---

[1] It is possible for the deformation profile of a structure to change shape when an increase of static loads produces large deflection effects, or geometry changes. In this case, the same change of profile would be incorporated into a quasi-static analysis for dynamic loads, as noted in §2.4, but any changes in the profile due to inertia effects are still ignored.

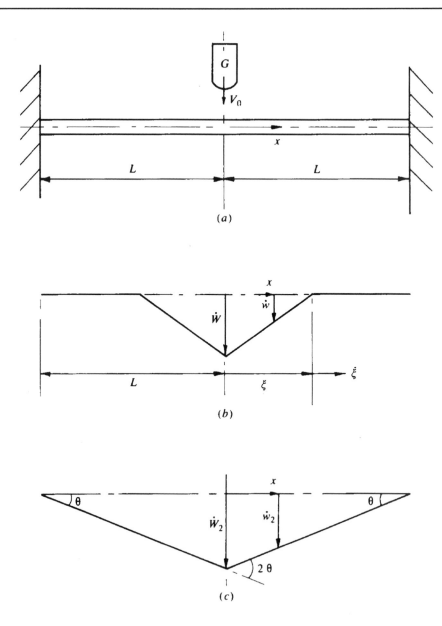

Figure 2.1 (a) Impact of a fully clamped beam with a mass G travelling with a velocity V_o.
(b) Transverse velocity field during the first phase of motion ($0 \le t \le t_l$). (c)
Transverse velocity field during the second phase of motion ($t_l \le t \le T$).

2.2   Quasi-static method of analysis

A quasi-static analysis for a rigid, perfectly plastic structural member struck by a mass having an initial kinetic energy, K_e, is obtained from the energy balance

$$P_c W_q = K_e , \tag{2.1}$$

where W_q is a quasi-static estimate of the maximum permanent transverse displacement, and P_c, is a plastic collapse force for the same structure when loaded statically at the same location as the striking mass. It is evident that the external work, $P_c W_q$, in equation (2.1), is also equal to the internal energy dissipated at the plastic hinges and zones which develop in a structural member during plastic deformation. In this case, equation (2.1) becomes

$$\sum_{i=1}^{n} D_i = K_e , \tag{2.2}$$

where D_i is the total energy absorbed at each plastic hinge and zone during the plastic deformation required to absorb K_e, and n is the total number of such sites in a structural member.

If only plastic bending[2] hinges develop in a structure during plastic collapse, then equation (2.2) takes the form

$$\sum_{i=1}^{n} M_i \theta_i = K_e \tag{2.3}$$

where M_i is the plastic collapse moment for the cross-section at the location i and θ_i is the corresponding total rotation across that plastic hinge during collapse. In the particular case of a uniform structural cross-section with a plastic collapse moment, M_o, then equation (2.3) becomes

$$M_O \sum_{i=1}^{n} \theta_i = K_e \tag{2.4}$$

The left-hand sides of equations (2.2) to (2.4) are time-independent unlike the corresponding expressions in §1.2 and §1.4. It is assumed for a quasi-static analysis that the shapes of the velocity and displacement fields are the same

---

[2] No membrane or transverse shear forces contribute to the total energy absorbed.

and time-independent so that there is no propagation of any plastic hinges or movements at the boundaries of plastic zones[3].

2.3   Impact of a mass on a fully clamped beam

2.3.1 Introduction

Parkes [2.4] has shown that an exact theoretical solution for the beam shown in Figure 2.1(a), when made from a rigid, perfectly plastic material, has two phases of motion; a transient phase with travelling plastic hinges in Figure 2.1(b) which is followed by a modal phase of motion with stationary plastic hinges, as shown in Figure 2.1(c). The corresponding maximum permanent transverse displacement at the mid-span is [2.4, 2.5]

$$W_f = \frac{G^2 V_o^2}{24 m M_o} \left\{ \frac{\bar{\alpha}}{1+\bar{\alpha}} + 2 \log e(1+\bar{\alpha}) \right\}, \tag{2.5}$$

where

$$\bar{\alpha} = mL/G. \tag{2.6}$$

2.3.2 Quasi-static analysis

The static plastic collapse force for the particular fully clamped beam in Figure 2.2, is $P_c = 4M_o/L$ when the beam has a uniform cross-section with a plastic collapse moment, M_o. Thus, a quasi-static analysis using equation (2.1) predicts a maximum permanent transverse displacement

$$W_q = G V_o^2 L/8 M_O, \tag{2.7}$$

for the beam in Figure 2.1(a) struck at the mid-span by a mass, G, having $K_e = G V_O^2/2$. This analysis uses the transverse displacement profile in Figure 2.2(b)

---

[3] Martin [2.1] derived a theorem to predict an upper bound to the exact permanent displacement of a rigid, perfectly plastic continuum subjected to impulsive loading. This approach might be used for mass impact loadings by neglecting the structural mass. In this case [2.2], it leads to an expression having the same form as equation (2.1). Martin's theorem proves also that the predictions of a quasi-static method of analysis provides an upper bound to the exact value of the permanent displacement. Martin and Symonds[2.3] have also studied the requirements of an optimal mode for the dynamic plastic analysis of structures subjected to impulsive loadings.

which has the same form as Figure 2.1(c) employed in the exact solution[4]. The permanent transverse displacement profile of the beam is, therefore, triangular and can be expressed in the form

$$w_q = GV_o^2L(1-x/L)/8M_o, \quad 0 \leq x \leq L \tag{2.8}[5]$$

Equation (2.8) with $x = 0$ is identical to the exact solution given by equation (2.5) when $G/mL >> 1$, or $\bar{\alpha} << 0$, for heavy strikers. It is evident, therefore, that a quasi-static method of analysis predicts accurately the maximum permanent transverse displacement and the permanent displacement profile[6] throughout the beam for a striker mass, G, which is much heavier than one-half of the beam mass, mL.

The collapse mechanism in Figure 2.2(b), which is associated with the impact loading of the beam in Figure 2.1(a) with a uniform cross-section, has three plastic hinges which absorb the energy $M_o\theta + M_o(2\theta) + M_o(\theta)$, where $\theta \cong W_q/L$. It is straightforward to show that equating this internal energy to $K_e = GV_o^2/2$, as demanded by equation (2.4), again leads to equations (2.7) and (2.8).

2.3.3 Accuracy of quasi-static analysis

Now, the difference between the maximum permanent transverse displacement predicted by a quasi-static method of analysis, W_q, and an exact theoretical solution, W_f, could be expressed as a percentage error [2.6].

$$e_w = (W_q - W_f)100/W_f. \, (\%) \tag{2.9}$$

Equation (2.9) when using W_q from equation (2.7) and W_f from equation (2.5), gives the variation of e_w with the mass ratio $G/2mL$ in Figure 2.3. These results reveal that the error in the prediction of a quasi-static analysis is less than one percent when the mass ratio is greater than 33. The error is about ten per cent for a mass ratio as small as 3.25. However, it is evident from Figure 2.3 that the error is significant for smaller mass ratios and is 31% when a striker mass is equal to the beam mass.

---

[4] Velocity and not displacement fields are used in chapter 1 because of the path-dependence of an inelastic material, as noted in Appendix 2 of Reference [2.5]. However, the shapes of the velocity and displacement fields are the same in a quasi-static analysis because any plastic hinges remain stationary.

[5] For the beam in Figure 2.1(a) and other symmetric cases, the right hand side, $0 \leq x \leq L$, only is considered.

[6] See also §3.8.4 in Reference [2.5]

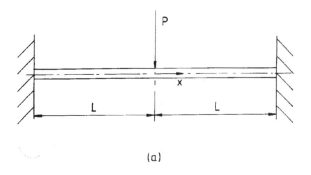

(a)

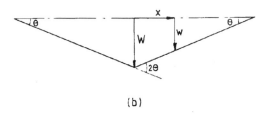

(b)

Figure 2.2 Static plastic collapse load for a fully clamped beam.

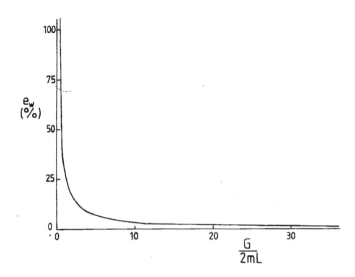

Figure 2.3 Error of quasi-static analysis as a function of the ratio between the striker and the beam masses.

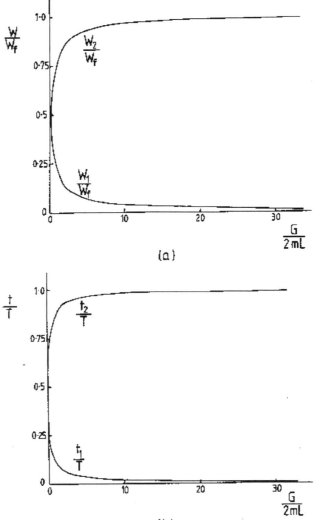

Figure 2.4   (a) Dimensionless transverse displacements at the end of the first phase of motion (W$_1$) and acquired during the second phase of motion (W$_2$) for the exact dynamic rigid-plastic analysis, and (b) dimensionless time durations for the two phases of motion according to the exact analysis as functions of the ratio between the striker and the beam masses.

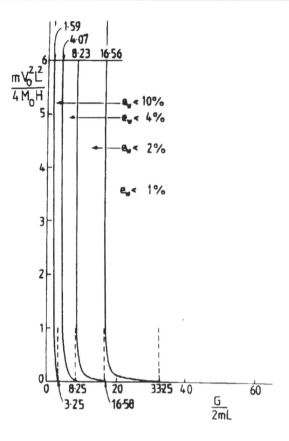

Figure 2.5   Percentage difference between the theoretical predictions of a quasi-static
analysis and a dynamic rigid-plastic analysis [2.8] for the beam impact problem
in Figure 2.1(a). ———: finite-displacement effects retained in analysis. - - -:
infinitesimal displacements [2.6].

2.3.4 Comments

The maximum transverse displacements acquired during the transient
(W_1) and modal (W_2) phases[7] of motion according to the exact theoretical
analysis, are plotted in Figure 2.4(a) when expressed as a ratio of W_f. The
corresponding dimensionless time durations of these two phases of motion are
plotted in Figure 2.4(b). It is evident from the results in Figure 2.4 that the
modal phase of motion dominates the response for large mass ratios, while the
transient phase is dominant only for small mass ratios.

---

[7] See Figures 2.1(b,c).

The impact force, $P = -G\ddot{W}$, between the mass, G, and the beam in Figure 2.1, is infinitely large immediately after impact [2.6]. In fact, the magnitude of P cannot exceed the transverse shear strength, $2Q_o$, in a rigid, perfectly plastic beam. Nevertheless, transverse shear effects are not retained in this section. However, the impact force, P, decreases rapidly with time as the plastic hinges in Figure 2.1(b) propagate away from the impact position at the mid-span. It is similar to the static collapse force, P_c, throughout the duration of the second phase of motion for mass ratios larger than about ten [2.6].

2.4   Quasi-static method of analysis with finite-deflection effects

The static and dynamic behaviour of axially restrained beams and plates is influenced significantly by finite-transverse deflections, particularly when the maximum values are larger than the structural thickness, approximately [2.5]. This phenomenon is related to geometry changes of the structural members and causes the development of in-plane, or membrane forces. These forces become increasingly important with increase in deflection and, for sufficiently large deflections, they dominate the influence of bending moments, which alone develop in infinitesimal deflection, or first order, theory, examined in §2.2-2.3. In these circumstances, a quasi-static analysis is obtained using the equation

$$\int_0^{W_q} P \, dW = K_e , \qquad (2.10)$$

where the magnitude of the concentrated force, P, is a function of the associated transverse deflection, W. Equation (2.10) reduces to equation (2.1) for infinitesimal deflections with $P = P_c$.

Equation (2.10) might be expressed in terms of the internal energy absorbed by a structure, as shown in equation (2.2) for infinitesimal deflections. Thus, equation (2.10) becomes

$$\int_0^{W_q} \dot{D}_T (W) \, dW = K_e , \qquad (2.11)$$

where the increment of external work, PdW, is replaced by the rate of internal energy absorption with respect to W at a deflection, W.

2.5   Influence of finite-deflections in beams

2.5.1 Impact loading of a fully clamped beam

Equation (2.10) is used to obtain a quasi-static estimate for the maximum permanent transverse deflection of the beam impact problem which is illustrated in Figure 2.1(a). This equation requires the force (P) - deflection (W) relation for a static concentrated force acting at the mid-span, as shown in Figure 2.2(a). This particular case was studied by Haythornthwaite [2.7] and in Reference [2.5] and gave

$$P = 4M_0 (1 + W^2/H^2)/L, \quad 0 \le W/H \le 1, \tag{2.12a}$$

and

$$P = 8M_0 W/H L, \quad W/H \ge 1, \tag{2.12b}$$

where H is the thickness of a beam having a solid rectangular cross-section.

Now, substituting equations (2.12a,b) into equation (2.10) with $K_e = GV_0^2/2$ gives

$$(4M_0/L) \int_0^{W_q} (1+W^2/H^2)\, dW = GV_0^2/2 \tag{2.13a}$$

if $W_q \le H$, and

$$(4M_0/L) \left\{ \int_0^H (1+W^2/H^2)\, dW + \int_H^{W_q} 2WdW/H \right\} = GV_0^2/2 \tag{2.13b}$$

if $W_q \ge H$.   Equations (2.13a,b) predict the quasi-static estimates for the maximum permanent transverse deflections at the mid-span

$$W_q/H + (W_q/H)^3/3 = \Omega \tag{2.14a}$$

when $W_q \le H$, and

$$W_q/H = (\Omega - 1/3)^{1/2} \tag{2.14b}$$

if $W_q \ge H$, where

$$\Omega = GV_0^2 L/8M_0 H. \tag{2.14c}$$

Equation (2.14a) with $W_q \to 0$ reduces to equation (2.7) for infinitesimal deflections, as expected. The cubic term in equation (2.14a) arises from the strengthening influence of membrane forces produced by finite transverse deflections. It is evident from equation (2.14a) that this phenomenon leads to a one-third increase in the dimensionless external energy absorbed by a beam with $W_q = H$, while it is more than doubled when $W_q/H = 2$ according to equation (2.14b).

2.5.2 Accuracy of a quasi-static analysis

Now, a theoretical rigid, perfectly plastic analysis is presented in Reference [2.8] for the impact beam problem in Figure 2.1(a) which retains the influence of finite-deflection effects in the basic equations. This analysis has a first phase of motion with a stationary plastic hinge in the beam which forms immediately underneath the striker mass, G, and two travelling plastic hinges which propagate away from the mid-span towards the fully clamped supports. This phase of motion is completed when the travelling plastic hinges reach the supports which occurs simultaneously for an impact at the mid-span. The kinetic energy remaining in the beam and striker at the end of this phase of motion is absorbed in a final phase of motion with stationary plastic hinges at both supports and at the mid-span, which is identical to the single modal form used to obtain equations (2.12a) and (2.12b) for static loads.

The percentage difference (e_w) between the maximum permanent transverse displacements of a quasi-static theoretical procedure (W_q, equations (2.14a) and (2.14(b)) and the dynamic analysis (W_f) in reference (2.8) for the problem in Figure 2.1(a), is defined by equation (2.9). The comparisons between the two analyses in Figure 2.5 reveals that the error introduced by a quasi-static procedure is less than ten per cent for mass ratios $(G/2mL)$ larger than 1.59-3.25, and that the error is less than one per cent for mass ratios larger than about 16-33, depending on the dimensionless impact velocity.

The theoretical predictions in Figures 2.3 and 2.5 for infinitesimal deflections are independent of the dimensionless impact velocity. The curves associated with the finite-deflection case in Figure 2.5 are virtually independent of the dimensionless impact velocity except for small dimensionless values when the beam behaviour changes from an initial bending response to a membrane one for larger deflections. It is interesting to observe that the influence of finite-deflections, or geometry changes, expands the range of validity of a quasi-static method of analysis. This is particularly noticeable for the larger mass ratios. Thus, an error of less than one per cent is associated with a quasi-static method of analysis for infinitesimal deflections and mass ratios larger than about 33. To maintain an error of less than one per cent, the mass ratio reduces to 16.56 for large dimensionless impact velocities when the influence of finite-deflections is retained in the analysis.

Experimental test results are reported in References [2.9] and [2.10] for fully clamped aluminium alloy beams struck at the mid-span by a mass, G, travelling with an initial velocity V_o, as shown in Figure 2.1(a). The dimensionless maximum permanent transverse displacements at the mid-span are plotted in Figure 2.6 against the dimensionless initial kinetic energy and compared with the theoretical quasi-static predictions of equations (2.14a,b). The agreement between the theoretical predictions and both sets of experimental results in Figure 2.6 illustrates the accuracy which is possible with quasi-static methods of analysis.

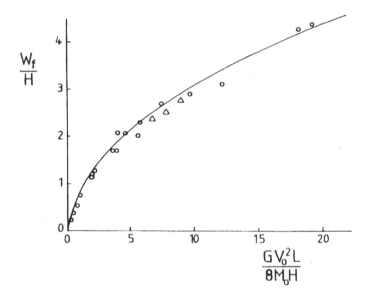

Figure 2.6   Comparison between the theoretical quasi-static predictions (equations (2.14a,b)), for the maximum permanent transverse deflections (W_f) of the beam impact problem in Figure 2.1(a), and the corresponding experimental values on aluminium alloy beams. ———: equations (2.14a,b); O: experimental results on flat beams [2.9]; Δ: experimental results [2.10] with $\sigma_o = (\sigma_y + \sigma_m)/2$, where σ_y and σ_m are the yield and ultimate tensile stresses, respectively.

REFERENCES

[2.1]   Martin, J. B., Impulsive loading theorems for rigid-plastic continua, *Proceedings of the ASCE, Journal of the Engineering Mechanics Division,* **90**(EM5), 27-42, 1964.

[2.2]   Jones, N., Bounds on the dynamic plastic behaviour of structures including transverse shear effects, *International Journal of Impact Engineering,* **3**(4), 273-291, 1985.

[2.3]   Martin, J. B. and Symonds, P. S., Mode approximations for impulsively-loaded rigid-plastic structures, *Proceedings of the ASCE, Journal of the Engineering Mechanics Division,* **92**(EM5), 43-66, 1966.

[2.4]   Parkes, E. W., The permanent deformation of an encastré beam struck transversely at any point in its span, *Proceedings of the Institution of Civil Engineers,* **10**, 277-304, 1958.

[2.5]   Jones, N., *Structural Impact,* Cambridge University Press, Cambridge, England, 1989, (Paperback edition (1997))

[2.6]   Jones, N. Quasi-static analysis of structural impact damage, *Journal of Constructional Steel Research,* **33**(3), 151-177, 1995.

[2.7]   Haythornthwaite, R. M., Beams with full end fixity, *Engineering,* **183**, 110-112, 1957.

[2.8]   Shen, W. Q. and Jones, N., A comment on the low speed impact of a clamped beam by a heavy striker, *Mechanics, Structures and Machines,* **19**(4), 527-549, 1991.

[2.9]   Liu, J. and Jones, N., Experimental investigation of clamped beams struck transversely by a mass, *International Journal of Impact Engineering,* **6**(4), 303-335, 1987.

[2.10]   Yu, J. L. and Jones, N., Further experimental investigations on the failure of clamped beams under impact loads, *International Journal of Solids and Structures,* **27**(9), 1113-1137, 1991.

3. MATERIAL STRAIN RATE SENSITIVITY

3.1   Introduction

The theoretical analyses developed in chapter 1 have examined the influence of inertia on the response of an elementary structure subjected to dynamic loads which cause plastic behaviour. The yield criterion, which governs the plastic flow, was assumed to be independent of the rate of strain ($\dot{\varepsilon}$). However, the plastic flow of some materials is sensitive to strain rate, which is known as material strain rate sensitivity, or viscoplasticity [3.1]. This phenomenon is illustrated in Figure 3.1 for mild steel specimens which were tested at various uniaxial compressive strain rates [3.2].

It is evident from Figure 3.1 that the plastic behaviour of mild steel is highly sensitive to strain rate. The strain rates in Figure 3.1 are realistic and are encountered in practical engineering problems. For example, consider a vertical 1 m long mild steel bar ($L = 1$ m) which is struck at one end with a large mass dropped from a height of 5 m and having an axial velocity of 10 m/s on impact.[1] An average axial strain rate of $\dot{\varepsilon} = \varepsilon/t = (\delta L)/t = V/L = 10 \, \text{s}^{-1}$, approximately, is generated in the bar when neglecting any stress wave effects. The plastic flow stress corresponding to a strain rate of $10 \, \text{s}^{-1}$ is approximately double the static flow stress according to the experimental results in Figure 3.1 for strains up to about 0.02.

---

[1] Impact velocity $V = \sqrt{2g \times 5} \cong 10 \, \text{m/s}$. Average axial strain ($\varepsilon$) in bar $\cong \delta/L$, where δ is the axial deformation at the struck end. Axial velocity at struck end $(V) \cong \delta/t$, where t is time.

The theoretical rigid, perfectly plastic predictions in Figure 2.6 for the permanent transverse displacements of fully clamped beams subjected to impact loads, agree reasonably well with the experimental results on aluminium 6061-T6 beams. However, the same theoretical predictions are larger than the corresponding maximum permanent transverse displacements of mild steel beams as illustrated in Figure 3.2 for the impulsively loaded case [3.3]. It is evident, though, that the dimensionless term $\lambda/4 = \rho V_0^2 L^2 / \sigma_0 H^2$ in the abscissa contains the strain-rate-independent, or static, yield stress (σ_0) in the denominator. Thus, any increase in the plastic flow stress of the test specimens due to strain rate effects, as shown in Figure 3.1 for mild steel, means that the value of λ for the experimental test points should be smaller, thereby giving better agreement with the theoretical predictions.

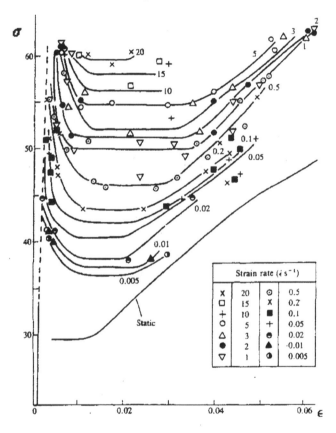

Figure 3.1   Stress (σ)-strain (ε) curves for mild steel at various uniaxial compressive strain rates according to Marsh and Campbell [3.2]. 1 unit of ordinate is 10^3 lbf/in^2 or 6.895 MN/m^2.

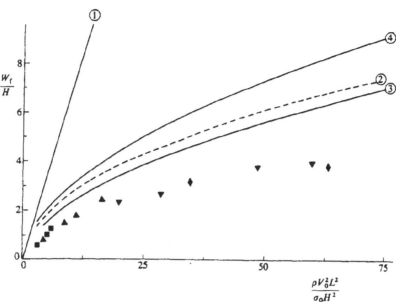

Figure 3.2   Maximum permanent transverse displacements (W_f) of a fully clamped beam
subjected to an impulsive velocity of magnitude V_o distributed uniformly across
the entire span 2L.   1: infinitesimal analysis (bending only). 2, 3, 4:
theoretical analyses which include the influence of finite deflections. ■, ▲, ♦, ▼:
experimental results on mild steel beams.[3.3]

It is evident from Figure 3.2 that the influence of material strain rate
sensitivity manifests itself as a strengthening effect in a structure.[2] This might
suggest that it is a beneficial phenomenon since it provides an additional safety
factor.[3] However, Perrone [3.5] remarked that energy absorption systems for
enhancing the structural crashworthiness of vehicles, for example, could impart
unacceptable forces on the human body which might otherwise have been
acceptable   for   an   identical   material   (and   structure)   but   with
strain-rate-independent material properties.

Material strain rate sensitivity is a material effect and is independent of
the structural geometry. It is impossible to review here the vast amount of
literature, which is available on the strain-rate-sensitive behaviour of materials.

---

[2] Sometimes a structural mode change occurs which causes larger and not smaller associated
permanent deformations, as discussed by Bodner and Symonds [3.4].

[3] In fact, the associated fracture strain might decrease with increase in strain rate, as
indicated in Figure 3.6.

However, several excellent surveys on various aspects of the subject have been published [3.1, 3.6-3.10]. This chapter focuses, primarily, on the behaviour of metals which undergo moderate strain rates.

An introduction to the strain rate characteristics of some metals subjected to various loads is presented in §3.2, while §3.3 contains an elementary constitutive equation, which is particularly valuable for dynamic structural problems.

3.2   Material characteristics

3.2.1   Introduction

It is the purpose of this section to introduce some of the strain rate characteristics of materials which play an important role in the dynamic plastic behaviour of structures. This is achieved by examining the behaviour of materials under various simple dynamic loads. No attempt is made to describe the underlying materials science. Nor is any attention given to the considerable practical difficulties encountered in high strain rate tests (e.g., stress wave effects and inertia of the test rigs and recording equipment [3.11].

3.2.2   Compression

The uniaxial compressive results of Marsh and Campbell [3.2] shown in Figure 3.1 are assembled in Figure 3.3 together with additional test values on mild steel. This figure shows clearly a significant increase in the upper yield stress with increase of strain rate. The stress at a strain of 0.05 also increases, but less markedly. Material strain hardening, therefore, decreases with increase of strain rate for this type of mild steel with strains up to at least 0.05.

The compressive strain rate behaviour of titanium and aluminium 6061-T6 according to Maiden and Green [3.12] is shown in Figure 3.4. The test results for the titanium in Figure 3.4(a) do not indicate the presence of an upper yield stress and are less sensitive to strain rate than the mild steel in Figures 3.1 and 3.3.

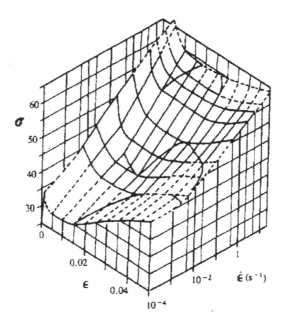

Figure 3.3   Isometric projection of stress (σ), strain (ε), strain rate ($\dot{\varepsilon}$) surface for mild steel subjected to dynamic uniaxial compression [3.2]. 1 unit of ordinate is 10^3 lbf/in^2 or 6.895 MN/m^2.

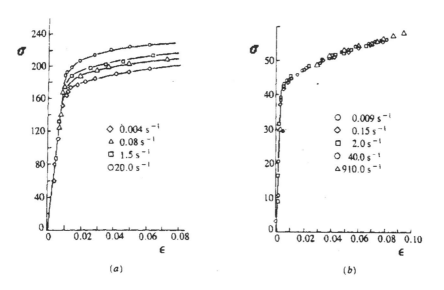

Figure 3.4   Dynamic uniaxial compression tests [3.12].   (a) titanium 6Al-4V. (b) aluminium 6061-T6, 1 unit of ordinate is 10^3 Ibf/in^2 or 6.895 MN/m^2.

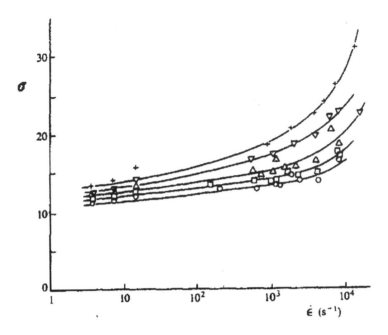

Figure 3.5   Dynamic uniaxial compressive tests on work-hardened aluminium [3.13] at various constant strains (○: $\varepsilon = 0.01$; □: $\varepsilon = 0.02$; △: $\varepsilon = 0.04$; ◊: $\varepsilon = 0.08$; +: $\varepsilon = 0.16$). 1 unit of ordinate is 10^3 lbf/in^2 or 6.895 MN/m^2.

It is evident that the behaviour of the aluminium 6061-T6 in Figure 3.4(*b*) is essentially strain-rate-insensitive, which possibly accounts for the good agreement between rigid, perfectly plastic theoretical methods and the corresponding experimental results for beams (e.g., Figure 2.6) and other structures. Nevertheless, the experimental compressive test results on a work-hardened aluminium by Hauser [3.13] presented in Figure 3.5 do show a dependence on strain rate, which, however, is noticeably less for a given change of strain rate than the mild steel in Figures 3.1 and 3.3.

3.2.3   Tension

Manjoine [3.14] reported in 1944 on some tensile tests which he conducted on a low carbon steel using a high-speed tension machine. His experimental results indicated that the lower yield stress and the ultimate tensile stress increased with increase in strain rate, the increase being more significant for the lower yield stress.

Campbell and Cooper [3.15] have examined the dynamic tensile behaviour of low-carbon mild steel specimens up to fracture, as shown in Figure 3.6. A summary of their results is presented in Figure 3.7. The upper and lower yield stresses increase with increase in strain rate, as observed by Manjoine [3.14]. However, the ultimate tensile stress also increases, but more slowly. Thus, the reduction in the importance of material strain hardening with increase in strain rate observed in Figure 3.3 for dynamic compression is also found in mild steel with large tensile strains and large strain rates. Indeed, it appears, apart from the upper yield stress, that this material behaves as a perfectly plastic material with little or no strain hardening at high strain rates.

It is interesting to note from the results of Campbell and Cooper [3.15] shown in Figure 3.6 that the fracture strain decreases with increase in strain rate. In other words, the material becomes more brittle at higher strain rates [3.16].

Many authors have conducted dynamic tensile tests since the early experiments of Manjoine [3.14]. Symonds [3.17] has gathered together the dynamic lower yield or flow stresses for mild steel which have been recorded over a thirty-year period in a number of laboratories. These results are presented

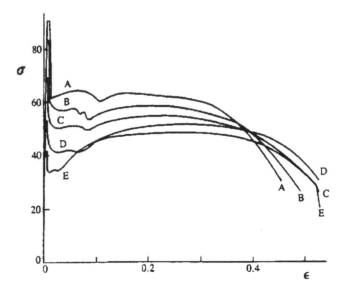

Figure 3.6   Dynamic uniaxial tensile tests on mild steel[3.15] at various mean plastic strain rates. A: $\dot{\varepsilon}$ = 106 s^{-1}; B: $\dot{\varepsilon}$ = 55 s^{-1}, C: $\dot{\varepsilon}$ = 2 s^{-1}; D: $\dot{\varepsilon}$ = 0.22 s^{-1}; E: $\dot{\varepsilon}$ = 0.001 s^{-1}. 1 unit of ordinate is 10^3 lbf/in^2 or 6.895 MN/m^2.

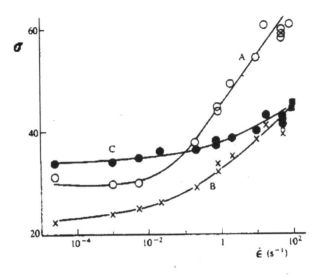

Figure 3.7 Variation of strength with strain rate for the dynamic uniaxial tensile behaviour of mild steel.[3.15] A: upper yield stress; B: lower yield stress; C: ultimate tensile stress. 1 unit of ordinate is 1 kg/mm^2 or 9.807 MN/m^2.

in Figure 3.8 and reveal a trend of increasing flow stress with increase of strain rate over a wide range of strain rates. The data have considerable scatter which is, undoubtedly, related to the range of different mild steels having different grain sizes and heat treatments, and the variety of testing machines and data recording equipment.

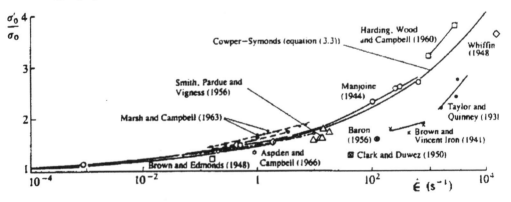

Figure 3.8 Variation of dynamic uniaxial lower yield stress of mild steel with strain-rate.[3.17] All test results were obtained in dynamic tension except the dynamic compression tests of Marsh and Campbell (1963), Aspden and Campbell (1966) and Whiffin (1948). Equation (3.3) is the Cowper-Symonds empirical relation which is introduced in §3.3.2 ($D = 40.4$ s^{-1} and q = 5).

3.3   Constitutive equations

3.3.1   Introduction

Many different constitutive equations for the strain-rate-sensitive behaviour of materials have been proposed in the literature [3.6-3.10]. Careful experimental work is required in order to generate the various coefficients in these constitutive equations. Many authors have cast light on the characteristics of constitutive equations which is indispensable for guiding experimental test programmes. However, there is still considerable uncertainty and lack of reliable data even for some common materials. For example, some authors have observed that aluminium 6061-T6 is strain-rate-sensitive [3.18,3.19], while others have not [3.12, 3.20, 3.21]. In addition, insufficient data are available for the behaviour of materials under dynamic biaxial loadings and for the influence of generalised stresses (i.e., bending moments, membrane forces, etc.). Thus, a constitutive relation which places relatively small demands on experimental test programmes, yet gives reasonable agreement with the available experimental data,is discussed briefly in this chapter

3.3.2   Cowper-Symonds constitutive equation

Cowper and Symonds[3.22] suggested the constitutive equation

$$\dot{\varepsilon} = D\left(\frac{\sigma_0'}{\sigma_0} - 1\right)^q, \quad \sigma_0' \geq \sigma_0, \tag{3.1}$$

where σ_0' is the dynamic flow stress at a uniaxial plastic strain rate $\dot{\varepsilon}$, σ_0 is the associated static flow stress and D and q are constants for a particular material. Now, equation (3.1) .may be written as

$$\log_e \dot{\varepsilon} = q \log_e\left(\frac{\sigma_0'}{\sigma_0} - 1\right) + \log_e D \tag{3.2}$$

which is the equation of a straight line with $\log_e(\sigma_0'/\sigma_0 - 1)$ plotted against $\log_e \dot{\varepsilon}$. The parameter q is the slope of this straight line, while the intercept on the ordinate is $\log_e D$.

Equation (3.1) may be recast in the form

$$\frac{\sigma_0'}{\sigma_0} = 1 + \left(\frac{\dot{\varepsilon}}{D}\right)^{1/q},$$ (3.3)

which, with $D = 40.4$ s^{-1} and $q = 5$, gives reasonable agreement with the experimental data for mild steel assembled by Symonds[3.17] and shown in Figure 3.8. There is considerable scatter of the experimental data in Figure 3.8, as remarked in §3.2.3. However, from an engineering viewpoint, equations (3.1) to (3.3) do provide a reasonable estimate of the strain rate sensitive uniaxial behaviour of mild steel.

Table 3.1   Coefficients of equation (3.3) for various materials

Material	D (s^{-1})	q	Reference
Mild steel	40.4	5	Cowper and Symonds [3.22]
Aluminium alloy	6500	4	Bodner and Symonds[3.4]
a-Titanium(Ti 50A)	120	9	Symonds and Chon[3.23]
Stainless steel 304	100	10	Forrestal and Sagartz[3.24]

The coefficients D and q in the Cowper-Symonds equation have also been determined for the materials listed in Table 3.1.

It is interesting to note from equation (3.3), that $\sigma_0' = 2\sigma_0$ when $\dot{\varepsilon} = D$, regardless of the value of q, as indicated in Figure 3.9. Thus, the dynamic flow stress of mild steel is doubled at a strain rate of 40.4 s^{-1}, while a strain rate of 6500 s^{-1} is required to double the dynamic flow stress of an aluminium alloy. This large strain rate for an aluminium alloy contributes to the difficulty of detecting material strain-rate-sensitive effects and is partly responsible for the lack of agreement between experimental tests which was noted in §3.3.1. In fact, it was observed[3.21] that equation (3.3), with $D = 1288\,000$ s^{-1} and $q = 4$, passes through the average of the widely scattered experimental data on aluminium 6061-T6. However, it is evident from Figure 3.10, which shows the strain rate sensitivity of various aluminium alloys, that it is vital to know the exact material specification.

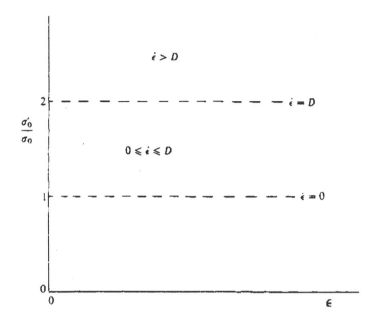

Figure 3.9   Cowper-Symonds rigid, perfectly plastic strain-rate-sensitive relation according to equation (3.3) with any values of D and q.

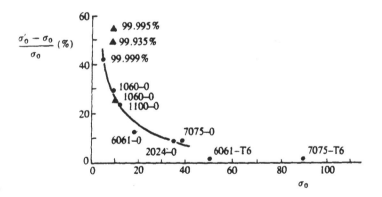

Figure 3.10   Percentage increase in the plastic flow stress at $\varepsilon = 0.06$ between $\dot{\varepsilon} = 10^{-3}\ \text{s}^{-1}$ and $\dot{\varepsilon} = 10^{3}\ \text{s}^{-1}$ for various aluminium alloys [3.7, 3.11]. σ_0 is the static plastic flow stress at $\dot{\varepsilon} = 10^{-3}\text{s}^{-1}$ and σ_0' is the dynamic flow stress at $\dot{\varepsilon} = 10^{3}\text{s}^{-1}$. 1 unit on abscissa is $10^{3}\ \text{lbf/in}^{2}$ or 6.895 MN/m^{2}.

The Cowper-Symonds constitutive equation (3.3) may be expressed in the form

$$\frac{\sigma'_e}{\sigma_0} = 1 + \left(\frac{\dot{\varepsilon}_e}{D}\right)^{1/q} \tag{3.4}$$

where

$$\sigma'_e = \left[\left(\sigma'_x - \sigma'_y\right)^2 + \left(\sigma'_y - \sigma'_z\right)^2 + \left(\sigma'_z - \sigma'_x\right)^2 + 6\left(\tau'^2_{xy} + \tau'^2_{yz} + \tau'^2_{zx}\right)\right]^{1/2} \Bigg/ \sqrt{2} \tag{3.5}^4$$

is the equivalent, or effective, dynamic flow stress and

$$\dot{\varepsilon}_e = \sqrt{2}\left[\left(\dot{\varepsilon}_x - \dot{\varepsilon}_y\right)^2 + \left(\dot{\varepsilon}_y - \dot{\varepsilon}_z\right)^2 + \left(\dot{\varepsilon}_z - \dot{\varepsilon}_x\right)^2 + 6\left(\dot{\varepsilon}^2_{xy} + \dot{\varepsilon}^2_{yz} + \dot{\varepsilon}^2_{zx}\right)\right]^{1/2} \Bigg/ 3 \tag{3.6}^4$$

is the associated equivalent, or effective, strain rate. The constants D and q are obtained from dynamic uniaxial, or pure shear, tests on the material and σ_0 is the corresponding static uniaxial flow stress.

Equation (3.5) reduces to $\sigma'_e = \sigma'_x$ in the uniaxial case, when all the stress components are zero except σ'_x. If the material in a dynamic uniaxial test with $\dot{\varepsilon}_x > 0$ obeys the incompressibility relation $(\dot{\varepsilon}_x + \dot{\varepsilon}_y + \dot{\varepsilon}_z = 0)$, then $\dot{\varepsilon}_x = \dot{\varepsilon}_z = -\dot{\varepsilon}_x/2$ and $\dot{\varepsilon}_e = \dot{\varepsilon}_x$.

Reference [3.25] contains some further discussion of the phenomenon of material strain rate sensitivity largely from the perspective of the dynamic plastic behaviour of structures. There is a vast literature now available on this general topic much of which has been cited over the years in the Applied Mechanics Reviews.

---

[4] The equivalent or effective stress and strain rate may also be written with the aid of the summation convention[3.26] in the tensorial forms $\sigma'_e = \left(3 S'_{ij} S'_{ij}/2\right)^{1/2}$ and $\dot{\varepsilon}_e = \left(2\dot{\varepsilon}_{ij}\dot{\varepsilon}_{ij}/3\right)^{1/2}$, respectively, where S'_{ij} is the dynamic deviatoric flow stress tensor, i and j have the range 1 to 3 when 1, 2 and 3 are identified with the cartesian coordinates, x, y and z, respectively. Note that $\sigma'_e = (3J_2)^{1/2}$, where the second invariant of the deviatoric stress tensor $J_2 = S'_{ij} S'_{ij}/2$.

REFERENCES

[3.1]   Campbell, J. D., *Dynamic Plasticity of Metals,* Springer, Vienna and New York, 1972.

[3.2]   Marsh, K. J. and Campbell, J. D., The effect of strain rate on the post-yield flow of mild steel, *Journal of the Mechanics and Physics of Solids,* **11**, 49-63, 1963.

[3.3]   Symonds, P. S. and Jones, N., Impulsive loading of fully clamped beams with finite plastic deflections and strain rate sensitivity, *International Journal of Mechanical Sciences,* **14**, 49-69, 1972.

[3.4]   Bodner, S. R. and Symonds, P. S., Experimental and theoretical investigation of the plastic deformation of cantilever beams subjected to impulsive loading, *Journal of Applied Mechanics,* **29**, 719-28, 1962.

[3.5]   Perrone, N., Crashworthiness and biomechanics of vehicle impact, *Dynamic Response of Biomechanical Systems,* N. Perroneed, Ed., ASME, 1-22, 1970.

[3.6]   Perzyna, P., Fundamental problems in viscoplasticity, *Advances in Applied Mechanics,* Academic Press, vol. 9, 243-377, 1966.

[3.7]   Campbell, J. D., Dynamic plasticity: macroscopic and microscopic aspects, *Materials Science and Engineering,* **12**, 3-21, 1973.

[3.8]   Duffy, J., Testing techniques and material behavior at high rates of strain, *Mechanical Properties at High Rates of Strain,* J. Harding, Ed., Institute of Physics Conference Series No. 47, 1-15, 1979.

[3.9]   Malvern, L. E., Experimental and theoretical approaches to characterisation of material behaviour at high rates of deformation, *Mechanical Properties at High Rates of Strain,* J. Harding, Ed., Institute of Physics Conference Series No. 70, 1-20, 1984.

[3.10] Goldsmith, W., *Impact,* Edward Arnold, London, 1960.

[3.11]   Nicholas, T., Material behavior at high strain rates, *Impact Dynamics,* J. A. Zukas, *et al.,* Ed., John Wiley, New York, ch. 8, 277-332, 1982.

[3.12]   Maiden, C. J. and Green, S. J., Compressive strain-rate tests on six selected materials at strain rates from 10^{-3} to 10^4 in/in/sec, *Journal of Applied Mechanics,* **33**, 496-504, 1966.

[3.13]   Hauser, F. E., Techniques for measuring stress-strain relations at high strain rates, *Experimental Mechanics,* **6**, 395-402, 1966.

[3.14]   Manjoine, M. J., Influence of rate of strain and temperature on yield stresses of mild steel, *J. of Applied Mechanics,* **11**, 211-18, 1944.

[3.15]   Campbell, J. D. and Cooper, R. H., Yield and flow of low-carbon steel at medium strain rates, In *Proceedings of the Conference on the Physical Basis of Yield and Fracture,* Institute of Physics and Physical Society, London, 77-87, 1966.

[3.16]   Jones, N., Some comments on the modelling of material properties for dynamic structural plasticity, In *Proceedings* 4th *International Conference on the Mechanical Properties of Materials at High Rates of Strain*, Oxford, Institute of Physics Conference Series No. 102, 435-445, 1989.

[3.17]   Symonds, P. S., *Survey of Methods of Analysis for Plastic Deformation of Structures under Dynamic Loading,* Brown University, Division of Engineering Report BU/NSRDC/1-67, 1967.

[3.18]   Ng, D. H. Y., Delich, M. and Lee, L. H. N., Yielding of 6061-T6 aluminium tubings under dynamic biaxial loadings, *Experimental Mechanics,* **19**, 200-6, 1979.

[3.19]   Hoge, K. G., Influence of strain rate on mechanical properties of 6061-T6 aluminium under uniaxial and biaxial states of stress, *Experimental Mechanics,* **6**, 204-11, 1966.

[3.20]   Gerard, G. and Papirno, R., Dynamic biaxial stress-strain characteristics of aluminium and mild steel, *Transactions of the American Society for Metals,* **49**, 132-48, 1957.

[3.21]   Jones, N., Some remarks on the strain-rate sensitive behaviour of shells, *Problems of Plasticity,* A. Sawczuk, Ed., Noordhoff, Groningen, vol. 2, 403-7, 1974.

[3.22]   Cowper, G. R. and Symonds, P. S., *Strain Hardening And Strain-Rate Effects In The Impact Loading Of Cantilever Beams*, Brown University Division of Applied Mathematics Report No. 28, September, 1957.

[3.23]   Symonds, P. S. and Chon, C. T., Approximation techniques for impulsive loading of structures of time-dependent plastic behaviour with finite-deflections, *Mechanical Properties of Materiais at High Strain Rates,* Institute of Physics Conference Series No.21, 299-316, 1974.

[3.24]   Forrestal, M. J. and Sagartz, M. J., Elastic-plastic response of 304 stainless steel beams to impulse loads, *Journal of Applied Mechanics,* **45**, 685-7, 1978.

[3.25]   Jones, N., *Structural Impact*, Cambridge University Press, Cambridge, 1989, paperback edition (1997).

[3.26]   Fung, Y. C., *A First Course in Continuum Mechanics*, Prentice-Hall, Englewood Cliffs, N.J., 1969.

4. DYNAMIC AXIAL CRUSHING

4.1   Introduction

The structural members examined in the previous chapters responded in a stable manner when subjected to dynamic loads. However, in practice, dynamic loads may cause an unstable response. This chapter examines the dynamic progressive buckling phenomenon [4.1], which is illustrated in figure 4.1 for a circular tube.

Figure 4.1   Static and dynamic axial crushing of a thin-walled mild steel cylindrical shell with a mean radius (R) of 27.98 mm, mean wall thickness (H) of 1.2 mm and an initial axial length (L) of 178 mm. The top three wrinkles developed as a result of static loading, while the remaining wrinkles were produced when the tube was struck by a mass (G) of 70 kg travelling at an impact velocity of 6.51 m/s.

A thin-walled cylindrical shell, or tube, when subjected to a static axial load, as shown in figure 4.2, may have force-axial displacement characteristics similar to those in figure 4.3(a). It is evident that the tube exhibits an unstable behaviour after reaching the first peak load at point A. Most structural designs are based on a load equal to this peak load divided by a safety factor. The magnitude of this safety factor is selected by taking into account the slope AB of the load-deflection behaviour (post-buckling characteristics). However, thin-walled circular tubes are used in many practical situations to absorb impact energy. Indeed, Pugsley [4.2] examined the axial impact of thin-walled circular and square tubes in order to study the structural crashworthiness of railway coaches. In this circumstance, the total axial displacement of a tube exceeds considerably the displacement associated with the load at B in figure 4.3(a). Thus, an entirely different approach is required from that employed normally to examine the plastic buckling of structures.

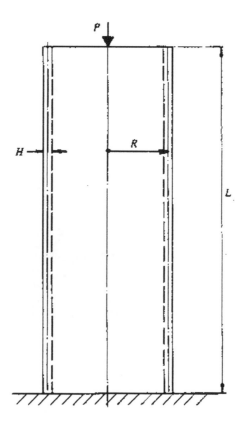

Figure 4.2   A thin-walled cylindrical shell subjected to an axial crushing force P.

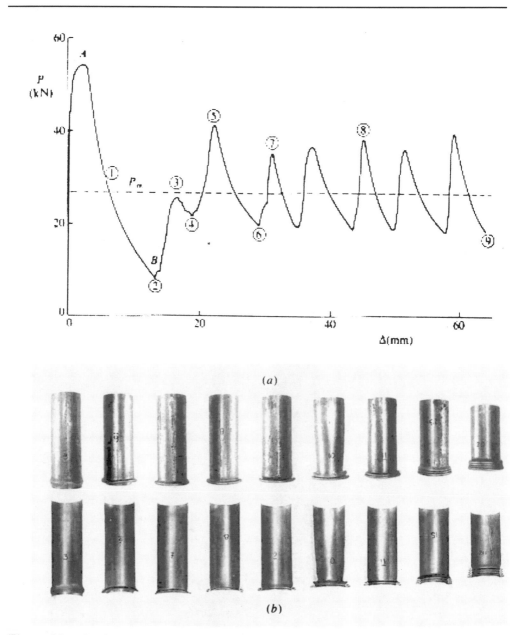

(a)

(b)

Figure 4.3   Static axial crushing behaviour of a thin-walled mild steel circular tube with a mean radius (*R*) of 27.98 mm, mean wall thickness (*H*) of 1.2 mm and an initial axial length (*L*) of 178 mm. (a) Axial force versus axial crushing distance. (b) Photographic record of the development of wrinkles during axial crushing. The photographs (from left to right) refer to the numbers 1 to 9 in figure 4.3(a). The upper row gives the outside views, while the lower row shows the specimens cut open across a diameter.

It is evident from figure 4.3(a) that the load-displacement behaviour exhibits a repeated pattern. In fact, each pair of peaks in figure 4.3(a) is associated with the development of a wrinkle or buckle in figure 4.3(b). Usually, these wrinkles, or buckles, develop sequentially from one end of a tube so that the phenomenon is known as progressive buckling. The most efficient use of the tube material occurs when as much as possible is crushed, as indicated in figure 4.4 for a thin-walled tube with a square cross-section. For convenience, designers often ignore the fluctuations in the load-displacement characteristics and use a mean value (P_m), as indicated in figure 4.3(a). Incidentally, an ideal energy-absorbing device is defined, for some purposes, as one which has a constant resistance and, therefore, offers a constant deceleration throughout the entire stroke [4.4].

Figure 4.4   A thin-walled mild steel tube with a square cross-section before and after static axial crushing [4.3].

The axial impact behaviour of circular tubes with low velocities (up to tens of metres per second for metal tubes) is taken [4.1] as quasi-static, and the influence of inertia forces is, therefore, ignored as discussed in Chapter 2. This is a reasonable simplification when the striking mass *(G)* is much larger than the mass of a tube *(m)*. The axial inertia force of a striking mass is G$\ddot{u}$, where $\ddot{u}$ *is* the axial deceleration during the impact event. If the axial velocity-time history is continuous at the interface between a striking mass and the end of a tube, then the axial inertia force in the tube is of order *m*$\ddot{u}$, which is negligible compared with G$\ddot{u}$ when *m* << *G*.

The response of the impact problem described above is controlled by the phenomenon associated with static progressive buckling. Nevertheless, it is called dynamic progressive buckling in this chapter because material strain rate effects are important for a strain-rate-sensitive material [4.1]. If the influence of inertia effects of the tube are important in a practical problem, with larger axial impact velocities, then the phenomenon is known as dynamic plastic buckling which is discussed in Chapter 9 of Reference [4.5].

Figure 4.5   Axially crushed circular tube test specimens [4.1] Axisymmetric, or concertina, deformation mode on left and non- axisymmetric, or diamond deformation modes at centre and right.

4.2 Static axial crushing of a circular tube

4.2.1 Introduction

A thin-walled circular tube of mean radius R and thickness H, when subjected to an axial force, as shown in figure 4.2, may develop either axisymmetric buckles similar to those in figure 4.1, or a non-axisymmetric (diamond) pattern, which is indicated in figure 4.5. Various theoretical methods predict that thicker tubes with R/H < 40-45, approximately, deform axisymmetrically, while the thinner tubes, with larger values of R/H, buckle into a non-axisymmetric mode [4.1]. However, some tubes may switch, during a test, from an axisymmetric deformation mode into a diamond pattern [4.6].

Alexander [4.7] and Pugsley and Macaulay [4.8] presented static theoretical analyses for the axisymmetric and non-axisymmetric behaviour of circular tubes, respectively. The axisymmetric solution of Alexander [4.7] is outlined in the next section.

4.2.2 Axisymmetric crushing

4.2.2.1   Introduction

An approximate theoretical analysis for the axially loaded thin-walled circular tube in figure 4.2 was obtained by Alexander [4.7]. He assumed that the tube was made from a rigid, perfectly plastic material and used the simplified axisymmetric deformation pattern with plastic hinges which is illustrated in figure 4.6[1]. This collapse pattern is an idealisation of the actual behaviour, since it is evident from figure 4.3(b) that the profiles of the wrinkles are curved rather than straight. Nevertheless, the theoretical predictions have some value for design purposes and have been extended in recent years to cater for various effects, giving improved agreement with experimental results. Moreover, this analysis illustrates the general approach which is employed for some dynamic progressive buckling problems and, therefore, has value as an introduction to the topic.

It is evident from figure 4.6 that the work $P_m \times 2l$, which is expended by a constant or mean external force P_m to develop and flatten completely one axisymmetric wrinkle or buckle, is equal to the internal energy dissipated due to plastic deformation in the tube. The internal energy dissipation rate is derived in the next section.

---

[1] The dimension l in Figure 4.6 is taken as constant, regardless of the number of wrinkles.

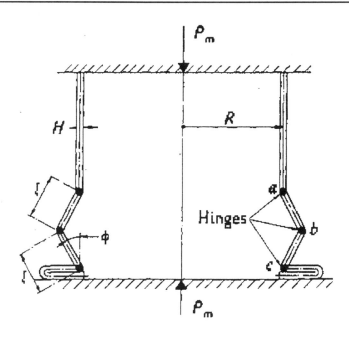

Figure 4.6   Idealised axisymmetric, or concentric, crushing mode for an axially compressed cylindrical shell.

4.2.2.2   Internal energy dissipation

The total plastic energy absorbed by the two stationary axisymmetric plastic hinges at a and c in figure 4.6 during the formation of one wrinkle is

$$D_1 = 2 \times 2\pi R M_o/2 \qquad (4.1)$$

where the plastic collapse moment for the cross-section (per unit circumferential length) is

$$M_o = (2\sigma_o \sqrt{3})H^2/4. \qquad (4.2)^2$$

---

[2] Alexander [4.7] used the von Mises yield condition and assumed that the tube was in a state of plane strain. This is strictly true only for the plastic hinges at a and c in Figure 4.6 because $\varepsilon_\theta \geq 0$ for the axisymmetric hinge at b.

The radial position of the axisymmetric plastic hinge at b increases from R to $R + l$ during the formation of one complete wrinkle. Thus, the energy absorbed by the axisymmetric central hinge b in figure 4.6 during an incremental change $d\phi$ is

$$dD_2 = 2\pi(R + l\sin\phi)M_0\,(2d\phi),$$

which gives the total energy dissipated as

$$D_2 = \int_0^{\pi/2} 4\pi(R + l\sin\phi)M_0\,d\phi$$

or

$$D_2 = 4\pi M_0(R\pi/2 + l)\,. \tag{4.3}$$

It is evident from figure 4.6 that the axisymmetric portions ab and bc of a wrinkle are stretched circumferentially between ϕ and $\phi+d\phi$ with a mean[3] engineering strain increment

$$d\varepsilon_\theta = \frac{2\pi\{(l/2)\sin(\phi + d\phi)\} - 2\pi\{(l/2)\sin\phi\}}{2\pi R},$$

which, expanding the $sin(\phi+d\phi)$ term using standard trigonometric relations, becomes

$$d\varepsilon_\theta = l\cos\phi\,d\phi/2R \tag{4.4}$$

when $sind\phi{\to}d\phi$ and $cosd\phi{\to}1$. Thus, the energy absorbed in circumferential stretching during an incremental change from ϕ to $\phi+d\phi$ is

$$dD_3 = \sigma_0\,d\varepsilon_\theta 2lH2\pi R,$$

which, using equation (4.4), may be written

$$D_3 = \int_0^{\pi/2} \sigma_0(l\cos\phi\,d\phi)\,2lH\pi$$

or

$$D_3 = 2\sigma_0 l^2 H\pi\,. \tag{4.5}[3]$$

---

[3] The assumption of a mean circumferential strain is removed in reference [4.1] and equation (4.5) is then multiplied by $(1 + l/3R)$.

The total energy absorbed during the development of one complete wrinkle in a thin-walled circular tube is

$$D_T = D_1 + D_2 + D_3 \tag{4.6}$$

which, using equations (4.1), (4.3) and (4.5), becomes

$$D_T = 4\pi M_o(\pi R + l) + 2\pi \sigma_o l^2 H$$

or

$$D_T = 2\pi \sigma_o H^2(\pi R + l)/\sqrt{3} + 2\pi \sigma_o l^2 H \tag{4.7}$$

when eliminating M_o with equation (4.2).

<u>4.2.2.3   Axial crushing force</u>

Now, in order to conserve energy, the total external work done by a constant axial force during the formation and complete flattening of one wrinkle $(P_m \times 2l)$ equals the internal work according to equation (4.7). Thus,

$$P_m 2l = 2\pi \sigma_o H\{H(\pi R + l)/\sqrt{3} + l^2\},$$

or

$$P_m/\sigma_o = \pi H\{H(\pi R/l + 1)/\sqrt{3} + l\} \tag{4.8}$$

The axial length $l=ab=bc$ between the axisymmetric hinges in figure 4.6 is unknown but may be obtained by minimising the axial crushing force, or $dP_m/dl=0$, which gives

$$H(-\pi R/l^2)/\sqrt{3} + 1 = 0,$$

or

$$l = (\pi RH/\sqrt{3})^{\frac{1}{2}}. \tag{4.9}$$

Substituting equation (4.9) into equation (4.8) gives the axial crushing force

$$P_m/M_o = 4(3)^{1/4} \pi^{3/2} (R/H)^{1/2} + 2\pi, \tag{4.10a}$$

or

$$P_m/M_o = 29.31 (R/H)^{1/2} + 6.31, \tag{4.10b}$$

which is identical to Alexander's theoretical prediction [4.7] when M_o is defined by equation (4.2).

Equation (4.l0b) was obtained using the collapse mechanism in figure 4.6 which assumes that the convolutions, or wrinkles, form on the outside of a tube. Alexander [4.7] repeated the theoretical analysis and found that the mean crushing force for a tube with convolutions which form internally instead of externally is

$$P_m/M_o = 4(3)^{1/4} \pi^{3/2} (R/H)^{1/2} - 2\pi, \qquad (4.11)$$

Alexander assumed that the average of equations (4.10a) and (4.11)

$$P_m/M_o = 4(3)^{1/4} \pi^{3/2} (R/H)^{1/2} \qquad (4.12a)^4$$

or

$$P_m = 2(\pi H)^{3/2} R^{1/2} \sigma_o /3^{1/4}, \qquad (4.12b)$$

offers a reasonable approximation to the actual crushing force.

4.2.2.4   Comment

The theoretical procedure in §§ 4.2.2.2 and 4.2.2.3 does not satisfy the upper bound theorem of plasticity which is discussed in Chapter 1. The collapse mechanism in figure 4.6 requires finite displacements and finite rotations, whereas the static plastic collapse theorems for a perfectly plastic material were developed using the principle of virtual velocities for infinitesimal displacements.

4.3   Dynamic axial crushing of a circular tube

4.3.1   Introduction

The theoretical analysis, which is outlined in § 4.2, was developed for the axisymmetric crushing of a cylindrical tube when subjected to a static axial load. However, it is noted in § 4.1 that this analysis also describes the dynamic progressive buckling of a circular tube which may be considered as a quasi-static problem (see Chapter 2). Although inertia effects may be neglected,

---

[4] Alexander [4.7] further modifies the coefficient in equation (4.12a) by assuming that the crushing force is the average of the values predicted by equation (4.12a) and equation (4.8) without the $\pi H^2/\sqrt{3}$ term and with $l = \pi(2RH)^{1/2}/2\{3(1 - v^2)\}^{1/4}$ for the linear elastic case.

the influence of material strain rate sensitivity must be retained for many materials (as discussed in Chapter 3). Thus, if a circular tube is made from a strain-rate-sensitive material, then it is necessary to modify the plastic flow stress in equation (4.12b) in order to cater for the enhancement of the flow stress with strain rate.

4.3.2   Influence of strain rate sensitivity

Equation (3.3), known as the Cowper-Symonds constitutive equation, gives dynamic flow stresses which agree reasonably well with the dynamic uniaxial tension and compression test results on several materials. Now, if the static flow stress (σ_0) in equation (4.12b) is replaced by the dynamic flow stress according to equation (3.3), then

$$P_m = 2(\pi H)^{3/2} R^{1/2} \, \sigma_0 \, [1+(\dot{\varepsilon}/D)^{1/q}]/3^{1/4}, \tag{4.13}$$

where D and q are constants for the tube material and are given in Table 3.1.

The strain rate ($\dot{\varepsilon}$) in equation (4.13) is taken as constant, although it varies spatially and temporally during an impact event. An estimate is now made for $\dot{\varepsilon}$ using an approximate procedure which is suggested in Reference 4.1.

Equation (4.4) predicts that the mean circumferential strain in a completely flattened buckle of the circular tube in figure 4.6 is

$$\varepsilon_\theta \cong l/2R, \tag{4.14}$$

which may also be obtained by inspection. The time T to flatten completely a single buckle is now required in order to estimate the mean circumferential strain rate as $\dot{\varepsilon}_\theta = \varepsilon_\theta / T$. It is assumed that the axial velocity at the struck end of a tube varies linearly with time from the impact velocity V_o at $t = 0$ until motion ceases when $t > T$ after the development of several wrinkles. This gives a constant deceleration and a constant axial force which is consistent with a mean crushing force P_m. Now the time required to form the first wrinkle, or convolution, is $T = 2l/V_o$ when taking the impact velocity V_o to remain constant which would be reasonable when a large number of wrinkles are formed in the tube. Therefore, the average strain rate $\dot{\varepsilon}_\theta = \varepsilon_\theta / T$ is

$$\dot{\varepsilon}_\theta = V_0/4R, \tag{4.15}$$

which is used as an approximation for $\dot{\varepsilon}$ in equation (4.13). Equation (4.13) may now be written

$$P_m = 2(\pi H)^{3/2} R^{1/2} \, \sigma_o \, [1+(V_o/4RD)^{1/q}]/3^{1/4}, \tag{4.16}$$

where V_o is the axial impact velocity at the end of a circular tube. If $D \to \infty$ for a strain-rate-insensitive material, then equation (4.16) reduces to equation (4.12b) as expected.

The lack of accuracy in the simplified estimates for the mean strain rate $\dot{\varepsilon}_\theta = V_o/4R$ is not as important as appears at first sight because the term $(V_o/4RD)^{1/q}$ is highly non-linear owing to the relatively large values for q in Table 3.1. For example, if the actual mean strain rate in a steel tube with $q = 5$ were twice as large as estimated by equation (4.15), then the actual strain-rate-sensitive term in equation (4.16) would be $(2)^{1/5}(V_o/4RD)^{1/5}$ or $1.15(V_o/4RD)^{1/5}$, which is 15 per cent larger.

4.3.3   Structural effectiveness and solidity ratio

In order to assist in the presentation of experimental results and theoretical predictions for the axial crushing of thin-walled sections, Pugsley [4.2] introduced the two dimensionless ratios known as structural effectiveness and solidity ratio.

The structural effectiveness is defined as

$$\eta = P_m/A\sigma_1, \tag{4.17}$$

where P_m is the mean axial crushing force, A is the cross-sectional area of the thin-walled cross-section and σ_1 is a characteristic stress. If $\sigma_1 = \sigma_o$, where σ_o is the plastic flow stress, then $A\sigma_o$ is the squash load required to cause uniform plastic flow due to an axial force. The structural effectiveness $\eta = P_m/A\sigma_o$ is, therefore, a ratio between the mean crushing force and the squash load. For the particular case of a thin-walled circular cylindrical tube, $A = 2\pi RH$ and

$$\eta = P_m/2\pi RH\sigma_o. \tag{4.18}$$

The solidity ratio, or relative density, is defined as

$$\phi = A/A_c \tag{4.19}$$

where A_c is the cross-sectional area which is enclosed by the cross-section. Clearly, $\phi \to 0$ represents a section with very thin walls. For the particular case of a thin-walled circular cylindrical tube, $A = 2\pi RH$ and $A_c = \pi R^2$ so that equation (4.19) becomes

$$\phi = 2\pi RH/\pi R^2,$$

or

$$\phi = 2H/R. \tag{4.20}$$

The dimensionless parameters (4.18) and (4.20) may be used to write equations (4.12b) and (4.16) in the form

$$\eta = (\pi \phi/2\sqrt{3})^{1/2} \tag{4.21}$$

and

$$\eta = (\pi \phi/2\sqrt{3})^{1/2}[1+(V_o/4RD)^{1/q}], \tag{4.22}$$

respectively. The additional dimensionless parameter $V_o/4RD$ in equation (4.22) caters for the phenomenon of material strain rate sensitivity. If $V_o/4RD = 1$, then the correction factor for material strain rate sensitivity equals 2 and the mean dynamic progressive buckling force is double the corresponding static value regardless of the value of q.

4.4   Comparison with Experimental Results on Static Crushing

Some experimental results for the static axial crushing of thin-walled cylindrical shells are presented in figure 4.7 in terms of the dimensionless parameters η and ϕ which were introduced in the last section. The considerable scatter in this figure is due to any initial imperfections in the tubes, different testing arrangements, different material properties, different static loading rates, and varying numbers of fully developed wrinkles, etc. Nevertheless, the theoretical predictions of equation (4.21) tends to form a lower bound on the dimensionless mean crushing force.

A more realistic collapse profile than that shown in figure 4.6 was introduced in References [4.1] and [4.9] by recognising an effective crushing distance.   Incorporating this effect in the basic equations leads to a dimensionless crushing force

$$\eta = 3.36 \left(1 + 0.29 \, \phi^{1/2}\right)/\left(3.03\phi^{-1/2} - 1\right). \tag{4.23}$$

which tends towards an upper bound of the experimental results in figure 4.7.

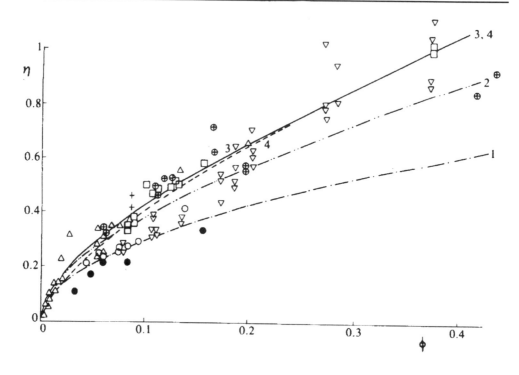

Figure 4.7   Static axial crushing of thin-walled cylindrical shells. — · — 1:equation (4.21).
———— 3: equation (4.23). — ·· — ·· 2 and - - - - 4 are defined in figure 9.8 of
Reference [4.5]. Experimental results: +: Reference [4.1] (mild steel). •:
Mamalis and Johnson [4.10] (aluminium 6061 T6). ⊕: Reference [4.6] (mild
steel). O: Alexander [4.7] (mild steel). Δ: Macaulay and Redwood [4.11].
Γ, ∇: taken from figure 4.4 of Reference [4.12].

4.5   Comparison with Experimental Results on Dynamic Crushing

Some experimental results for the dynamic progressive buckling of thin-
walled cylindrical shells subjected to axial impact loads are presented in figure
4.8 and reveal the importance of material strain rate sensitivity. Equation (4.22)
shows that the enhancement of the crushing load due to this phenomenon is

$$P_m^d/P_m^s = 1 + (V_o/4RD)^{1/q}, \qquad (4.24)$$

where P_m^d and P_m^s are the dynamic and static progressive buckling forces,
respectively.

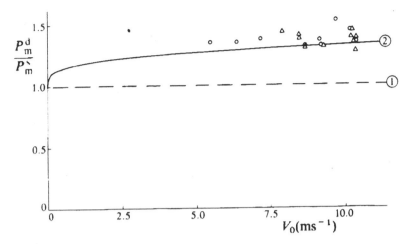

Figure 4.8 Ratio of dynamic axial crushing forces to static axial crushing forces for cylindrical shells versus axial impact velocity, V_o. - - - - 1: equation (4.24) for a strain rate insensitive material. ———— 2: equation (4.24) with D = 6844 s^{-1} and q = 3.91. Experimental results [4.1] with axisymmetric (O) and non-axisymmetric (Δ) deformations.

The parameters D and q in equation (4.24) are listed in table 3.1 for several materials. However, these constants were obtained from dynamic tests on materials with strains up to a magnitude of only a few per cent. The average strains in the specimens in figure 4.8 were of order 12 per cent [4.1] with an associated plastic flow stress similar to the ultimate tensile stress. The experimental results in figure 3.1 reveal that the material strain rate characteristics for mild steel are sensitive to the magnitude of strains larger than about 2-3 per cent. It transpires that $D = 6844\ s^{-1}$ and $q = 3.91$ [4.3] when Campbell Cooper's [3.15] experimental data for the variation of the ultimate tensile stress with strain rate was replotted using the method in §3.3.2. Equation (4.24), with these coefficients, gives fair agreement with the experimental results in figure 4.8.

4.6   Final comments

A large literature has been published on the static and dynamic progressive buckling and dynamic plastic buckling of thin-walled structural members having various shaped cross-sections. Many of these studies assume a quasi-static response, as discussed in References [4.5] and [4.13]. Wierzbicki and Abramowicz [4.14, 4.15] have studied the crushing mechanics of thin-walled sections and some of this important work is reported in Part II of this book. These methods have been used, for example, to predict the response of top-hat and double-hat sections of interest in the automobile industry [4.16, 4.17]. If the impact velocities are larger than those which give rise to dynamic progressive buckling, or a quasi-static response (say 10-15 m/s, approximately), then transverse inertia effects become important and the phenomenon of dynamic plastic buckling develops [4.5, 4.13, 4.18]. Moreover, if the axial length of a thin-walled structural member exceeds a certain value, then the less efficient overall or global buckling mode might intervene during the response [4.19].

REFERENCES

[4.1]   Abramowicz, W. and Jones, N., Dynamic axial crushing of circular tubes, *International Journal of Impact Engineering*, **2**, 263-281, 1984.

[4.2]   Pugsley, Sir A., The crumpling of tubular structures under impact conditions, In *Proceedings of the Symposium on The Use of Aluminium in Railway Rolling Stock*, Institute of Locomotive Engineers, The Aluminum Development Association, London, England, 33-41, 1960.

[4.3]   Abramowicz, W. and Jones, N., Dynamic axial crushing of square tubes, *International Journal of Impact Engineering*, **2**, 179-208, 1984.

[4.4]   Ezra, A. A. and Fay, R. J., An assessment of energy absorbing devices for prospective use in aircraft impact situations, In *Dynamic Response of Structures*, (G. Herrmann and N. Perroneed, Eds.), Pergamon, New York, 225-246, 1972.

[4.5]   Jones, N., *Structural Impact*, Cambridge University Press, 1989; paperback edition, 1997.

[4.6]   Abramowicz, W. and Jones, N., Dynamic progressive buckling of circular and square tubes, *International Journal of Impact Engineering*, **4**, 243-270, 1986.

[4.7]   Alexander, J. M., An approximate analysis of the collapse of thin cylindrical shells under axial loading, *Quarterly Journal of Mechanics and Applied Mathematics*, **13**, 10-15, 1960.

[4.8]   Pugsley, Sir A. and Macaulay, M., The large scale crumpling of thin cylindrical columns, *Quarterly Journal of Mechanics and Applied Mathematics*, **13**, 1-9, 1960.

[4.9]   Abramowicz, W., The effective crushing distance in axially compressed thin walled metal columns, *International Journal of Impact Engineering*, **1**, 309-317, 1983.

[4.10] Mamalis, A. G. and Johnson, W., The quasi-static crumpling of thin-walled circular cylinders and frusta under axial compression, *International Journal of Mechanical Sciences*, **25**, 713-732, 1983.

[4.11] Macaulay, M. A. and Redwood, R. G., Small scale model railway coaches under impact, *The Engineer*, 1041-6, 25 Dec., 1964.

[4.12] Thornton, P. H., Mahmood, H. F. and Magee, C. L., Energy absorption by structural collapse, In *Structural Crashworthiness*, (N. Jones and T. Wierzbicki, ed.), Butterworths, London, England, 96-117, 1983.

[4.13] Jones, N., Some phenomena in the structural crashworthiness field, *International Journal of Crashworthiness*, **4**(4), 335-350, 1999.

[4.14] Wierzbicki, T. and Abramowicz, W., On the crushing mechanics of thin-walled structures, *Journal of Applied Mechanics*, **50**, 727-734, 1983.

[4.15] Wierzbicki, T., Crushing behaviour of plate intersections, In *Structural Crashworthiness*, (N. Jones and T. Wierzbicki, ed.), Butterworths, London, England, 66-95, 1983.

[4.16] White, M. D., Jones, N. and Abramowicz, W., A theoretical analysis for the quasi-static axial crushing of top-hat and double-hat thin-walled sections, *International Journal of Mechanical Sciences*, **41**, 209-233, 1999.

[4.17] White, M. D. and Jones, N., Experimental study into the energy absorbing characteristics of top-hat and double-hat sections subjected to dynamic axial crushing, *Proceedings Institution of Mechanical Engineers*, **213**(D), 259-278, 1999.

[4.18] Karagiozova, D., Alves, M and Jones, N., Inertia effects in axisymmetrically deformed cylindrical shells under axial impact, *International Journal of Impact Engineering*, **24**(10), 1083-1115, 2000.

[4.19] Abramowicz, W. and Jones, N., Transition from initial global bending to progressive buckling of tubes loaded statically and dynamically, *International Journal of Impact Engineering*, **19**(5-6), 415-437, 1997.

5. GENERAL INTRODUCTION TO STRUCTURAL CRASHWORTHINESS

5.1   Introduction

The term 'structural crashworthiness' is used to describe the impact performance of a structure when it collides with another object. A study into the structural crashworthiness characteristics of a system is required in order to calculate the forces during a collision which are needed to assess the damage to structures and the survivability of passengers in vehicles, for example. This topic embraces the collision protection of aircraft, buses, cars, trains, ships and offshore platforms, etc. [5.1-5.7] and even spacecraft [5.8]. No attempt is made to review the entire field and only that part which is related to dynamic progressive buckling introduced in the previous chapter is discussed briefly.

5.2   Elementary aspects of inelastic impact

Consider a stationary mass M_1 which is struck by a mass M_2 travelling with an initial velocity V_2, as shown in figure 5.1(a). Conservation of linear momentum demands that

$$M_2V_2 = (M_1 + M_2)V_3. \tag{5.1}$$

where V_3 is the common velocity of both masses immediately after an inelastic impact.[1] The loss of kinetic energy is, therefore,

$$K_l = M_2V_2^2/2 - (M_1 + M_2)V_3^2/2, \tag{5.2}$$

---

[1] The coefficient of restitution is taken as zero *(e = 0)*.

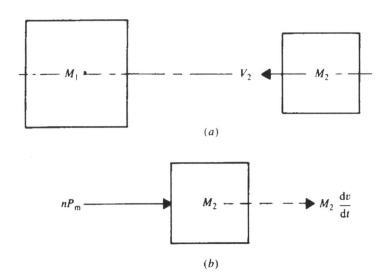

Figure 5.1   (a) A mass M_2 travelling with a velocity V_2 towards a stationary mass M_1.
(b) Horizontal forces acting on mass M_2 during the impact event.

which, when using equation (5.1) for V_3, may be recast into the form

$$K_1 = (M_2 V_2^2/2) / (1+M_2 / M_1),\tag{5.3}$$

where $M_2 V_2^2/2$ is the initial kinetic energy of the mass M_2.

Equation (5.3) gives the energy which must be absorbed by an energy-absorbing system which is interposed between the two masses M_1 and M_2, in figure 5.1(a). If the striking mass M_2 is much larger than the struck mass M_1 (i.e., $M_2/M_1 >> 1$), then $K_1 \cong 0$, and no kinetic energy is lost during the impact event. In the other extreme case of a striking mass M_2 which is much smaller than the struck mass M_1 (i.e., $M_2/M_1 << 1$), then $K_1 \cong M_2 V_2^2/2$ and all of the initial kinetic energy of the mass M_2 must be absorbed during the impact. The loss of kinetic energy for an impact between two equal masses is $K_1 = M_2 V_2^2/4$, which is one-half of the initial kinetic energy of the striking mass M_2. The variation of the dimensionless kinetic energy loss $K_1/(M_2 V_2^2/2)$ with the mass ratio M_2/M_1 is shown in figure 5.2.

Frequently, the struck mass M_1 in figure 5.1(a) is constrained to remain stationary throughout a practical impact event. In other words, $M_1/M_2 >> 1$ and equation (5.3) gives $K_1 = M_2 V_2^2/2$, as expected.

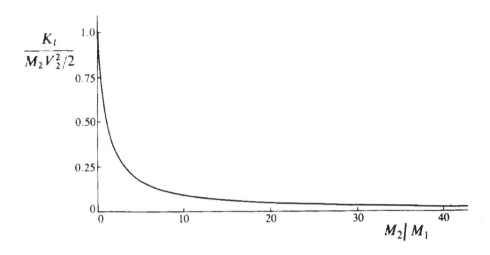

Figure 5.2   Variation of the dimensionless kinetic energy loss given by equation (5.3) with the mass ratio M_2/M_1.

Now, it is evident from figure 4.3(a) that $D_a = P_m \Delta$ is the energy which is absorbed in an axially crushed circular tube and indeed is the energy absorbed in any thin-walled structure which undergoes dynamic progressive buckling with a mean force P_m (e.g., honeycomb structure [5.9]). A nest of n axially loaded thin-walled tubes, for example, may be used to absorb the energy in the impact scenario shown in figure 5.1(a), and, therefore,

$$D_a = n\, P_m\, \Delta = K_l \qquad (5.4)$$

provided $\Delta \le \Delta_b$, where Δ_b is the bottoming-out displacement when no undeformed tube material remains to form further wrinkles. Thus,

$$n\, P_m\, \Delta = M_2\, V_2^2/2 \qquad (5.5)$$

when one end of the energy-absorbing device remains stationary throughout the impact event.

The mean dynamic crushing force for n tubes is

$$n P_m = -M_2\, dv/dt \qquad (5.6)^2$$

---

[2] A dynamic compressive crushing force P_m, which is defined as shown in Figure 5.1(b), is taken as positive throughout this chapter.

according to the free body diagram in Figure 5.1(b), with $v=V_2$ at $t=0$ and $v=0$ when the striking mass M_2 has moved a distance Δ at $t = T$, where T is the response duration. Clearly, the deceleration is

$$a = dv/dt = -nP_m/M_2 \qquad (5.7)$$

which is constant throughout motion and, therefore, the velocity-time history is

$$v = -nP_m t/M_2 + V_2 \qquad (5.8)$$

when satisfying the initial condition. Motion ceases when $v = 0$, which gives the response duration

$$T = M_2 V_2/nP_m \qquad (5.9a)$$

or

$$T = -V_2/a. \qquad (5.9b)$$

A further integration of equation (5.8) gives the displacement-time history

$$\delta = -nP_m t^2/2M_2 + V_2 t, \qquad (5.10)$$

which, substituting equation (5.9a), predicts that the total crushed distance is

$$\Delta = M_2 V_2^2/2nP_m, \qquad (5.11a)$$

or

$$\Delta = V_2^2/2a, \qquad (5.11b)$$

when motion ceases at $t = T$. Equation (5.11a) may be obtained directly from the energy balance equation (5.5). The mean crushing force is often estimated in experimental test programmes and design calculations by dividing the initial kinetic energy $(M_2 V_2^2/2)$ by the total axial crushed distance Δ. This agrees with equations (5.5) or (5.11a).

The impact scenario in figure 5.1(a) lies in a horizontal plane and is related to situations which arise during bus, car, train and ship collisions, for example. However, another important class of impacts are caused by a mass M_2 which drops vertically onto a mass M_1 with an impact velocity V_2 as shown in figure 5.3(a). In this case, equation (5.5) for the conservation of energy is replaced by

$$nP_m\Delta = M_2 V_2^2/2 + M_2 g\Delta, \qquad (5.12)$$

where $M_2 g\Delta$ is the additional potential energy of the mass M_2 which crushes the energy absorber by an amount Δ during the impact event.

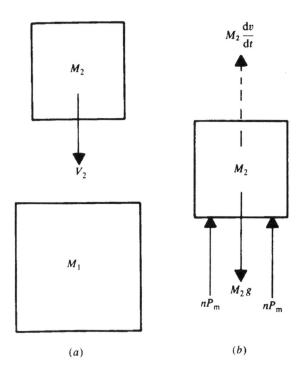

Figure 5.3   (a) A mass M_2 dropping vertically onto a stationary mass M_1 with an impact velocity V_2. (b) Vertical forces acting on a mass M_2 during the impact event.

The vertical equation of motion for the mass M_2 in figure 5.3(b) during impact is

$$nP_m + M_2 \, dv/dt - M_2 \, g = 0. \tag{5.13}$$

which predicts a constant deceleration

$$a = dv/dt = -nP_m/M_2 + g. \tag{5.14}$$

Now, integrating equation (5.14) with respect to time and introducing the initial and terminal conditions gives a response time

$$T = V_2/(nP_m/M_2 - g) \tag{5.15}$$

and a crushed distance

$$\Delta = V_2^2/[2(nP_m/M_2 - g)] \tag{5.16}$$

which may also be obtained directly from the conservation-of-energy equation (5.12).

In many practical impacts, the decelerations $|a|>>g$ so that equation (5.14) gives $a \cong -nP_m/M_2$, and equations (5.12) to (5.16) then reduce to equations (5.5) to (5.7), (5.9a) and (5.11a), respectively.

5.3   Thin-walled circular tubes

Thin-walled circular tubes have been proposed as energy absorbers behind car bumpers, train buffers and at the bottom of lift shafts to absorb the energy of a runaway lift. They are inexpensive, efficient and versatile and it is, therefore, worthwhile to examine the crushing behaviour of an energy-absorbing system which consists of thin-walled circular tubes.

The constant deceleration during an impact event is predicted by equation (5.7), with the mean dynamic crushing force P_m given by equations (4.23) and (4.24), or

$$P_m = \frac{21.1\,n\,\sigma_0 RH\,[1\,+\,0.41(H/R)^{1/2}][1\,+\,(V_2/4RD)^{1/q}]}{2.14\,(R/H)^{1/2}\,-\,1} \tag{5.17}$$

when using equations (4.18) and (4.20). Thus,

$$a = -\,\frac{21.1\,n\,\sigma_0 RH\,[1\,+\,0.41(H/R)^{1/2}][1\,+\,(V_2/4RD)^{1/q}]}{M_2(2.14\,(R/H)^{1/2}\,-\,1)} \tag{5.18}$$

Similarly, the response time T and axially crushed distance Δ are predicted by equations (5.9b) and (5.11b), respectively, or

$$T = \frac{M_2 V_2[2.14\,(R/H)^{1/2}\,-\,1]}{21.1\,n\,\sigma_0 RH\,[1\,+\,0.41(H/R)^{1/2}][1\,+\,(V_2/4RD)^{1/q}]} \tag{5.19}$$

and

$$\Delta = \frac{M_2 V_2^2[2.14\,(R/H)^{1/2}\,-\,1]}{42.2\,n\,\sigma_0 RH\,[1\,+\,0.41(H/R)^{1/2}][1\,+\,(V_2/4RD)^{1/q}]}. \tag{5.20}$$

Equations (5.17) to (5.20) have been developed for an energy absorbing system which does not bottom-out. Thus, equation (5.20) must satisfy the inequality $\Delta \leq \Delta_b$, where Δ_b is the bottoming-out displacement which is defined after equation (5.4).

It is evident that equations (5.18) to (5.20) have been developed for horizontal impact as illustrated in figure 5.1(a). However, the results are also valid for the vertical impact situation in figure 5.3(a) provided $|a|>>g$.

The foregoing equations are now used to examine the dynamic progressive buckling and energy-absorbing characteristics of a thin-walled cylindrical shell. The shell is made from a ductile material with a uniaxial static flow stress (σ_o) of 300 MN/m^2, and has a mean radius $R = 30$ mm and mean wall thickness $H = 1.2$ mm. One end of the shell is fixed while the other end is struck by a mass $M_2 = 100$ kg travelling with an initial velocity[3] $V_2 = 10$ m/s. Thus, substituting these quantities with $n = 1$ for a single tube into equations (5.18) to (5.20) gives $a = -254$ m/s^2, $T = 39.3$ ms and $\Delta = 196.7$ mm, respectively, provided the tube is made from a strain-rate-insensitive material (i.e., $D \rightarrow \infty$).

These calculations reveal that the deceleration of the striking mass M_2 is $254/9.81 = 25.9$ times larger than the gravitational acccleration[4] (9.81m/s^2). This suggests that the gravitational terms in equations (5.13) to (5.16) may be neglected for this particular circular tube when impacted vertically. The additional potential energy in equation (5.12) is $M_2g\Delta=193J$, which is about 3.9 percent of the initial kinetic energy for the mass M_2 (5 kJ).

The response duration is very short (39.3 ms) and the event happens too quickly for the human eye to follow the deformation. Nevertheless, it is long compared with the time required for an elastic stress wave to travel along the length of a tube. Uniaxial tensile or compressive elastic stress waves travel [5.10] at 5150 m/s and 5100 m/s in mild steel and aluminium, respectively, and,

---

[3] In the case of a vertical impact, a mass dropped from a height $h = 5.1$ m would give an impact velocity $V_2 = 10$ m/s according to the well-known formula $V_2 =(2gh)^{1/2}$, with $g = 9.81$ m/s^2

[4] It is evident that the gravitational force of the 100 kg striking mass is 981 N, while the mean dynamic crushing force on the tube is 25.9 times larger (i.e. P_m=25.4 kN). Alternatively, the mean dynamic crushing force may be calculated from equation (5.17) or by noting that the initial kinetic energy = 5 kJ = $P_m \times \Delta$, or P_m=5x10^3/0.1967 = 25.4 kN. The static uniaxial compressive squash load for this shell is $2\pi RH\sigma_o = 67.9$ kN. Thus, P_m is only 37 per cent of the squash load. However, the initial peak force is larger than the mean crushing force as shown in figure 4.3(a).

therefore, would take about 58 μs to travel along a 300 mm long mild steel or aluminium bar. Thus, the response time T of 39.3 ms is 678 times longer than the time taken for a uniaxial elastic stress wave to traverse a 300 mm long bar.

Now, $\Delta_b/L=0.78$ for a circular tube [5.11] so that it must be at least 196.7/0.78=252 mm long in order to prevent the phenomenon of bottoming-out.

The foregoing calculations are associated with a strain- rate-insensitive tube. If a tube is made from mild steel with the coefficients $D = 6844$ s^{-1} and $q = 3.91$ from §4.5, then the term $1 + (V_2/4RD)^{1/q}$, which appears in equations (5.17) to (5.20), equals 1.32. In this circumstance, $a = -335$ m/s^2, $T = 29.8$ ms and $\Delta = 149$ mm. The influence of material strain rate sensitivity on the values of P_m, a, T and Δ for the above circular tube is shown in figure 5.4.

5.4   Thin-walled square tubes

The general procedure, which is outlined in § 5.2 and illustrated for a circular tube in §5.3, may also be used for square tubes having the mean dynamic crushing force given by [5.11, 5.12]

$$P_m = 13.05 \ \sigma_o H^2 (C/H)^{1/3} \ [1 + (0.33V_2/CD)^{1/q}] , \qquad (5.21)$$

where V_2 is the impact velocity of a striking mass M_2.

5.5   Impact injury

Theoretical calculations such as those in § 5.2 to 5.4 enable an engineer to assess whether a structural design is capable of withstanding the forces generated during an impact and absorbing the impact energy without excessive damage. However, in the case of passenger transportation systems, it is also necessary to ensure that the passengers can tolerate an impact. Clearly, designers require guidelines and criteria which, however, are difficult to obtain on human beings and are further complicated by the observation that people's response and tolerance to impact varies with size, age and sex, etc. Nevertheless, many studies have been conducted in this field over the years and References 5.1, 5.2, 5.13-5.18, together with other articles cited therein, contain some valuable data.

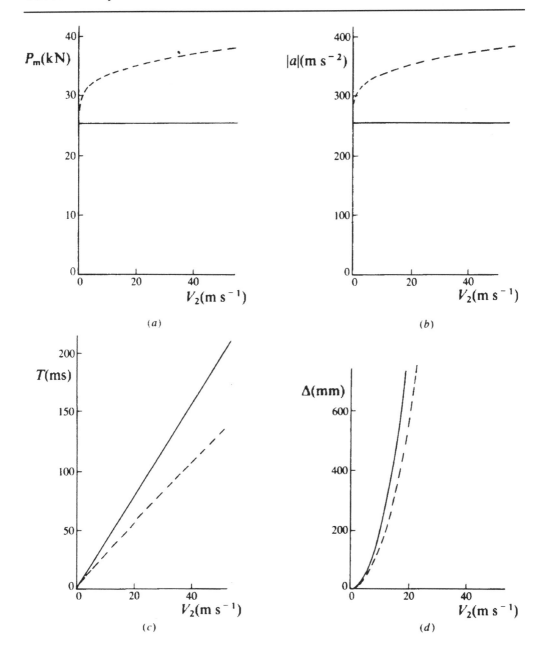

Figure 5.4   (a)-(d) Variation of mean crushing force, deceleration, response time and total crushing distance with impact velocity, respectively, for a thin-walled circular tube having R = 30 mm, H = 1.2 mm, σ_o = 300 MN/m² and M_2 = 100 kg. ————: equations (5.17) to (5.20) with D → ∞ for a strain-rate-insensitive material. - - - - -: equations (5.17) to (5.20) with D = 6844 s⁻¹ and q = 3.91.

Some typical values [5.13,5.14] of whole body acceleration (and deceleration), which are associated with various types of impact, are presented in figure 5.5. However, the influence of pulse length plays an important role in the severity of injury, as indicated in figure 5.6, which is taken from Macaulay [5.13]. A line through the middle of the moderate injury band in figure 5.6 is approximated by the equation

$$TA^{2.5} = 1000, \qquad\qquad (5.22)$$

where T is the pulse duration (s) and $A = a/g$ when a is the whole body acceleration (or deceleration).

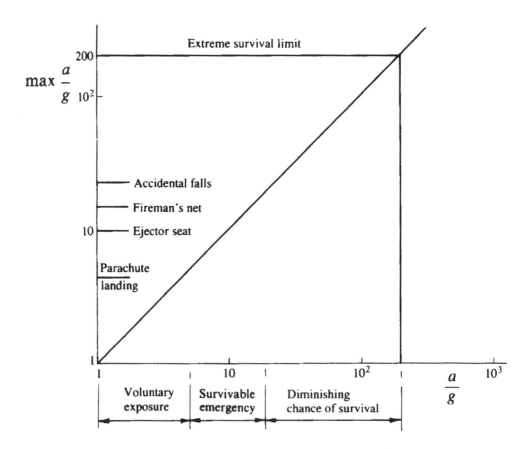

Figure 5.5   Whole-body tolerance of impact [5.13, 5.14]

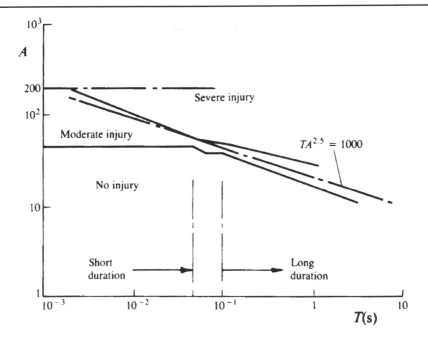

Figure 5.6   Influence of the pulse duration on whole-body tolerance of impact [5.13]

Equation (5.22) only applies to impacts which involve whole-body accelerations. Other impact situations occur in which specific body components are struck and, therefore, data is required for many parts of the body. For example, figure 5.7 presents data [5.13, 5.19] on the strength of fresh femur bones. The strength decreases significantly with a person's age and is less for impact loads. Data for other parts of the body under various static and dynamic loads may be found in Reference 5.13 and in the articles cited therein.

Head injuries are responsible for a significant loss of life and serious injuries in transportation accidents [5.20]. The Wayne State tolerance curve in figure 5.8 was obtained by dropping embalmed cadaver heads onto hard flat surfaces to determine incipient skull fracture [5.16]. This work has been re-examined by Gadd and others to develop criteria for head injuries which are caused when the front of the head strikes a hard object, or when the head is loaded through the neck by a decelerating body [5.13, 5.21]. Gadd introduced the severity index [5.14-5.16, 5.20, 5.21].

$$SI = \int_0^T A_v^{2.5} dt .$$

(5.23)

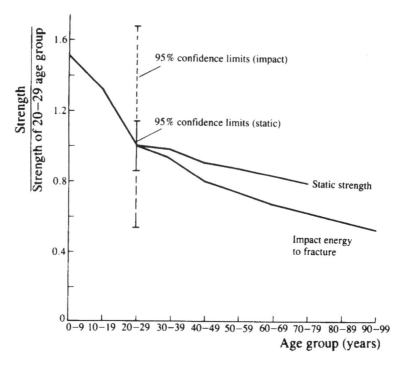

Figure 5.7   Variation in static and impact strength of fresh femur bone with age [5.13, 5.19]

where $A_v = a_v/g$, a_v is the average head acceleration (or deceleration) which may vary throughout the loading pulse having a duration $T(s)$ with $2.5ms < T < 50ms$. Gadd suggested that $SI = 1000$ marks the threshold conditions between fatal and non-fatal head injuries, although there is a continuing debate on the choice of a threshold value and about the relevance of this criterion [5.22].

The Wayne State tolerance curve in figure 5.8 was constructed using an average acceleration for the entire impact event, whereas the acceleration is allowed to vary with time in equation (5.23). This inconsistency was recognised and eliminated by replacing the dimensionless acceleration A_v in equation (5.23) by an average value to produce the Head Injury Criterion [5.1, 5.16].

$$HIC = (T_2 - T_1)\left\{ \int_0^T A_v \, dt \, /(T_2 - T_1) \right\}^{2.5} \qquad (5.24)$$

where any time interval $T_2 - T_1$ is selected to maximise the right-hand side of equation (5.24) and a value of HIC = 1000 is considered [5.22] as life-threatening.

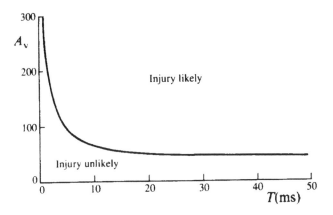

Figure 5.8   Wayne State tolerance curve for head impact [5.13]

Now, consider the impact situation in § 5.2 where a mass M_2, which is travelling with a velocity V_2, strikes an infinitely large stationary mass (i.e., $M_1 \rightarrow \infty$) in figure 5.1(a). The deceleration during the impact event when an energy absorber with a constant resistance P_m is interposed between the two masses M_1 and M_2 is

$$a = - V_2^2/2\Delta \tag{5.25}$$

according to equation (5.11b), where Δ is the total stopping distance. The associated duration of response from equation (5.9b) is

$$T = 2\Delta/V_2 \tag{5.26}$$

when eliminating a with equation (5.25). Thus, substituting equations (5.25) and (5.26) into equation (5.22) for whole body accelerations gives

$$V_2^4/\{g(2g\Delta)^{1.5}\} = 1000. \tag{5.27}$$

The calculations for the strain-rate-insensitive circular tube examined in § 5.3 predict a crushing displacement $\Delta = 196.7$ mm for an initial impact velocity $V_2 = 10$ m/s. Thus, the left-hand side of equation (5.27) equals 134. This value is well below the threshold limit of 1000 which is required for serious injury during whole body impact.

The deceleration a in equation (5.25) is constant so that equation (5.23) predicts $SI = 134$ for the thin-walled cylindrical tube examined in §5.3. It should be noted that the response time $T = 39.3$ ms lies within the range of validity of equation (5.23). Similarly, HIC = 134 according to equation (5.24) for the same cylindrical tube.

Now, substituting equations (5.7) and (5.9b) with n = 1 into equation (5.24) gives

$$HIC = (P_m / M_2 g)^{1.5} V_2 / g \qquad (5.28)$$

or

$$HIC = V_2 a^{1.5} / g^{2.5} \qquad (5.29)$$

which relates the Head Injury Criterion to the impact velocity and acceleration (or deceleration) as shown in figure 5.9.

The field of impact injury is an active area of research and Parts V and VI of this book discuss important new developments in crashworthiness biomechanics and injury biomechanics, respectively.   Other recent review articles have been published by the International Crashworthiness Conferences [5.23, 5.24] and elsewhere.

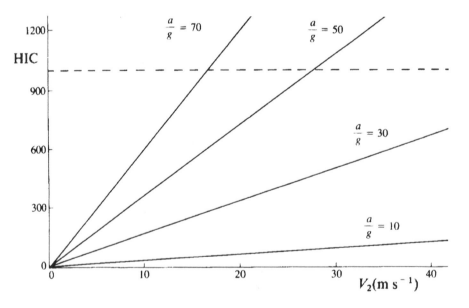

Figure 5.9   Variation of Head Injury Criterion (HIC) with impact velocity V_2 and dimensionless acceleration (a/g) according to equation (5.29) for a constant impact force and acceleration (or deceleration). - - -: threshold between potentially fatal and non-fatal impacts.

REFERENCES

[5.1]   Johnson, W. and Mamalis, A. G., *Crashworthiness of Vehicles*, MEP, London, 1978.

[5.2]   Singley, G. T., Survey of rotary-wing aircraft crashworthiness, In *Dynamic Response of Structures*, G. Herrmann and N. Perrone (Eds.), Pergamon, Oxford, 179-223, 1972.

[5.3]   Pugsley, Sir A. and Macaulay, M. A., Cars in collision-safe structures, *New Scientist*, **78**(1105), 596-598, 1978.

[5.4]   Thornton, P. H., Energy absorption in composite structures, *Journal of Composite Materials*, **13**, 247-262, 1979.

[5.5]   Jones, N. and Wierzbicki, T. (Eds) *Structural Crashworthiness*, Butterworths, London, 1983.

[5.6]   Scholes, A., Railway passenger vehicle design loads and structural crashworthiness, *Proceedings of the Institution of Mechanical Engineers*, **201**(D3), 201-207, 1987.

[5.7]   Miles, J. C., Molyneaux, T. C. K. and Dowler, H. J., Analysis of the forces on a nuclear fuel transport flask in an impact by a train, *Proceedings of the Institution of Mechanical Engineers*, **201**(A1), 55-68, 1987.

[5.8]   Jones, N. and Wierzbicki, T., Dynamic plastic failure of a free-free beam, *Int. Journal of Impact Engineering*, **6**(3), 225-240 (1987).

[5.9]   Wierzbicki, T., Crushing analysis of metal honeycombs, International *Journal of Impact Engineering*, **1**(2), 157-174, 1983.

[5.10] Johnson, W., *Impact Strength of Materials*, Edward Arnold, London and Crane, Russak, New York, 1972.

[5.11] Jones, N., *Structural Impact*, Cambridge University Press, 1989, paperback edition (1997).

[5.12] Abramowicz, W. and Jones, N., Dynamic axial crushing of square tubes, *Int. Journal of Impact Engineering*, **2**, 179-208, 1984.

[5.13] Macaulay, M. A., *Introduction to Impact Engineering*, Chapman & Hall, London, 1987.

[5.14]   Snyder, R.G., Human impact tolerance, SAE paper N. 700398, In *International Automobile Safety Conference Compendium*, SAE, 712-782, 1970.

[5.15]   King, A. L, Human tolerance limitations related to aircraft crashworthiness, In *Dynamic Response of Structures*, G. Herrmann and N. Perrone (Eds.), Pergamon, New York, 247-263, 1972.

[5.16]   Huston, R. L. and Perrone, N., Dynamic response and protection of the human body and skull in impact situations, In *Perspectives in Biomechanics*, H. Reul, D. N. Ghista and G. Rau, Harwood (Eds.), vol. 1, pt B, 531-571, 1978.

[5.17]   Perrone,N., Biomechanical and structural aspects of design for vehicle impact, In *Human Body Dynamics, Impact, Occupational and Athletic Aspects*, D.N.Ghista (Ed.), Clarendon, Oxford, 181-200, 1982.

[5.18]   Johnson,W., Mamalis, A.G. and Reid, S.R., Aspects of car design and human injury, In *Human Body Dynamics, Impact, Occupational and Athletic Aspects*, D.N.Ghista (Ed.), Clarendon, Oxford, 164-180, 1982.

[5.19]   Currey, J. D.. Changes in the impact energy absorption of bone with age, *Journal of Biomechanics*, **12**, 459-469, 1979.

[5.20]   King, A. L, Crash course, Mechanical Engineering, ASME, **108**(6), 58-61, 1986.

[5.21]   Perrone, N., Biomechanical problems related to vehicle impact, In *Biomechanics: Its Foundations and Objectives*, Y.C. Fung, N.Perrone and M. Anlikered (Eds.), Prentice Hall, Englewood Cliffs, N.J., 567-583, 1972.

[5.22]   Goldsmith, W., Current controversies in the stipulation of head injury criteria, *Journal of Biomechanics*, **14**, 883-884, 1981.

[5.23]   Chirwa, E. C. and Viano, D. C. (Eds.), *Proceedings of International Crashworthiness Conference*, Detroit, September 1998, Woodhead Publishing Limited, Cambridge, U.K., 1998.

[5.24]   Chirwa, E. C. and Otte, D. (Eds.), *Proceedings of International Crashworthiness Conference*, London, September 2000, ICrash 2000, Bolton BL3 5AB, U.K., 2000.

Part II

MACRO ELEMENT METHOD IN CRASHWORTHINESS OF VEHICLES

W. Abramowicz

Impact Design Europe, Michalowice, Poland

The objective of this part is to present a complete introduction to the macro element method employed in modern crashworthiness. The characteristic feature that distinguishes the macro element approach from all other classical formulations in nonlinear mechanics is that trial deformation functions are postulated on the basis of experimental observations rather then in the form of elementary algebraic functions. It is shown how experimentally determined shape functions are incorporated into a consistent, mathematically tractable engineering method. Apart from basic formulation it is also shown how the macro element concept is implemented in standard CAD/CAE programs.

6. INTRODUCTION TO THE MACRO ELEMENT MODELLING CONCEPT

6.1 Introduction

Crashworthiness emerged as an extensively explored engineering field in early 60 following introduction of the safety standards in the US automotive industry. The new regulations posed a serious challenge to both practicing engineers responsible for the development of safe vehicles as well as to scientists and researchers developing predictive tools. Crash response of structures involves a number of highly nonlinear phenomena such as localization of plastic flow, interaction of local and global buckling modes, large deformations and tearing of material. The industry driven requirements for adequate computational tools capable of capturing all the above phenomena triggered the development of a number of dedicated tools and techniques. In the early days of crashworthiness experimentation was the main design tools in the automotive industry. At the same time numerical techniques, especially FE methods were progressing rapidly towards highly reliable simulation tools. However, it has been soon recognized that precise numerical analysis is not always the only 'thing' the engineer needs. In many cases less accurate, qualitative answers might suffice or at least help in better understanding of a given crash phenomenon and assist in planning of simulation and experimental program.

One of the most successful predictive techniques that emerged during early days of crashworthiness is the so-called kinematic approach to large plastic deformations of structures. The kinematic approach originates from the famous work of Alexander published in 1960 [6.1]. During following decades the kinematic approach, frequently referred to as a macro element method, has grown to a consistent, mathematically tractable method of analyzing large shape distortions of sheet metal structures.

In the macro element method kinematic variables such as displacement, velocity and acceleration fields are postulated as a set of trial solutions in variational or extremal formulations of solid and structural mechanics. The characteristic feature that distinguishes the present approach from all other classical formulations in nonlinear mechanics is that trial deformation functions are postulated on the basis of experimental observations rather then in the form of elementary algebraic functions. In most cases a successful formulation of an individual macro element imposes strict limitations onto the type of a structure that can be modeled by means of such an element as well as certain restrictions onto the admissible boundary and loading conditions. Therefore, the range of applicability of a single macro element is much more narrow then general formulations of engineering theories like e.g. plate theory or FE method. However, the advantage of such a restricted formulation of an element is its simplicity. Frequently solution to the crushing problem is obtained in a closed form while computer programs based on the macro element method are fast and do not require complex input data.

This chapter focuses on the basic formulation and example applications of the macro element method. Short literature review pertinent to subjects not covered in this chapter is given in the conclusions of chapter 10.

6.2   The Macro Element modeling concept-illustrative example

The concept of macro elements in crash calculations is explained on an example of a frontal ship collision shown schematically in Figure 6.1.

The basic idea of the macro element method is to divide the structure into a number of representative structural elements and then determine crushing response of each element separately and/or an assemblage of interacting elements. In the case of a typical bow structure of a ship the representative elements are referred to as 'X', 'T' and 'Y' sections. Each element models the intersection of two or more major structural plate elements along a common intersection line. The motivation for dividing the bow structure into 'X' 'Y' and 'T' elements instead of plate elements is that the material in the vicinity of a plate intersection line absorbs most of the impact energy. In an actual accident the discussed sections are loaded in a variety of ways. The present illustrative example focuses on an axial crushing response of 'X' elements when the crushing force is parallel to the plate intersection line. This type of loading is typical for a side collision and grounding of ships on a wide rock when response of a ship structure is governed mostly by crushing of 'X' and 'T' sections. In addition during an axial loading there is generally a weak coupling between deformation of distant 'X' and/or 'T' elements so that in an approximate analysis the deformation of each element can be treated separately.

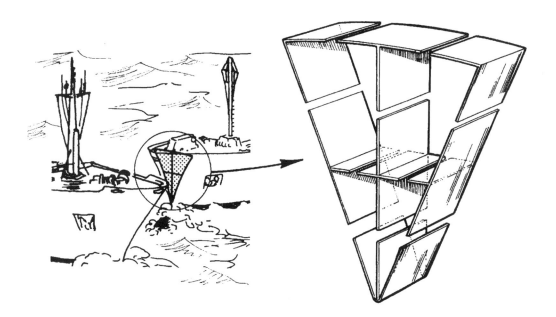

Figure 6.1   Ship impact analysis procedure based on dividing the structure into T X and Y macro elements containing plate intersection lines.

6.3   Folding modes of cruciforms ('X' elements)

Representative folding modes of 'X' elements subjected to axial loading are shown in Figure 6.2.

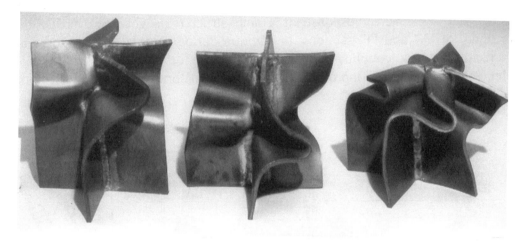

Figure 6.2   Different folding modes of mild steel 'X' sections tested under quasi-static and dynamic loading. (Courtesy of the Impact Research Centre at the Univ. Liverpool)

Depending on initial imperfections the crushed section may collapse in a variety of folding modes characterized by distinct levels of energy absorption.   However, a careful examination of contributing folding mechanisms reveals that all the experimentally observed deformation patterns can be assembled out of a single, characteristic deformation mode of a corner element, referred to as a Superfolding Element (SE).   For example the 'natural folding mode' most frequently observed in experiments, compare Figure 6.2, consists of four Superfolding Elements deforming in a symmetric mode, shown in Figure 6.3a.

Building models of collapsed elements out of construction paper, Figure 6.3b, frequently simplifies identification of experimentally observed folding modes.   The paper sheets are inextensible–so that any deformed pattern that can be made out of a flat paper by means of bending along local bend lines without cutting the paper involves only bending deformations of the shell.   Such a deformation is refereed in the literature to as quasi-isometric deformation, [6.3].   On the other hand all the cutouts and openings made in the paper model in order to reproduce an actual folding pattern correspond to localized membrane deformations. It transpires form the Figure 6.3b that natural folding mode of an X element involves both bending as well as membrane deformations.

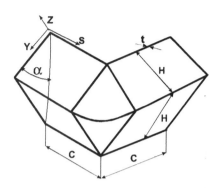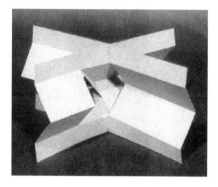

Figure 6.3   (a) Symmetric folding mode of a Superfolding Element of the length L=2c and thickness t. The height of the element 2H corresponds to the length of plastic folding wave. (b) A structural model, made of construction paper, illustrates an exact location of contributing Superfolding Elements within a deformed 'X' section, [6.2].

6.4   Internal energy dissipation

Internal energy dissipation in computational models made of construction paper can be conveniently calculated by using the concept of localized plastic hinge lines introduced in the classical theory of plasticity. In the hinge line method it is assumed that all the plastic deformations, either bending, tensile or shear are localized within the discontinuity lines (regions). For the rigid perfectly plastic idealization of the material the rate of energy dissipation is then given as

$$\dot{E}_b = M_o \dot{\theta} \quad and \quad \dot{E}_m = N_o \dot{u} \qquad (6.1)$$

for bending and tensile/compression deformations, respectively. In Eq. (6.1) $M_o = \sigma_o t^2 / 4$ denotes fully plastic bending moment, $N_o = \sigma_o t$ is the fully plastic membrane force (both quantities are given for unit length of a plastic hinge), t is the material thickness and σ_o is the level of plastic flow stress. The doted parameters, $\dot{\theta}$ and $\dot{u}$, denote rates of rotation and tension/compression, respectively. Since in the present model the bending and membrane plastic deformations are assigned to separate regions there is no interaction of generalized forces. In addition, it follows from the kinematics of the computational model that loadings in horizontal bending hinge lines and localized tensile zones are proportional refer to Figure 6.3a. Consequently, the rate equations, Eq. (6.1), can be effectively integrated over the complete crushing process.

The total bending angle in upper and lower hinge lines in Figure 6.3a is $\pi/2$ while the corresponding angle in the middle hinge is π. The total length of each hinge is $2C$, so that, the total dissipation due to bending is

$$E_b = 8C\pi M_o \tag{6.2}$$

Horizontal fibres in the conical zones of plastic deformations are elongated proportionally to the distance from the cone apex (coordinate s in Figure 6.3a). In the final, completely squeezed (flat) configuration the apex angle of the conical surface is $\pi/2$, so that, total membrane energy dissipated in four conical deformation zones is

$$E_m = 4\int_0^H \frac{\pi}{2} yN_o dy = \pi N_o H^2 = 4\pi M_o \frac{H^2}{t} \tag{6.3}$$

where H denotes half length of the local plastic folding wave. It should be noted that parameter H is unknown and must be determined as a part of the solution procedure.

6.5 Energy balance equation

Energy dissipated in the generalized hinge lines equals the energy supplied to the system by the axial crushing force acting on axial shortening of the section, δ. A convenient measure of the energy absorption capacity of an element, referred to as a mean crushing force, P_m, is defined as

$$E_{ext} = \int_0^{\delta_{eff}} P(\delta)d\delta = P_m \int_0^{\delta_{eff}} d\delta = P_m \delta_{eff} \implies P_m = \frac{E_{ext}}{\delta_{eff}} \tag{6.4}$$

In Eq. (6.4) the mean crushing force P_m is given as a mean value of the external energy functional. It can also be interpreted as average or specific energy absorption per unit crush of a structure. In Eq. (6.4) δ_{eff} denotes the so-called effective crushing distance. Separate studies show that the aspect ratio of the effective crushing distance to the length of the plastic folding wave, *2H*, equals $\delta_{eff}/2H = 0.73$, approximately, and is constant for a wide range of crushed structures, [6.3]. The effective crushing distance reflects the fact that an actual structure cannot be folded completely like a paper model. There always remains a stack of completely squeezed plastic lobes that have at least one-order-of-magnitude higher compression stiffness then an active lobe. Consequently an actual crushing distance corresponding to the creation of one plastic lobe is smaller then the length of the plastic folding wave, *2H* . This phenomenon is illustrated in Figure 6.4 for the case of completely squeezed square column that developed over a dozen of plastic lobes.

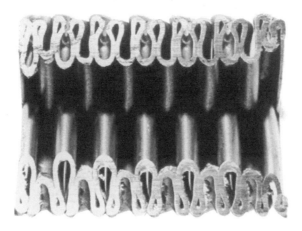

Figure 6.4   Section through a completely squeezed square column illustrates a residual stack of completed plastic folds.

Equating internal and external energies in Eq. (6.3) and Eq. (6.4) renders expression for the mean crushing force as a function of unknown length of the plastic folding wave, *2H.*

$$P_m(H) = \frac{2H}{\delta_{eff}}\left(2\pi\, M_o \frac{H}{t} + \frac{4\pi\, c\, M_o}{H\, t}\right) \tag{6.5}$$

The function $P_m(H)$ has a minimum $(\partial P_m/\partial H = 0)$ that provides for a conservative estimate of the energy absorption capacity of a cruciform element.

$$\frac{P_m}{M_o} = \frac{4\pi\sqrt{2}}{0.73}\sqrt{\frac{C}{t}} \quad ; \quad \frac{H}{t} = \sqrt{2}\sqrt{\frac{C}{t}} \tag{6.6}$$

6.6   Conclusions

The introductory example of the macro element approach reveals most of the distinctive features of the crushing response of thin-walled structures. The plastic crushing process develops following the elastic or elastic/plastic buckling and is characterized by localization of plastic deformations in relatively small parts of a structure. Plastic deformations are localized in narrow hinge lines where most of the plastic deformation takes place while the global deformation of the structure results from rigid body motion of undeformed or slightly deformed segments. A typical folding pattern observed in a variety of structural elements can be approximated by few folding modes of a corner element. Deformation of a corner element is described by using the concept of a specialized macro element referred to as Superfolding Element (SE).

The simplified calculation procedure presented in this section leaves several 'loose ends' in the formulation of the method. A formal formulation of the macro element method is presented in the next section.

REFERENCES

[6.1]   Alexander, J. M., An approximate analysis of the collapse of thin cylindrical shells under axial loading, *Q. J. Mech. Appl. Math.*, **13**(1), 10-15, 1960.

[6.2]   Abramowicz, W., *Crush Resistance of 'T' 'Y' and 'X' Sections*, Joint MIT-Industry Program on Tanker Safety, Massachusetts Institute of Technology, Report 24, 1994.

[6.3]   Wierzbicki, T. and Abramowicz, W. *The Manual of Crashworthiness Engineering Vol. I - IV*, Center for Transportation Studies, Massachusetts Institute of Technology, 1987 - 1989.

7. GENERAL FORMULATION OF THE MACRO ELEMENT APPROACH

This chapter shows how the formulation of the macro element approach is built on first principles of structural mechanics.

7.1   Prerequisites

In the Continuum Mechanics *deformation* of a body in a time interval $[t_o, t_f]$ is defined as a one-parameter family of configurations $\chi(\cdot,\tau) \in K(V)$; $\tau \in [t_o, t_f]$, in such a way that for each material point $\mathbf{X} \in V$ the function

$$\mathbf{x} = \chi(\mathbf{X},\tau) \quad \tau \in [t_o, t_f] \tag{7.1}$$

is continuous and has a continuous first and second order time derivatives, see e.g. [7.1].   Each invertible and continuously differentiable[1] mapping $\mathbf{x} = \chi(\mathbf{X},\tau)$ is called a position of the material point X, $X \in V$, at time instant t, $t \in [t_o, t_f]$, in the deformation $\mathbf{x} = \chi(\cdot,\tau)$ $\tau \in [t_o, t_f]$.   A region of Euclidean space corresponding to a *current configuration* of the body, $v \in K(V)$, is denoted as $v = \chi(V,t)$

---

[1] The term *continuously differentiable* which will be used frequently in this chapter should be understood, unless otherwise stated, as: *continuously differentiable as many times as required except at some surfaces, lines or points*.

Any mapping $\mathbf{x} = \chi(\mathbf{X})$ $\mathbf{X} \in V$ can be taken as a configuration (deformed body). In structural mechanics the class of admissible configurations is typically limited to same special sub-classes by imposing the internal kinematic constrains. A specific example of such constrained class of deformation is the Love-Kirchoff shell theory in which it is assumed that the material fiber that is normal to the midsurface in one configuration remains normal in any other configuration. Another example is the problem of plane deformations. In the formulation of any problem in the field of structural mechanics it is assumed, as a rule, that the region V and the set $K(V)$ are known a priori.

Once the set of admissible configurations is defined a generic form of the non-linear problem of structural mechanics is formulated by specifying the governing equations. These are:

- equations of equilibrium
- constitutive relations
- compatibility conditions
- kinematic boundary constraints, boundary loading and initial conditions

These equations constitute a set of governing relations for the unknown deformation $\chi(\mathbf{X}, \tau)$ $\tau \in [t_o, t_f]$. The function $\chi(\mathbf{X}, \tau)$, which constitutes the solution to the generic problem, will be referred to as a fundamental solution and denoted as $\chi^o(\mathbf{X}, \tau)$ [2]. Likewise, all other fields corresponding to the fundamental solution are identified by superscript 'o'.

7.2   General formulation of the macro element method

The general methods of structural mechanics (e.g. beam or shell theories) are formulated in such a way that the solution to the generic problem can be obtained for a wide class of reference configurations V (undeformed body), boundary and initial conditions and specific constitutive relations.

---

[2]   In the following considerations it is assumed that such a solution exists and is unique.

In contrast to these methods the macro element formulation is dedicated to narrow classes of structural elements or even to a single type of a structural element (or its representative part). Consequently, rigorous restrictions are imposed onto the admissible reference configuration V that, in the remaining of this chapter, will be identified with a reference configuration of a macro element. Typical examples of structural elements modeled by dedicated macro elements are thin-walled prismatic members (such as the cruciform sections discussed in the preceding section). Furthermore, the class of boundary, loading and initial conditions, for which a given method is designed, is also restricted and in several cases applies to only one type of boundary conditions. For example, a macro element that models axial crushing response of an X section will typically require that one end of the section is clamped while the opposite end moves with a constant velocity and remains parallel to the clamped end throughout the entire deformation. In terms of boundary conditions it means that the spatial position $\mathbf{x}_u = \chi(\mathbf{X}, \tau); \ \mathbf{X} \in S_u, \ \tau \in [t_o, t_f]$ of the surface $S_u \subset S$ on which displacements are prescribed is known for each time instant. Also, the velocity of each point on the surface S_u is known a priori (kinematic loading). When the boundary conditions change typically a new macro element must be formulated.

Under such strict limitations the deformation of a macro element, Eq. (7.1), resulting from a strictly defined loading and boundary conditions can be determined on the basis of experimental observations as discussed in the previous section. As a rule, the deformation $\chi(\mathbf{X}, \tau)$, is postulated with an accuracy to a vector of free scalar parameters $\beta = [\beta_1, \ldots \beta_N]$ and constitutes a set of trial deformations.

$$\{\mathbf{x}^* \in K(V): \ \mathbf{x}^* = \chi^*(V, \beta, \tau); \ \tau \in [t_o, t_f]; \ \beta \in \mathbf{R}^3\} \qquad (7.2)$$

In order to avoid needless complexity it is further assumed that the vector β does not depend on time. All variables pertinent to trial deformations, Eq. (7.2), will be denoted by a superscript '*'.

Once a class of admissible deformations is determined the general form of the macro element method is formulated in the following way. First, we note that due to limitations imposed onto the admissible set of deformations, Eq. (7.2), the boundary, loading and initial conditions are automatically satisfied for each $t \in [t_o, t_f]$. Second, the kinematic variables i.e. the material and spatial velocity fields

$$\mathbf{v}^* = \mathbf{G}(\mathbf{X}, \tau) \equiv \mathbf{g}[\mathbf{x}(\mathbf{X}, \tau), \tau] \ where \ \mathbf{G}(\mathbf{X}, \tau) \equiv \frac{\partial}{\partial \tau} \chi(\mathbf{X}, \tau); \tau \in [t_o, t_f]$$

gradient of deformation, $F^* = \nabla x^*$, and its rate, $\dot{F}^*$, are determined from the admissible deformation fields, Eq. (7.2), by time and spatial differentiation, respectively. Therefore, the compatibility conditions are also automatically satisfied.

All that remains to be done is to solve the equations of equilibrium. In the remaining of this section we shall consider only quasi-static deformations. In this case the equilibrium of a macro element, v, at each time instant t, $t \in [t_o, t_f]$, can be conveniently expressed via the following form of the principle of virtual velocities (weak formulation)

$$\dot{E}^o_{ext} = \dot{E}^o_{int} \tag{7.3}$$

where $\dot{E}^o_{ext} = \int_{\partial v} \mathbf{T} v d(\partial v)$ denotes the power input while $\dot{E}^o_{int} = \int_v \sigma \mathbf{d} \, dv$ is the stress power (rate of total internal dissipation in the case of plastic solids) corresponding to a fundamental solution. In the preceding expressions T denotes surface traction, σ is the Cauchy stress tensor while d is the rate of deformation tensor (symmetric part of the deformation gradient). In following calculations the spatial description is used consistently, so that, the integration in expressions for the total rate of internal and external work is performed over the current configuration $x = \chi(V, t)$. In the next step of the solution procedure a set of trial functions for the stress power is established.

$$\left\{ \dot{E}^*_{int}(\beta, \tau) : \dot{E}^*_{int} = \int_v \sigma^* \mathbf{d}^* dv^*; d_{ij} = \frac{1}{2}(v^*_{i,j} + v^*_{j,i}) \right\} \tag{7.4}$$

Each trial solution in Eq. (7.4) corresponds to a trial deformation $x^* = \chi^*(\mathbf{X}, \beta, \tau)$. Since deformations are finite the configurations $v \equiv \chi(V, \tau)$ and $v^* \equiv \chi * (V, \beta, \tau)$ may occupy different regions in space and therefore the limits of integration in Eq. (7.3) and Eq. (7.4) are, in general, different. It should also be noted that Eq. (7.4) is not a statement of equilibrium. It is simply an analogue of the internal dissipation coupled with the trial deformation $\mathbf{x}^* = \chi^*(\mathbf{X}, \beta, \tau)$. The 'discrepancy' between the exact and trial rates of internal dissipation is then expressed as a residual

$$\hat{R}(\beta, \tau) = \dot{E}_{int}^*(\beta, \tau) - \dot{E}_{int}^o(\tau); \ \tau \in [t_o, t_f] \tag{7.5}$$

which is a function of time and a vector of free parameters β only. Since the set of admissible deformations is known a global residual, $R(\beta)$, which describes 'global accuracy' of each trial deformation, Eq. (7.4) can be determined as

$$R(\beta) = \int_o^{t_f} \dot{E}_{int}^*(\beta, \tau) d\tau - \int_o^{t_f} \dot{E}_{int}^o(\tau) d\tau \tag{7.6}$$

The global residual $R(\beta)$ is a function of the vector β only. Consequently, the solution to the problem is reduced to the determination of an optimal vector of free parameters, β, which renders the global residual $R(\beta)$ an absolute minimum and thus, corresponds to the best approximation in the entire space-time subdomain. Once the vector β is known the unknown external loading can be determined directly from Eq. (7.3). A more detailed discussion of this problem is presented in the next section.

7.3   Solution Procedure

This section shows how a solution to the governing equation of the macro element method, Eq. (7.6), is constructed. First, the general procedure for a class of steady state and quasi-steady state processes is discussed. It is shown that there exists a class of deformation processes where the upper bound theorem of the Theory of Plasticity can be applied rendering the condition of convergence of the kinematicaly admissible solution to the fundamental solution. Secondly the simple minimum criterion for the mean crushing force is derived for a class of progressive axial crushing of prismatic members.

7.3.1   Solution procedure for a steady-state processes

In the energy-time space, $\{E, \tau\}$, function

$$E_{int}^{o}(\tau) = \int_{o}^{\tau} \dot{E}_{int}^{o}(\tau)d\tau; \quad \tau \in [t_o, t_f]$$

defines a single curve referred to as a fundamental energy trajectory. Likewise the set of trial functionals

$$\left\{ E_{int}^{*}(\beta,\tau): \; E_{int}^{*}(\beta,\tau) = \int_{o}^{\tau} \dot{E}_{int}^{*}(\beta,\tau)d\tau; \quad \tau \in [t_o, t_f], \; \beta \in \mathbf{R}^3 \right\}$$

define a set of curves referred to as trial trajectories. All trajectories start from a common origin, corresponding to the reference configuration $v_o = \chi(V, t_o)$, Figure 7.1. Typically, the reference configuration is identified with the initiation of the crushing process.

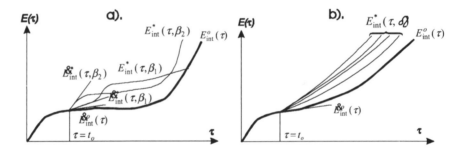

Figure 7.1   a). Fundamental energy trajectory, $E_{int}^{o}(\tau)$, and trial trajectories, $E_{int}^{*}(\tau, \beta)$, with a common origin corresponding to the reference configuration at $t = t_o$. b). A family of non-intersecting trial trajectories.

In the following we shall restrict the class of admissible constitutive laws to rigid/perfectly plastic isotropic materials with convex yield surface and the associated flow rule. In this case all trajectories are non-decreasing functions of time τ, or a time-like parameter, δ. In the vicinity of a reference configuration, $\tau = t_o$, the fundamental and trial trajectories can be expanded into a Taylor series

$$E_{int}^{*}(t_o + \Delta t) = E_{int}^{*}(t_o) + \dot{E}_{int}^{*}(t_o)\Delta t + O(\Delta t^2) \tag{7.7}$$

where the rate, $\dot{E}_{int}^*$, defines the slope of a given trajectory at the reference configuration. The upper bound theorem of the classical Theory of Plasticity for infinitesimal deformations states that out of all kinematically admissible velocity fields, defined over the constant reference configuration, $v_o = \chi(V, t_o)$ an actual velocity field minimizes the rate of energy dissipation.

In terms of energy trajectories defined in the energy space it means that the trial trajectory with the smallest slope at the reference configuration coincides with fundamental trajectory at least in the immediate vicinity of that configuration and with an accuracy to at least first order terms[3]. In other words the upper bound theorem defines certain property of *the state* of the system at the reference configuration and does not provide any definite clues as to *the impending deformation process*. In particular the principle of minimum work cannot be derived from the upper bound theorem, at least in the general case. This conclusion is illustrated schematically in Figure 7.1, which shows that a trial trajectory that coincides with the fundamental trajectory at the reference configuration can intersect other trajectories later in the deformation process. Thus, the total expenditure of work predicted by the corresponding solution might be larger or smaller then an actual work input. Furthermore, a trajectory that is optimal within the immediate vicinity of a reference configuration may diverge from the fundamental solution and vice versa a trajectory, which overestimates an incremental response, may constitute a better approximation later in the process or may be an optimal approximation in an average sense.

There are, however, certain sets of trial trajectories for which the upper bound theorem can still be used as an effective tool in selecting optimal solution. These are the non-intersecting sets of trial trajectories illustrated schematically in Figure 7.1b. In this case a trajectory selected on the basis of a minimum slope condition at the reference configuration remains the best solution later in the deformation process. The necessary condition for a *local minimum* of the slope is

$$\left(\frac{\partial \dot{E}_{int}^*(\tau, \beta)}{\beta} \right)_{\tau = t_o} = 0$$

---

[3] It can be shown that for a stable deformation process the curvature of the trajectory at the reference configuration is also positive.

This condition constitutes a set of N algebraic equations for N unknown components of the optimal vector β. A *global* character of the minimum is usually demonstrated by referring to a particular form of the function $\dot{E}_{int}^{*}$. In an abbreviated notation a set of necessary and sufficient conditions for the global minimum is denoted as

$$\beta^{o} = \min_{\beta}\left[\dot{E}_{int}^{*}\right] \tag{7.8}$$

An example of non-intersecting trajectories is a set of straight lines with a common origin, which describes trial solutions for steady-state deformations. Typical example of a steady-state deformation is propagation of a buckle in pipeline or flow of the material over a toroidal surface during plastic inversion of a cylinder. In steady deformation, all the process parameters are constant in time *at each spatial location*. Therefore, the rate of internal dissipation calculated, as a volume integral over the instantaneous configuration of the body, is also constant in time and the corresponding energy trajectory is a straight line.

All kinematically admissible steady-state processes that originate from the same reference configuration also result in a set of straight lines and quite obviously the solution selected on the basis of the upper bound theorem, Eq. (7.8) provides for the best solution in a global sense. So that, the optimal solution to a steady-state process, β^{o}, can be defined from either of the conditions

$$\beta^{o} = \min_{\beta}\left[\frac{\Delta E_{int}^{*}}{\Delta \tau}\right] \;\; or \;\; \beta^{o} = \min_{\beta}\left[\lim_{\tau \to 0}\frac{\Delta E_{int}^{*}}{\Delta \tau}\right] = \min_{\beta}\left[\dot{E}_{int}^{*}\right] \tag{7.9}$$

Another example of a deformation that can be solved by applying the upper bound theorem is the quasi-steady deformation process. This process is defined as a 'perturbed' steady-state deformation where trial trajectories oscillate periodically around a straight line that represents a stationary process, Figure 7.2.

In Figure 7.2 the period of oscillations is denoted as T_{0}. The optimal solution to the quasi-steady deformation is found in a straightforward way by minimizing the slope at the reference configuration, Figure 7.2. In practical

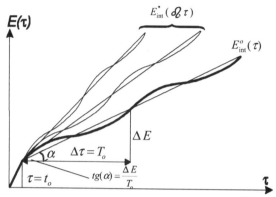

Figure 7.2   The quasi steady-state deformation process and a corresponding family of trial
trajectories. An average slope, α, is defined for the full oscillation cycle, T_o.

calculations, the optimal solution β^o is usually found from an average slope for
the full cycle of oscillations

$$\beta^o = \min_{\beta}\left[\Delta E_{int}^*/T_o\right] \tag{7.10}$$

In Eq. (7.10) ΔE is an increment of internal dissipation corresponding to the
full cycle. Obviously, a number of other approximations based, for example,
on the weighted residuals method can be applied here. Quite surprisingly,
however, this possibility seems not to have received due mention in the
relevant literature.

7.3.2   Progressive crushing and global minimum condition

Progressive crushing of prismatic members is a special case of the
quasi-steady deformation.   It has received a significant attention in the
literature due to its applicability to energy absorbing devices. The crushing of
prismatic members is characterized by the presence of highly localized zones
of plastic flow. Outside these zones the shell is assumed to be rigid. The rigid
parts, however, can be subjected to an arbitrary rigid motion characterized by a
rigid-body translation vector, δ and rigid-body rotation vector θ. In the rigid-
body dynamics, the external loads are the global cross-sectional forces, P, and
moments, M. Consequently, the rate of external work is

$$\dot{E}_{ext}^o = P\dot{\delta} + M\dot{\theta} \tag{7.11}$$

and in general consists of six contributions.

In the next section progressive axial crushing of prismatic members will be discussed as a representative example of the application of the macro element method. In the case of axial crushing it is convenient to use an actual axial shortening of the member, δ as a time-like parameter of the process. Furthermore, for a perfectly/plastic material we can assume, without a loss of generality, that the rate of deformation $\dot{\delta}$ is constant throughout the deformation process. Consequently, multiplying the time coordinate by a constant velocity, $\dot{\delta}$, does the transformation from the energy/time space $\{E, \tau\}$ to the energy/time-like parameter space $\{E, \delta\}$. This transformation changes the shape of trajectories in Figure 7.2 but does not affect relations between their average slopes. So that, the minimum condition, Eq. (7.10), can be rewritten in the form

$$\beta^{\circ} = \min_{\beta} \left[\frac{\Delta E_{int}^*}{\Delta \delta} \right] \tag{7.12}$$

where $\Delta \delta$ is an axial shortening corresponding to a full cycle of oscillations or to the creation of a single plastic lobe. Hereinafter, $\Delta \delta$ will be identified as an effective crushing distance in axially crushed prismatic members. During an axial crushing the total power input to the system is due to the axial crushing force, P, acting on a conjugate rate of axial shortening $\dot{\delta}$, $\dot{E}_{ext}^* = P\dot{\delta}$. The total expenditure of work corresponding to a full cycle of oscillations $\Delta \delta = \delta(t_o + T_o) - \delta(t_o)$, is simply $E_{ext}^* = P_m \Delta \delta$ and equals to the total internal dissipation E_{int}^*. Substituting E_{ext}^* into Eq. (7.12) finally predicts

$$\beta^{\circ} = \min_{\beta} \left[\frac{\Delta E_{int}^*}{\Delta \delta} \right] = \min_{\beta} [P_m] \tag{7.13}$$

The above result is simply the criterion of an absolute minimum of the mean crushing force, P_m, per single plastic lobe. This criterion was conjectured rather then proved by Alexander in 1960, [7.2], and has been used ever since in virtually all solutions concerned with progressive crushing although up to now the conjecture has never been substantiated by a rigorous proof.

7.3.3   Possible generalizations of the global minimum condition

In the preceding sections it has been shown that the solution to the progressive crushing problem can be obtained from the upper bound theorem in the case of a particular set of non-intersecting energy trajectories, Figure 7.1. The question remains, however, whether an actual deformation is governed by the requirement of an absolute minimum of the total expenditure of work. Deeper insight into this problem is obtained from the analysis of the straining history of a representative material point.

In 1986 Hill, [7.3], proved that there exist optimal paths of homogenous deformation between different states of finite strains for which the total expenditure of work gives a global minimum. An optimal path is such that certain triad of material fibers is perpetually orthogonal, while the logarithms of their stretches vary monotonically in fixed ratio. In other words optimal paths correspond to a pure straining deformation which, moreover coincides with a path of proportional loading in the space of logarithmic strains

$$\frac{\log\lambda_1}{\dot{\varepsilon}_1} = \frac{\log\lambda_2}{\dot{\varepsilon}_2} = \frac{\log\lambda_3}{\dot{\varepsilon}_3} = t \tag{7.14}$$

In Eq. (7.14) λ_i denotes i-th component of the principal stretch while $\dot{\varepsilon}_i$ is the rate of logarithmic strain, $\dot{\varepsilon}_i = \dot{\lambda}_i/\lambda_i$. In a plane-strain deformation of incompressible material the principal Eulerian strain-rates are always automatically in fixed ratio, namely $\{1, -1, 0\}$. Hence, the condition of proportional loading, Eq. (7.14), is identically satisfied. Therefore, any (locally) plane pure straining deformation coincides with an absolute minimum of the total work expenditure (for given initial and final strains).

There is strong experimental evidence that in fact the above conditions are fulfilled in a majority of representative deformation mechanisms associated with local collapse of crushed shells. Indeed, the deformed pattern of thin-walled members can be assembled out of axisymmetric shells such as moving and/or stationary cylinders, cylindrical cones and toroids. Moreover, it is observed that plastic deformations are negligible in one of the principal directions and therefore majority of computational models assume inextensibility in this direction. Such an axisymmetric deformations are 'locally plane' in planes that are perpendicular to the direction of inextensibility. Since during such deformations the same material fibers are principal throughout the crushing process both optimum conditions are satisfied. This implies that an actual deformation of crushed shells follows the path of minimal work expenditure.

Even more importantly the above conclusion remains valid for a strain-hardening materials as well as for a purely geometrical state variable.

$$\int_0^f \sqrt{\dot{\varepsilon}_1^2 + \dot{\varepsilon}_2^2 + \dot{\varepsilon}_3^2} \, dt \qquad (7.15)$$

The path integral in Eq. (7.15) is frequently identified with an equivalent strain (or is proportional to such a strain measure). This corollary to the minimum condition has an important implication: if the requirement of an absolute minimum of work is an underlying law that governs crushing response of shells then elements made of highly strain-hardening as well as non-metallic materials should deform in a manner similar to elements made of plastic material. This supposition has strong experimental evidence. It is observed that typical deformation patterns of shell-like structures made of sufficiently ductile materials can be approximated, with a reasonable accuracy, by segments of axisymmetric shells. This simple geometrical observation suggests that regardless of a material property crushing deformation of shells tends to follow an optimal path defined by Eq. (7.14). Therefore the criterion of the minimum mean crushing force is also used in the case of shells made of strain hardening as well as non-metallic materials. The only limitation here seems to be a sufficient ductility of material, so that, large strains in localized deformation zones can be accommodated without damage of material or rupture of a shell.

REFERENCES

[7.1]   Malvern, L. E. *Introduction to the Mechanics of a Continuous Medium*, Prentice-Hall, New Jersey, 1969.

[7.2]   Alexander, J. M. An approximate analysis of the collapse of thin cylindrical shells under axial loading, *Q. J. Mech. Appl. Math.*, **13**(1), 10-15, 1960.

[7.3]   Hill, R. Extremal paths of plastic work and deformation, *J. Mech. Phys. Solids*, **34**(5), 511-523, 1986.

8. THE SUPERFOLDING ELEMENT METHOD

8.1   Introduction to the Superfolding Element (SE) method

The crushing deformation of prismatic shell structures results from a local loss of stability and creation of the so-called local plastic fold or plastic wave.   Once created, the plastic fold accommodates most of the plastic deformation in a shell. The local deformation process continues up to the point where local contacts prevent further deformation of an actual fold and induce development of subsequent fold. Such a deformation process is referred to as a progressive crushing or progressive folding process.

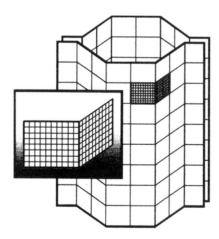

Figure 8.1   A checkerboard of SE in a spot welded hexagonal column illustrates the discretization procedure by using SE's. The insert reveals the 'macro-size' of a SE as compared to the standard FE mesh, [8.1].

An interesting feature of the progressive folding is its 'geometrical similitude'. It has been observed that most of actual deformation patterns can be assembled from a single, typical folding lobe. A single Superfolding Element (SE) models the crushing behavior of such a lobe. This section provides for an overview of underlying concepts of the Superfolding Element method. The presentation starts from general formulation of the SE used in computerized applications. Then, detailed discussions of possible simplifications to the general solution are presented. All results of this section are valid for shells made of plastic isotropic strain hardening materials with the convex yield surface and an associated flow rule.

8.2   Discretization into Superfolding Elements

In the initial undeformed configuration a SE represents the segment of a corner line of a prismatic column refer to Figure 8.1. It is cut off from a column by a set of two parallel horizontal planes. The distance between planes, *2H*, equals the length of the plastic folding wave of the column. The vertical boundaries of a SE are defined by a set of two vertical planes equally distanced from the neighboring corners and/or vertical edges of a column.

8.2.1   Dimensions of a SE

In the initial, undeformed configuration a single SE is defined by four parameters:

1. total length, C, of two arms of a SE, $C = a + b$,
2. central angle, Φ
3. wall thickness t_a of the arm of the length a
4. wall thickness t_b of the arm of the length b

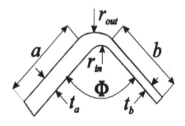

Figure 8.2   Basic cross-sectional dimension of a Superfolding Element shown on the insert in Figure 8.1.

In stamped sheet metal structures $t_a = t_b = t$ while in extruded aluminium components thickness of the two adjacent walls might be different.

It should be noted, at this point, that the height of a SE which corresponds to the length of a plastic folding wave, *2H*, is not given a priori and must be calculated as a part of the solution. Accordingly, a computerized implementation of the SE method requires an adaptive meshing algorithm.

8.2.2   An active layer of folds

A set of Superfolding Elements located between two horizontal planes defines a single layer of plastic folds (also referred to as a deformable cell, Figure 8.3).

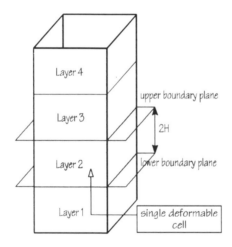

Figure 8.3   A deformable cell represents a single layer of active plastic folds in a progressively crushed prismatic column, [8.1].

The number of SE in a given layer corresponds to the number of corners in a column.   In progressive crushing of real columns plastic deformation are always spread over two neighboring layers.   However, consideration of a single layer at a time is a useful approximation, which leads to accurate results.   An example of possible deformation patterns of a single active layer for various regimes of loading is shown schematically in Figure 8.4, [8.1].

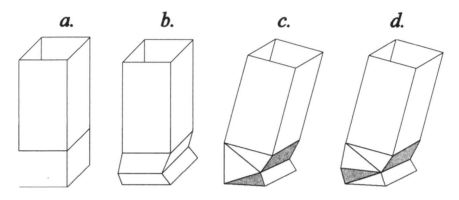

a.   *b.*   *c.*   *d.*

Figure 8.4   Various deformation modes of an active layer of folds in a square prismatic beam
subjected to different loading histories. The symmetric mode (b) corresponds to
the progressive axial crushing process, [8.1].

The present section is concerned with axial loading only. In this case
the deformation of all SE is symmetric and the boundary planes remain parallel
throughout the deformation process.

8.3   Folding modes of a SE

The most general deformation mode of a single SE is shown
schematically in Figure 8.4.   The plastic folding of the element involves
activation of five different deformation mechanisms. These are:

- continuous deformation of a section of the floating toroidal surface,
  1, at the so called corner point
- bending along horizontal stationary hinge lines, 2,
- rolling deformations along moving inclined hinge lines, 3,
- extensional deformations of the conical surface, 4, in the terminal
  phase of the deformation process and
- bending deformations along inclined hinge lines, 3, in the terminal
  phase of the folding process, when the moving hinge line is locked
  within the element.

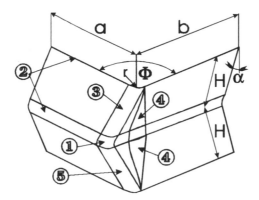

Figure 8.5   Basic folding mechanisms in a deformed Superfolding Element (SE)

The basic folding mechanisms shown in Figure 8.5 are:

①   Deformation of a floating toroidal surface.
②   Bending along stationary hinge lines.
③   Rolling deformations.
④   Opening of a conical surfaces.
⑤   Bending deformations along inclined, stationary, hinge lines
     following locking of the traveling hinge line 3.

The general, folding mechanism in Figure 8.5 is constructed from two simpler folding modes, illustrated in Figure 8.6. These modes are referred to as an asymmetric and symmetric deformation mode, respectively. The mode shown in Figure 8.5 is called an asymmetric mixed mode. A progress of the deformation process in each mode is controlled by a single process parameter α, $0 \le \alpha \le \alpha_f$, which defines the rotation of a side face of an element from the initial upright position, Figure 8.5. At the initiation of the folding process $\alpha = 0$. The process terminates when $\alpha = \alpha_f = \pi/2$. The asymmetric deformation mode does not have a conical surface 4 in Figure 8.5. Consequently, the propagating hinge line, 3, controls the entire folding process. The symmetric deformation mode, on the other hand, is short of the propagating hinge line 3 in Figure 8.5. In this case local extensional plastic deformations are confined to the conical surface 4 as discussed in Chapter 6.

The development of a particular folding mode in Figure 8.5 and Figure 8.6, is controlled by a single switching parameter α^*, $0 \le \alpha^* \le \alpha_f$. This parameter defines a configuration at which a symmetric mode takes over the control of the folding process. If $\alpha^* = \alpha_f$ the folding of a SE is controlled by an asymmetric mode alone while the case $\alpha^* = 0$ corresponds to a purely symmetric mode, see Figure 8.6. For $0 \le \alpha^* \le \alpha_f$ both mechanisms are involved in the folding process: folding starts as an asymmetric mode and continues up to the point where the moving hinge line 3 is locked within an element. At this point the conical surface 4 starts to grow.

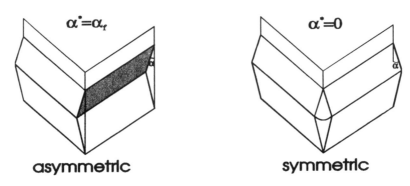

Figure 8.6   Two fundamental folding modes of a single Superfolding Element controlled by limit values of the switching parameter α^*.

An actual value of the switching parameter, α^*, depends on both the input parameters, $\{C, t_a, t_b, \Phi\}$, and constraints imposed onto deforming faces of a SE. In the case of an unconstrained or standing alone SE the asymmetric mode of deformation, $\alpha^* = 0$, is predominant for right angle and acute elements, $\Phi \le \pi/2$, while the symmetric mode controls the folding process of obtuse elements with the central angle, Φ larger then *120* degrees, approximately. In the intermediate range of central angles both modes coexist while the fractional contribution of each mode to the total energy dissipation depends on the central angle, Φ, and the width to thickness aspect ratio, C/t. The folding modes of a standing alone SE are referred to as natural folding modes.

A SE, which is a member of a deformable cell, is constrained by neighboring elements. Kinematic constraints are introduced into the element either by imposing the deformation of one or two arms of the element in a pre-defined direction or by constraining the deformation of the element's corner line. The former case is typical for an assemblage of elements, which model the deformation of a column with closed cross-section. In this case the requirement of the circumferential continuity of the deformation field may

force an element to the deformation mode different then the natural folding
mode. Similarly, constraints imposed onto the corner line can change the
natural deformation mode of an element. During the deformation of 'X' and 'Y'
sections, discussed earlier in this section, the continuity conditions imposed
onto the common corner line of all contributing flanges prevent the
development of asymmetric modes.

8.4   Trial functions

An example of a kinematicaly admissible configuration,
$x^* = \chi^*(X, \beta, t); t \in [t_o, t_f]$, of a single SE is shown schematically in Figure
8.5. In general vector β has four components.

$$\beta^o = \{r, H, \alpha^*, \delta_{eff}\} \tag{8.1}$$

These are (refer to Figure 8.5):

- rolling radius, r,
- length of the plastic folding wave, $2H$,
- switching parameter, α^*
- effective crushing distance, δ_{eff}.

It has been shown in chapter 6 that $\delta_{eff}/2H = 0.73$, approximately, for
all progressive crushing processes of practical importance. Consequently, the
effective number of free parameters of the process is reduced to three, $\beta = \{r,
H, \alpha^*\}$. The velocity field, v^*, and the corresponding rate of deformation
tensor, d^*, Eq. (7.4), could now be calculated from the postulated deformation,
$x^* = \chi^*(V, \beta, \tau); \tau \in [t_o, t_f]$, by an appropriate spatial and time differentiation.
The calculated trial fields are then used to construct trial solutions discussed in
section 7.3. In the next section the trial dissipation E_{int}^* for a SE will be
expressed in terms of the rate of curvature and rate of extensions defined at the
middle surface of the element. Particulars of the transition from a three-
dimensional description of continuum to a two-dimensional formulation of a
shell element (in Eulerian description) are not discussed here.

8.5   Trial solutions

The rate of internal energy dissipation in a deformed shell element results, in general, from the continuous and discontinuous velocity fields, [8.2].

$$\dot{E}_{int} = \int_S \left(M_{\alpha\beta} \dot{\kappa}_{\alpha\beta} + N_{\alpha\beta} \dot{\varepsilon}_{\alpha\beta} \right) dS + \sum_{i=1}^{n} \int_{L_i} M_o^i \left[\dot{\theta}_i \right] dl^i \qquad (8.2)$$

In Eq. (8.2) S denotes the current shell mid surface, n is the total number of plastic hinge lines, L_i is the length of the i-ht hinge while $[\dot{\theta}_i]$ denotes a jump of the rate of rotation across a moving hinge line. Components of the rate of curvature and rate of extensions tensors are denoted, respectively, as $\dot{\kappa}_{\alpha\beta}$ and $\dot{\varepsilon}_{\alpha\beta}$ while $M_{\alpha\beta\,\beta}$ and $N_{\alpha\beta}$ are the corresponding conjugate generalized stresses. For the sake of simplicity in the remaining part of this section the superscript '*' is omitted in all expressions for trial fields. The arguments τ and β will be specified whenever necessary.

Since the assumed deformation fields are axisymmetric and, in addition, inextensible in one of the principal directions, tangent to the shell mid surface, the expression for (trial) rate of internal energy dissipation, Eq. (8.2), reduces to

$$\dot{E}_{int} = \int_S N_o \dot{\varepsilon}_1 \, dS + \sum_{i=1}^{n} \int_{L_i} M_o^i \left[\dot{\theta}_i \right] dl^i \qquad (8.3)$$

where $\dot{\varepsilon}_1$ is the rate of straining in a principal direction, tangent to the shell's mid surface and equals to a corresponding component of the rate of deformation tensor $\dot{\varepsilon}_1 = d_{11}$. The strain rate component perpendicular to the mid surface does not contribute to the internal energy dissipation due to specific form of the Tresca yield condition used in the present calculations. Integrating Eq. (8.3) in the interval $0 \le \alpha \le \alpha_f$ renders the expression for (trial) energy dissipation per single plastic lobe

$$E_{int}(\beta) = \int_0^{\alpha^*} {}_{(1)}\dot{E}_{int}\, d\alpha + \int_{\alpha^*}^{\alpha_f} {}_{(2)}\dot{E}_{int}\, d\alpha \qquad (8.4)$$

Two integrals on the r.h.s of Eq. (8.4), defined through the switching parameter α^*, correspond to the contribution of asymmetric and symmetric modes, respectively.

In practical calculations it is convenient to define following expressions for the membrane and bending contributions to the internal energy dissipation, [8.3-8.5]

$$E_{int}^N \equiv \int_0^\alpha d\alpha \int_S N_0 \dot{\varepsilon} dS = \sigma_0^N(\bar{\varepsilon}) \int_0^\alpha d\alpha \int_S \dot{t} \dot{\varepsilon} dS$$

$$E_{int}^M \equiv \sum_{i=1}^n \int_{L^i} M_o^i \left[\dot{\theta}_i \right] dl^i = \sum_{i=1}^n \,_{(i)} \sigma_0^M(\bar{\varepsilon}) \frac{t}{4} \int_0^\alpha d\alpha \int_{L^i} \left[\dot{\theta}_i \right] dl^i$$

(8.5)

where σ_o^N and σ_o^M denote, respectively, an average level of the flow stress in the entire crushing process. This stress is referred to as an *energy equivalent stress*. The functional dependence of the energy equivalent stresses on the constitutive relation and characteristic deformations of the SE is discussed in the next section.

8.6   The energy equivalent stress measure

Definition of proper flow stress measures is crucial for the accuracy of all calculations based on the macro element approach. Although the stress state in each zone of localized plastic flow is typically one-dimensional the determination of the level of plastic flow is not a straightforward task and require right understanding of the global approach used in macro element method.   This section introduces the global or average stress measures compatible with measures of energy dissipation introduced in the preceding section.

8.6.1   *Quasi-static processes*

The material properties enter the energy relations, Eq. (8.4), via the equivalent stresses σ_o^M and σ_o^N. These stresses are defined, respectively, as

$$\sigma_0^N(\bar{\varepsilon}) = \frac{1}{\bar{\varepsilon}} \int_0^{\bar{\varepsilon}} \sigma(\varepsilon) d\varepsilon$$

$$\sigma_0^M(\bar{\varepsilon}) = \frac{1}{\bar{\varepsilon}^2} \int_0^{\bar{\varepsilon}} \sigma_0^N(\varepsilon) \varepsilon d\varepsilon$$

(8.6)

and correspond to an average level of the plastic flow stress in regions subjected to quasi-static uniaxial tension/compression or bending, characterized by a representative strain $\bar{\varepsilon}$ (in most cases $\bar{\varepsilon}$ corresponds to maximal strain in a given region).   In Eq. (8.6) $\sigma(\varepsilon)$ corresponds to a standard quasi-static tensile characteristic (detailed derivation of the above formulas is given in [8.3]).

8.6.2   Dynamic processes

The visco-plastic constitutive relation, $\sigma(\varepsilon,\dot{\varepsilon})$, is postulated as a product of two contributions

$$\sigma(\varepsilon,\dot{\varepsilon}\ )=\sigma_v(\varepsilon,\dot{\varepsilon}_v\ )\gamma(\varepsilon,\dot{\varepsilon}\ ) \tag{8.7}$$

where $\sigma_v(\cdot)$ corresponds to the standard quasi-static tensile characteristic of a given material, determined at the constant strain rate, $\dot{\varepsilon}_v$, $(10^{-3} \leq \dot{\varepsilon}_v \leq 10^{-4}$ [1/s]) while the dynamic factor $\gamma(\cdot)$ describes strain rate effects. As usual the stress and strain measures, used in the Eq. (8.6) and Eq. (8.7), correspond, respectively, to the Cauchy stress σ and logarithmic strain ε.

8.6.3   Evaluation of energy equivalent measures from standard material tests

Correct definitions of dynamic stress and strain rate measures require additional processing of standard experimental data.   The general processing procedures are explained here on an example of a simple power-type-hardening law. These methods are then applied to an actual stress-strain rate data obtained from laboratory experiments.

8.6.3.1 Quasi-static energy equivalent flow stress.

Consider a simple strain-hardening material defined by the power-type-hardening law:

$$\frac{\sigma(\varepsilon)}{\sigma_y} = \left(\frac{\varepsilon}{\varepsilon_o} \right)^n \tag{8.8}$$

In Eq. (8.8) σ_y corresponds to a 0.2% proof stress ($\varepsilon_o = 0.002$), ε is a current strain while n corresponds to stress hardening exponent. In a plastic deformation process the density of energy dissipation, e, corresponding to an actual level of monotonically increasing straining, $\bar{\varepsilon}$, is given as:

$$e= \int_0^\varepsilon \sigma(\varepsilon)\, d\varepsilon \tag{8.9}$$

which together with Eq. (8.8) predicts:

$$e = \frac{\varepsilon}{n+1} \sigma(\varepsilon) \quad or \quad e = \varepsilon \bar{\sigma}(\varepsilon) \tag{8.10}$$

In Eq. (8.10) the energy equivalent flow stress $\bar{\sigma}(\varepsilon)$ is defined as

$$\bar{\sigma}(\varepsilon) = \frac{\sigma_y}{n+1} \left(\frac{\varepsilon}{\varepsilon_o} \right)^n = \frac{1}{n+1} \sigma(\varepsilon) \tag{8.11}$$

A comparison of both above defined stress measures is shown in Figure 8.7.

8.6.3.2 Dynamic energy equivalent flow stress

In a logarithmic scale the graphs of energy density, Eq. (8.10) and energy equivalent flow stress, Eq. (8.11), correspond to straight lines. Assuming that this feature of the energy density and equivalent flow stress function holds also in the dynamic case using a Cowper–Symonds-type of constitutive relation, conveniently captures the strain-rate effect:

$$e^d(\varepsilon,\dot{\varepsilon}) = e(\varepsilon) f(\dot{\varepsilon}) \quad or \quad \frac{e^d(\varepsilon,\dot{\varepsilon})}{e(\varepsilon)} = 1 + \left(\frac{\dot{\varepsilon}}{D} \right)^{1/p} \tag{8.12}$$

where D and p are two new material constants responsible exclusively for the rate effects. In Eq. (8.12) $\dot{\varepsilon}$ is the rate of strain while the relation $e(\varepsilon)$ is a fictitious static energy density function corresponding to $\dot{\varepsilon} = 0$. In practical applications $e(\varepsilon)$ is identified as an energy density corresponding to the quasi-static tensile test. Once the energy density function is defined by means of the second of Eq. (8.12) the corresponding dynamic energy equivalent flow stress is calculated as:

$$\bar{\sigma}^d(\varepsilon,\dot{\varepsilon}) = \frac{e(\varepsilon)}{\varepsilon} \left[1 + \left(\frac{\dot{\varepsilon}}{D} \right)^{1/p} \right] = \bar{\sigma}(\varepsilon) \left[1 + \left(\frac{\dot{\varepsilon}}{D} \right)^{1/p} \right] \tag{8.13}$$

So that, in the case of linear approximation to the strain rate effect the dynamic factor $\gamma(\cdot)$ in Eq. (8.8) has a form

$$\gamma(\dot{\varepsilon}) = 1 + \left(\frac{\dot{\varepsilon}}{D} \right)^{1/p} \tag{8.14}$$

and does not depend on an actual strain level.

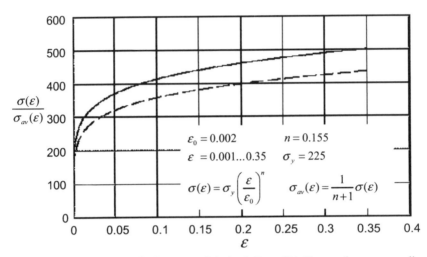

Figure 8.7 A power-type-hardening material (n=0.2), solid line and corresponding energy equivalent flow stress, broken line.

8.6.3.3 Illustrative example

In the case of an actual strain-stress relation deviations of various stress measures form a straight line can be quite significant. However, experimental observations show that typically the dissipation density function, Eq. (8.10) is close to a straight line (in a logarithmic scale), as illustrated in Figure 8.8 for typical mild steel.

Consequently the definition energy equivalent flow stress relations, Eq. (8.6), and material constants D and p in Eq. (8.13) is done in three steps:

- First, the energy density functions are calculated for all experimental dynamic and quasi-static tensile curves.
- Second, the linear approximations to these curves are defined by means of linear regression method. An example of the corresponding procedure is shown in Figure 8.8.
- Finally the D and p values are calculated by applying the linear regression method to the slopes of energy density functions at various deformation rates. This procedure is illustrated in Figure 8.9 for the reference strain equal to 20%.

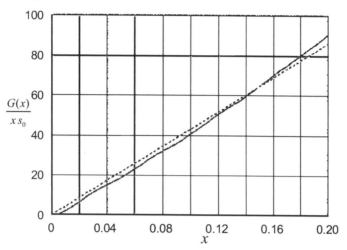

Figure 8.8   The experimental energy density function for a mild steel (solid) and corresponding linear approximation (quasi-static response).

It should be noted in passing that the strain rate constants D and p calculated by means of the above-discussed procedure are quite different then the values reported in the literature for the initial yield of material.   For example the typical values for the initial yield of a mild steel are $p=5$ and $D=40.4$ [1/s] while in the case of energy equivalent flow stress the corresponding data are $p=7.5$ and $D=5466000$ [1/s].

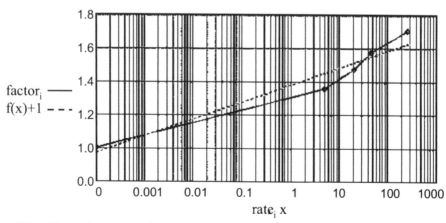

Figure 8.9   Dynamic energy density function normalized with respect to the quasi-static response as a function of the strain rate.   Energy density is calculated for a reference strain of 20%.   Beaded line corresponds to experimental findings while the broken line shows results of the linear regression method.

REFERENCES

[8.1] *Crash Cad User's Manual*, Impact Design, Europe, 1998.

[8.2] W. Abramowicz and T. Wierzbicki, Axial crushing of multi-corner sheet metal columns, *J. App. Mech.*, **56**, 1, 113-120 (1989).

[8.3] W. Abramowicz, *Crush Resistance of 'T' 'Y' and 'X' Sections*, Joint MIT-Industry Program on Tanker Safety, Massachusetts Institute of Technology, Report 24 (1994).

[8.4] Abramowicz, W., Jones, N., Dynamic Progressive Buckling of Circular and Square Tubes., *Int. J. Impact Engng.*, 4, 4, 243 - 270, 1986.

[8.5] Wierzbicki, T., Abramowicz, W. *The Manual of Crashworthiness Engineering*, Vol. I - IV, Centre for Transportation Studies, Massachusetts Institute of Technology, 1987 - 1989.

9. CRUSHING RESPONSE OF SIMPLE STRUCTURES

9.1   Introduction

The explicit expression for the mean crushing force, P_m, corresponding to a complete folding of a standing alone SE is derived by substituting expressions for contributing energy dissipation mechanisms, Eq. (8.5), into the governing minimum condition, Eq. (7.13).   The complete derivation of the following result is given in [9.1]. The final form of the governing expression is

$$P_m = \frac{t^2}{4}\{\sigma_o^{\ N}(\bar{\varepsilon}_1)A_1\frac{r}{t}+\sigma_o^{\ M}(\bar{\varepsilon}_2)A_2\frac{C}{H}+\sigma_o^{\ M}(\bar{\varepsilon}_3)A_3\frac{H}{r}+$$
$$+\sigma_o^{\ M}(\bar{\varepsilon}_4)A_4\frac{H}{t}+\sigma_o^{\ N}(\bar{\varepsilon}_5)A_5\}\frac{2H}{\delta_{\textit{eff}}} \tag{9.1}$$

where $A_i = A_i(\Phi, \alpha^*)$, $i = 1, 2...5$.   The five terms in parenthesis on the right hand side of Eq. (9.1) describe, respectively, fractional contributions to the total energy dissipation resulting from five elementary deformation mechanisms, identified in Figure 8.5. The five factors, A_i, $i = 1,2...5$ result from the surface--time integration.   Factors. A_2 and A_4, are easily calculated as a closed-form functions of geometrical parameters. The remaining factors A_1, A_3 and A_5, are functions of elliptic integrals and must be calculated numerically. The meaning of other variables appearing in Eq. (8.2) is explained in section 8.2.1.

Parameters corresponding to the equilibrium of a SE are determined from the set of three non-linear algebraic equations, resulting from the minimum condition, Eq. (7.13)

$$\frac{\partial P_m}{\partial H} = 0 \; ; \quad \frac{\partial P_m}{\partial r} = 0 \; ; \quad \frac{\partial P_m}{\partial \alpha^*} = 0 \tag{9.2}$$

In general the set of governing equations, Eq.'s (8.3), has no closed-form solution. Such a solution exists, however, for two fundamental folding modes of a Superfolding Element illustrated in Figure 8.6. Such a solution will be discussed later in this section in conjunction with the crushing behavior of square columns.

9.2   Single Superfolding Element

Equation (8.2), can be easily generalized to the case of a Superfolding Element with various thicknesses, t_a and t_b, of two arms, a and b respectively, refer to Figure 8.2.

$$P_m = \frac{1}{4}\left\{ t_1^2\sigma_o^{\ N}(\bar{\varepsilon}_1)A_1\frac{r}{t_1}+A_2\left(\sigma_o^{\ M}(\bar{\varepsilon}_2)\frac{at_a^2}{H}+\sigma_o^{\ M}(\bar{\varepsilon}_2)\frac{bt_b^2}{H}\right)+ \right.$$

$$\left. t_3^2\sigma_o^{\ M}(\bar{\varepsilon}_3)A_3\frac{H}{r}+t_4^2\sigma_o^{\ M}(\bar{\varepsilon}_4)A_4\frac{H}{t_4}+t_5^2\sigma_o^{\ N}(\bar{\varepsilon}_5)A_5\right\}\frac{2H}{\delta_{eff}} \tag{9.3}$$

Such a cross-section is typical for extruded aluminium elements where the difference in thickness of neighboring walls can be as large as one hundred percent or even more. The plastic energy dissipation in consecutive folding lobes depends now on the direction of folding of a SE. For example, in the case illustrated in Figure 8.5 the moving hinge line 3 sweeps the face a of the element and, thus, involves the deformation of a shell of the thickness t_a. At the some time the conical surface 4 is developed at the boundary between both faces and propagates into a thinner face. This feature of the folding process is reflected in the governing equation, Eq. (9.3), by specifying an appropriate thickness for each contributing mechanism. For example, the deformation pattern in Figure 8.5 corresponds to the following set of thickness, t_i $i = 1, ...5$:

- $t_1 = t_3 = t_5 = t_a$ and
- $t_4 = t_a$ for $t_a \le t_b$ or $t_4 = t_b$ for $t_b < t_a$.

Similarly as in the case of a uniform-thickness element the equilibrium solution to the Eq. (9.3) is obtained via the minimization procedure, Eq. (8.3). It should be noted, however, that in this case a useful approximation of the folding mode by pure asymmetric or symmetric folding modes does not apply, even in the case of a right-angle element, and consequently, the solution must be found by minimizing the complete governing equation, Eq. (9.2).

9.3 Crushing response of an assemblage of Superfolding elements

The crushing response of a single layer of folds, compare Figure 8.3 and Figure 8.4 is calculated by summing up fractional contributions of all SE in a given layer. Accordingly, the governing equation of the problem is

$$P_m = \sum_{i=1}^{J} \frac{t_i^2}{4} \{ \sigma_0^N(\bar{\varepsilon}_1) \, A_1^i \frac{r}{t_i} + \sigma_o^M(\bar{\varepsilon}_2) \, A_2^i \frac{C_i}{H} + \sigma_o^M(\bar{\varepsilon}_3) \, A_3^i \frac{H}{r} +$$

$$+ \sigma_o^M(\bar{\varepsilon}_4) \, A_4^i \frac{H}{t_i} + \sigma_o^N(\bar{\varepsilon}_5) \, A_5^i \} \frac{2H}{\delta_{eff}}$$

(9.4)

where the summation is extended over the J contributing SE. It is assumed here that the column is made of one material, so that, all average stresses are calculated on the basis of a single constitutive relation. Each element, however, may have different geometrical dimensions: C_i, Φ_i and t_i, $i = 1, 2, ... J$.

Since, all elements in a given layer of folds deform with the same length of the folding wave, $2H$, there is only one 'H' parameter in the governing equation. In order to simplify the calculation routines it is also assumed that values of the two other free parameters, i. e. the rolling radius r and the switching parameter α^*, are also the same in all contributing elements. Consequently, the solution procedure for an assemblage of SE parallels the corresponding procedure for a single SE.

9.4 Closed form solution for rectangular columns

Theoretical procedures developed in preceding sections are applied here to the crushing response of a rectangular column of uniform thickness, t, and cross - sectional dimensions $2a$ and $2b$, Figure 9.1.

Figure 9.1   Cross-sectional dimensions of a rectangular column.

The column is made of a strain hardening material defined by the power - type constitutive law, Eq. (8.8). The governing equation of the crushing problem, Eq. (9.4), can now be rewritten in a simplified form, applicable to right - angle elements, [9.2]

$$P_m = \frac{t^2}{4}(\sigma_o^1 A_1 \frac{r}{t} + \sigma_o^2 A_2 \frac{C}{H} + \sigma_o^3 A_3 \frac{H}{r}) \frac{2H}{\delta_{eff}}$$ (9.5)

where σ_o^i, $i = 1,2,3$, denotes equivalent stresses in three main regions of plastic deformations, refer Figure 8.5, while C is the length of a representative SE, $C = (a+b)$, refer to Figure 9.1. For a power-type constitutive relation, Eq. (8.8), energy equivalent stresses, σ_o^N and σ_o^M are given as

$$\frac{\sigma_o^N (\bar{\varepsilon})}{\sigma_y} = \frac{1}{1+n}\left(\frac{\bar{\varepsilon}}{\varepsilon_0}\right)^n$$ (9.6a),

$$\frac{\sigma_o^M (\bar{\varepsilon})}{\sigma_y} = \frac{2}{(n+1)(n+2)}\left(\frac{\bar{\varepsilon}}{\varepsilon_0}\right)^n$$ (9.6b)

where $\bar{\varepsilon}$ denotes the characteristic final strain. In the case of an uniform tensile-compression deformations $\bar{\varepsilon}$ corresponds to the final strain measured at any point of the deformed region whereas in the case of bending deformations $\bar{\varepsilon}$ equals to the final strain at an outer fibre

The relevant strains for three main mechanisms of plastic deformation in Eq. (9.5) are readily determined from the kinematics of the folding process. Consider first rolling deformations. The final strain at the outer fibre of a travelling hinge line 3 Figure 8.5 is inversely proportional to the rolling radius, r, and equals

$$\varepsilon_1 = \ln(1+\frac{t}{2r})$$ (9.7)

Similarly, the final strain at the outer surface of horizontal stationary hinge lines is

$$\varepsilon_2 = \ln(1+\frac{t}{2R})$$ (9.8)

where R is the large radius of the toroidal surface. It follows from the kinematics of the process and in particular from the formula for the effective crushing distance that radius R equals $R = 0.54H$ where $2H$ is the length of the plastic folding wave. Thus, the representative strain in the region 2 of Figure 8.5, equals

$$\varepsilon_2 = \ln(1 + 0.93\frac{t}{H}) \tag{9.9}$$

Determination of a compressive strain in the moving segment of a toroidal surface, 1, in Figure 8.5 involves complex calculation, which exclude the possibility of getting a desired closed-form solution. However, separate numerical and experimental studies show that strains in this region are of the same order as those in moving hinge lines 3, [9.3]. Therefore, in the following analytical calculations Eq. (9.9) is used as a convenient approximation of final strains at material points on the toroidal surface.

Further simplifications to the expression for a characteristic strain, $\bar{\varepsilon}$, essential in analytical calculations, are achieved by expanding Eq. (9.5) and Eq. (9.9) into power series

$$\varepsilon_1 \cong \frac{t}{2r} \ ; \ \varepsilon_2 \cong 0.93\frac{t}{H} \ ; \ \varepsilon_3 \cong \frac{t}{2r} \tag{9.10}$$

where only the linear terms are retained. Substituting Eq. (9.10) back into the expressions for energy equivalent flow stresses, Eq. (9.6), finally yields

$$\sigma_1 \equiv \sigma_o{}^N(\varepsilon_1) = \sigma_y a_1 \left(\frac{t}{r}\right)^n \ ; \ a_1 \equiv \frac{1}{n+1}\left(\frac{0.5}{\varepsilon_y}\right)^n \tag{9.11a}$$

$$\sigma_2 \equiv \sigma_o{}^M(\varepsilon_2) = \sigma_y a_2 \left(\frac{t}{H}\right)^n \ ; \ a_2 \equiv \frac{2}{(n+1)(n+2)}\left(\frac{0.93}{\varepsilon_y}\right)^n \tag{9,11b}$$

$$\sigma_3 \equiv \sigma_o{}^M(\varepsilon_3) = \sigma_y a_3 \left(\frac{t}{r}\right)^n \ ; \ a_3 \equiv \frac{2}{(n+1)(n+2)}\left(\frac{0.5}{\varepsilon_y}\right)^n \tag{9.11c}$$

Having determined energy equivalent stresses, σ_i, as functions of kinematical parameters, r and H, the governing equation, Eq. (9.5), can be rewritten in the form

$$P_m = \frac{t^2 \sigma_y}{4} \left\{ A_1 \left(\frac{r}{t}\right)^{1-n} + A_2 \frac{C}{t} \left(\frac{t}{H}\right)^{1+n} + A_3 \frac{H}{t} \left(\frac{t}{r}\right)^{1+n} \right\} \frac{2H}{\delta_{eff}} \qquad (9.12)$$

$$A_i \equiv A_i\, a_i$$

where constant coefficients A_i, $i=1,\ 2,\ 3$, are functions of the central angle of a SE while coefficients a_i, $i=1,\ 2,\ 3$, defined in Eq. (9.11), are functions of material parameters only. For example, a set of coefficients A_i pertinent to the crushing response of a SE with the central angle $\Phi=90^o$ is:

$$A_1 = 8\, I_1 \; ; I_1 = 0.555$$

$$A_2 = \pi \qquad (9.13)$$

$$A_3 = 2\, I_3 \; ; I_3 = 1.148$$

The equilibrium of the system, governed by Eq. (9.12) is defined by applying the minimum conditions, Eq. (9.2).   The final results for optimal free parameters of the process, i.e., length of the folding wave $2H$ and rolling radius r, are, [9.1]:

$$\frac{r}{t} = A_1^{\frac{2+n}{3+n}} A_2^{\frac{1}{3+n}} A_3^{\frac{1+n}{3+n}} K_1 \left(\frac{C}{t}\right)^{\frac{1}{3+n}} \qquad (9.14a)$$

$$K_1 = (1+n)(1-n)^{-\frac{2+n}{3+n}} \qquad (9.14b)$$

$$\frac{H}{t} = A_1^{\frac{1+n}{3+n}} A_2^{\frac{2}{3+n}} A_3^{\frac{1-n}{3+n}} K_2 \left(\frac{C}{t}\right)^{\frac{2}{3+n}} \qquad (9.14c)$$

$$K_2 = (1+n)(1-n)^{-\frac{1+n}{3+n}} \qquad (9.14d)$$

Substituting Eq. (9.14) back into Eq. (9.12) renders the closed-form solution for the average crushing force P_m.

$$\frac{P_m}{M_y} = A_1^{\frac{(1+n)^2}{3+n}} A_2^{\frac{(1-n)}{3+n}} A_3^{\frac{1-n^2}{3+n}} \left(\frac{C}{t}\right)^{\frac{1-n}{3+n}} *$$

$$* \left[K_1^{1-n} + K_2^{-(1+n)} + K_2 K_1^{-(1+n)} \right] \frac{2H}{\delta_{eff}} \qquad (9.15)$$

In Eq. (9.15) the fully plastic bending moment, M_y, is calculated with respect to the yield stress of material, σ_y, and equals, $M_y = \sigma_y t^2/4$. For the perfectly-plastic material, $n = 0$, and results given in Eq. (9.14) and Eq. (9.15) converge to the solution for a perfectly plastic material.

$$\frac{H}{t} = \sqrt[3]{\frac{A_2^2}{A_1 A_3}} \sqrt[3]{\left(\frac{C}{t}\right)^2} \; ; C = (a + b) \tag{9.16a}$$

$$\frac{r}{t} = \sqrt[3]{\frac{A_2 A_3}{A_1^2}} \sqrt[3]{\frac{C}{t}} \tag{9.16b}$$

$$\frac{P_m}{M_y} = 3\sqrt[3]{A_1 A_2 A_3} \sqrt[3]{\frac{C}{t}} \frac{2H}{\delta_{eff}} \tag{9.16c}$$

which for given values of constants A_i, Eq. (9.16), and a given non-dimensional ratio of the length of the plastic folding wave to the effective crushing distance, $2H/\delta_{eff} \cong 1.37$, finally yields.

$$\frac{H}{t} \cong \sqrt[3]{\left(\frac{C}{t}\right)^2} \; ; C = (a + b) \tag{9.17a}$$

$$\frac{r}{t} = 0.715 \sqrt[3]{\frac{C}{t}} \tag{9.17b}$$

$$\frac{P_m}{M_y} = 13.052 \sqrt[3]{\frac{C}{t}} \tag{9.17c}$$

The later result is valid for a single right - angle element and differs form the result for the entire column by the level of non - dimensional mean force, P_m/M_y. For four SE contributing to the crushing strength of a rectangular column the mean crushing force is simply four times larger then the resistance of a single SE and equals, [9.4]

$$\frac{P_m}{M_y} = 52.2 \sqrt[3]{\frac{C}{t}} \; ; \quad C = (a + b) \tag{9.18}$$

while the length of the plastic folding wave, $2H$, and the magnitude of rolling radius, r, are the same as in the case of a standing alone element.

9.5   Dynamic response

In the constitutive relation, Eq. (8.14), the dynamic energy equivalent flow stress is given as a product of quasi-static stress and dynamic factor γ that depends on the strain rate only.   Thus, in the present formulation the strain-hardening and strain-rate effects are effectively decoupled. This feature of the constitutive relation greatly simplifies calculations for dynamic crush. Indeed, for a dynamic crushing process characterized by a constant strain-rate $\dot{\varepsilon}$, the integration of energy equivalent stresses, Eq. (8.5), reduces to the multiplication of the quasi-static average stress by the dynamic factor $\gamma(\dot{\varepsilon})$

$$\sigma_0^{N}(\bar{\varepsilon},\dot{\varepsilon}) = \frac{1}{\bar{\varepsilon}} \int_0^{\bar{\varepsilon}} \sigma(\varepsilon)\gamma(\dot{\varepsilon})d\varepsilon = \gamma(\dot{\varepsilon})\sigma_0^{N}(\bar{\varepsilon})$$

$$\sigma_0^{M}(\bar{\varepsilon},\dot{\varepsilon}) = \frac{1}{\bar{\varepsilon}^2} \int_0^{\bar{\varepsilon}} \sigma_0^{N}(\varepsilon,\dot{\varepsilon})d\varepsilon = \gamma(\dot{\varepsilon})\sigma_0^{M}(\bar{\varepsilon})$$

(9.19)

Consequently, the dynamic crushing force, P_m^d, is also calculated by multiplying the quasi-static force, P_m, by the dynamic correction factor, $\gamma(\dot{\varepsilon})$.

The average rate of strain, $\dot{\varepsilon}$, is calculated from the known kinematics of the folding mechanism and assumed constant velocity, $\dot{\delta}$, of compression of a single lobe.   The average strain rates depend on the deformation mode of a SE and equal

$$\dot{\varepsilon} \cong 0.33\frac{\dot{\delta}}{C}$$

$$\dot{\varepsilon} \cong 0.43\frac{\dot{\delta}}{C}$$

(9.20)

$$\dot{\varepsilon} \cong 0.25\frac{\dot{\delta}}{C}$$

for the asymmetric, mixed and symmetric modes, respectively (details of the derivation procedure are given in [9.4]).   In Eq. (9.20) the velocity of axial compression $\dot{\delta}$ must be given in meters per second, *[m/s]*, while C (the total cross-sectional length of a representative SE) is given in meters, *[m]*. For a cross-section composed of more then one SE the representative length, C, should be calculated by dividing the entire cross-sectional length, L, by the number of contributing elements, J.

Having determined a representative level of the strain-rate the mean dynamic crushing force, P_m^d, is calculated from the formula

$$P_m^d = P_m \left\{ 1 + \left(\frac{\dot{\varepsilon}}{D} \right)^{1/p} \right\} \qquad (9.21)$$

where P_m is the quasi-static mean crushing force defined Eq. (9.15).

REFERENCES

[9.1]   W. Abramowicz, *Crush Resistance of 'T' 'Y' and 'X' Sections*, Joint MIT-Industry Program on Tanker Safety, Massachusetts Institute of Technology, Report 24, 1994.

[9.2]   W. Abramowicz and T. Wierzbicki, Axial crushing of multi-corner sheet metal columns, *J. App. Mech.*, **56**(1), 113-120, 1989.

[9.3]   Wierzbicki, T., Abramowicz, W. *The Manual of Crashworthiness Engineering*, Vol. I - IV, Center for Transportation Studies, Massachusetts Institute of Technology, 1987 - 1989.

[9.4]   Abramowicz, W., Jones, N., Dynamic progressive buckling of circular and square tubes, *Int. J. Impact Engng.*, **4**(4), 243 - 270, 1986.

10. DESIGN AND CALCULATION OF ENERGY ABSORBING SYSTEMS

10.1   Design of energy absorbing structures

Design of energy absorbing structures is achieved in several steps, usually in a highly iterative design/calculation loop. Objectives of each design step are secured by usage of dedicated computational tools typically different for each step. The selection of an appropriate tool for a given step depends on the complexity of the problem and availability of suitable software or other predictive techniques such as engineering formulas, empirical data or handbook - type of information.   For example, in the pre - design or pre - prototyping stages factors such as the specific energy absorption per unit length or the maximal moment capacity of a cross - section are of primary importance.   In the case of typical cross - sections such information is readily available in handbooks. In the case of more complex, real - word, members the desired parameters can be easily calculated by using specialized tools, such as CRASH CAD[1].

Using of commercial FE packages at early stages of design process are not justified due to the excessive modeling effort. On the other extreme, only professional FE packages such as PAM-CRASH, DYNA or RADIOSS guarantee the detailed, full crash simulation. In the intermediate stages of the design process usage of both computational tools is highly desirable and can

---

[1] CRASH CAD is a highly interactive CAD/CAE software based on the Superfolding Element approach introduced in the preceding sections.

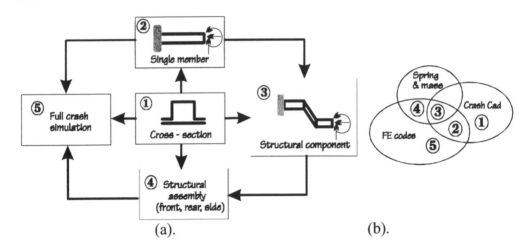

Figure 10.1   (a). Five stages of the design process leading to a crashworthy system. (b). Areas
of applicability of various tools in the simulation based design environment.

significantly increase the efficiency of the design process. Example of such a
process, typical for automotive industry, is shown schematically in Figure 10.1.
This figure also shows a chart illustrating possible application of various
predictive tools at different stages of the simulation based design process.

This section shows how the concept of Superfolding Element,
implemented in the program CRASH CAD, is used at various stages of the
design process. It is shown how synthetic approach to the design process
proceeds from selection of a cross - section and its basic dimensions, through
component design into the design of structural assemblages and culminates in
full FE crash simulation.

10.2   Crash analysis at the level of a cross - section.

The design of a structural member at the cross - sectional level is
especially important at the pre - design and early design stages when the proper
shape (topology) and optimal dimensions of a member are sought and the design
concept undergoes frequent modifications. Application of CRASH CAD at this
level is especially attractive as the program requires as input only overall
dimensions of the cross – section, and tensile characteristic of the material while
the calculation process takes only few seconds on a standard PC. Consequently,
designer can examine a wide range of cross - sectional topologies and run several
parametric studies within only few hours of work. The simplicity of CRASH CAD

input is illustrated in Figure 10.2 for a typical hat cross - section made of aluminium alloy.

An example of ranking of various cross - sectional topologies is shown in Figure 10.3 for a number of aluminium sections tested at Alcan laboratories, [10.1]. All sections were made of a 2-millimeter thick AA5754-0 aluminium sheets and tested under the axial crash loading. For the sake of comparison Figure 10.3 shows both experimental data and CRASH CAD predictions for the average crush force (specific energy absorption capacity per unit length) and length of the plastic folding wave – two; factors of primary importance for the optimal design of sections against crash.

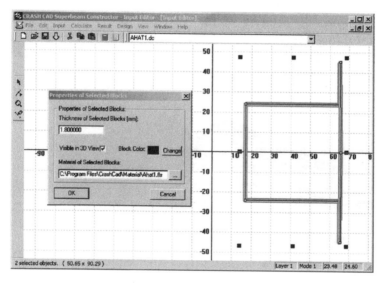

Figure 10.2   Input data screen of Crash Cad with a discretized cross - section of a hat member. Only overall dimensions of a section are needed to define the input data file.

Geometry 2 mm AA5754-O Spot-Welded					
Average Crush☐ Force (kN)⁻	CRASH CAD	28.1	53.2	41.9	66.6
	Experiment	26.0	49.6	39.8	70.1
Fold Half☐ Wavelength (mm)	CRASH CAD	35.9	32.4	36.0	46.1
	Experiment	37.5	33.5	36.5	41.5

Figure 10.3   Ranking of various cross-sectional topologies from the point of view of energy absorption capacity. The comparison between experimental and theoretical data emphasizes a typical accuracy of CRASH CAD in crash calculations, [10.1].

10.3   Crash analysis at the level of a single member.

Calculations at the cross - sectional level are the first essential step in all CRASH CAD calculations. This section shows how basic calculations, discussed in the previous section, are used to predict crushing response of a single prismatic member.

10.3.1   Design of a section for axial crush.

One of the most sensitive parts of the design at the level of a single prismatic section is concerned with progressive folding during a head - on collision. Among all possible deformation modes of a prismatic section the progressive folding absorbs most of the impact energy. At the same time, this folding mode is the one most difficult to obtain in a real - world design. Development of progressive folding requires simultaneous completion of several conditions. These are:

- The cross - section topology must be properly designed, so that, the local deformation of a section in each plastic lobe can be accommodated without internal contacts and penetrations.   In addition, the deformation of each plastic lobe must be compatible with the deformation of it's closest neighbour,
- Spot welds (rivets or laser weld - line) must not interfere with the local plastic deformation of a section,
- The section must be properly 'triggered' through the introduction of correctly designed hoop dents which guarantee the development of a proper 'natural' folding mode and reduce the peak load to such a level that the potentially unstable plastic deformations are induced only in the region of triggering dents and finally
- The boundary and loading conditions (stiffness of joints, loading direction) are kept in the range that guarantees the predominantly axial loading of the section.

The first three conditions, pertinent to the level of a single member, must be met at the design stage of a given member while the last condition must be checked at the level of full crash simulation of a car. This difficult task requires an interaction of a component level design tool with the full FE crash simulation model.

10.3.2 Design of the cross - sectional topology.

The design of a single member for axial crash is done in the CRASH CAD *design loop*, [10.2]. The program leads the user through several designs loops and suggests necessary corrections to the cross - sectional geometry up to the point when the section can collapse progressively without internal contacts and/or penetrations. This stage of the design requires fine-tuning of central angles, widths of side faces and appropriate geometry of spot welding. An example of initial, bad design of a corrugated panel and final, correct design is shown in Figure 10.4. It transpires from Figure 10.4 that both cross-sections are quite similar and the decision on the correctness of the design is impossible without a detailed numerical simulation.

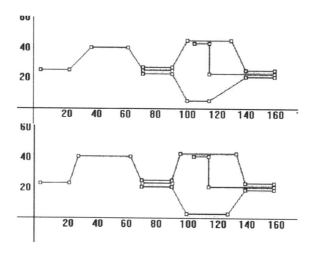

Figure 10.4   Initial (bad) and final (correct) topologies of a corrugated panel optimized fro axial crash through a proper selection of central angles and widths of side faces, [10.2].

10.3.3 Design of a triggering mechanism.

The final stage of cross-section design is concerned with appropriate triggering mechanism. An example of triggering dents, designed on the basis of SF modeling, are shown in Figure 10.5, respectively, for an octagonal (BMW) column with flanges.

Introduction of triggering dents in members designed for axial crash is necessary in order to promote a desired progressive folding pattern and reduce the peak force below the level, which is likely to induce a global, Euler - type

Figure 10.5 Paper model showing triggering dents in an octagonal (BMW) column with flanges (triggering dents designed on the basis of CRASH CAD calculations).

buckling of a column. Triggering of columns is especially important for complex cross - sections that develop a large number of natural folding modes. Usually only few of these modes are likely to converge to the desired progressive folding pattern while other modes lead to a premature column bending.

The importance of proper triggering is further clarified on an example of a laboratory experiment on a simple cross - sectional geometry, [10.3]. The first photograph in Figure 10.6 a shows a long square prismatic column made of mild steel. Such a column has a cross - sectional topology that guarantees proper folding without internal contacts and does not contain spot welds that may destroy progressive folding pattern. A properly triggered square column collapses progressively up to the point when the last plastic fold is completed.

For example, the 54x54x1.4 mm square column at the top of Figure 10.6a was squeezed over 700 mm and developed 32 symmetric plastic lobes with no sign of any global bending (the completely squeezed column in this figure was initially one-half of the size of the longer column and developed 16 lobes). The total energy absorbed was 25 kJ, which is sufficient to bring to the rest a 1000 kg car travelling with an initial velocity of 25 km/h. On the other hand much shorter, untriggered, columns in Figure 10.6b collapsed in bending or deteriorated from progressive folding despite of a carefully controlled loading and boundary conditions at both ends of a column.

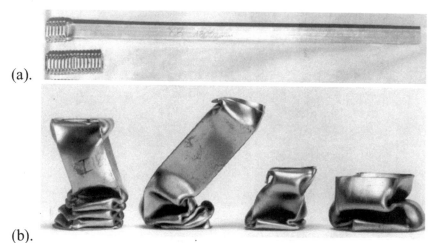

Figure 10.6   Progressive collapse of a properly triggered long square column, top, and global
bending or irregular folding of untriggered columns, bottom.

10.4   Structural components.

At the next level of a modeling process individual prismatic
components are assembled into more complex structural assemblies. While the
crushing characteristics as well as interaction curves for each individual
component are known from the previous step of calculations the main
difficulty at this stage of design/calculation process is concerned with the
construction of the appropriate mechanical model for a given configuration of
structural components. The modeling complexity at this stage is similar to that
encountered in the construction of a spring and mass or multibody system
models for given characteristics of *non - linear* springs. Simple models of
structural assemblies are also developed by using Superbeam Elements.
Example of such a model is presented in the next section.

10.4.1   The 2-D 'S'-Frame model

The 'S' frame model is used to determine the crushing response of
planar deformation of 'S' frame composed of prismatic segments with an
arbitrary cross - sectional topology. The entire 'S' frame is discretized into only
four 'Superbeam' elements (Superbeam is modeled by means of SF elements).
Consequently, the input procedure requires the specification of just overall
dimensions of a frame, as illustrated in Figure 10.7.

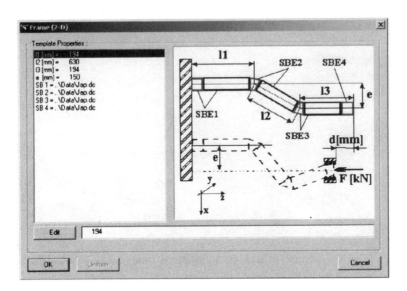

Figure 10.7 The input template of 'S' frame model. The whole frame is discretized into four
Superbeam Elements. The user is asked to specify the dimensions shown in the
figure.

The 'S' frame module calculates the peak force, energy absorption, as
well as the whole force - deflection characteristic for a given frame.   An
example of calculations is shown in Figure 10.8 together with corresponding
experimental data taken from the SAE paper by Y. Ohkami et al, [10.4].

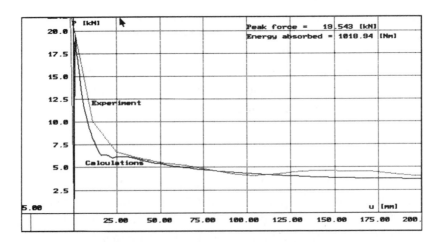

Figure 10.8 Comparison of Crash Cad calculations with experimental results reported in [10.5].
The energy absorption is calculated for the crush of 200 [mm].

10.5   Conclusions

Present chapter addresses only a limited number of issues concerned with macro element method. First of all it should be recognized that despite all the unquestionable successes of numerical simulation methods (e.g. FE) their limitations are obvious. Designing of cars and/or predicting crushing response of structures require intuitive judgment and qualitative assessment; processes that cannot be precisely quantified. Recognition of this fact lead to the development of the so-called *Qualitative Physics* which involves methods such as commonsense reasoning, expert-system-like techniques, diagrammatic reasoning or hybrid methods, see e.g. [10.6]. The Superfolding Element approach seems to be a method that embraces these two recognized areas of structural mechanics in a particular case of auto body design process. The brief overview of the synthetic approach to the simulation-based design process, discussed in this chapter, shows how both techniques could be combined into a single integrated environment for crash simulation. In fact such environments are currently under the development by leading car manufacturers. However, the related modeling methods and solution techniques are not yet available in the open literature.

Secondly, a limited content of the present chapter does not allow for the full exposition of the macro element method. The results presented in chapters 7 and 8 are limited to the basic formulation of the method and basic examples of calculation routines, [10.7], [10.5]. Problems such as bending and torsion crush response [10.8],[10.9], interaction of cross-sectional forces [10.10], or stability of collapse [10.3], [10.5], are not covered here. Interested readers are referred to a number of papers and books listed at the end of this chapter for more information regarding these subjects.

REFERENCES

[10.1]   McGregor, I., J., et al, Impact Performance of Aluminium Structures, In *Structural Crashworthiness and Failure*, N. Jones and T. Wierzbicki, Eds., Elsevier, 1993.

[10.2]   *Crash Cad User's Manual*, Impact Design, Europe, 1998.

[10.3]   Abramowicz, W. and Jones, N., Transition from static and dynamic progressive buckling to global bending of thin - walled tubes, *International Journal of Impact Engineering*, (to be published).

[10.4]   Ohkami, Y., et al., Collapse of thin - walled curved beam with closed - hat section - part. 1: study on collapse characteristic, SAE paper No 900460, 1990.

[10.5]   Wierzbicki, T., Abramowicz, W., *The Manual of Crashworthiness Engineering, Vol. I - IV*, Center for Transportation Studies, Massachusetts Institute of Technology, 1987 - 1989.

[10.6]   Kleiber, M. and Kulpa, Z., Computer-assisted hybrid reasoning and physical systems, *Computer Assisted Mechanics and Engineering Sciences*, **2,** 165-186, 1995.

[10.7]   Abramowicz, W. and Wierzbicki, T., Axial crushing of multi-corner sheet metal columns, *J. App. Mech.*, **56**(1), 113-120, 1989.

[10.8]   Wierzbicki, T., Recke L., Gholami, and Abramowicz, W., Stress profiles in thin-walled prismatic columns subjected to crush loading. part i. compression, *Computers & Structures*, **51**(6), 611-623, 1994.

[10.9]   Wierzbicki, T., Recke L., Gholami, and Abramowicz, W., Stress profiles in thin-walled prismatic columns subjected to crush loading. part ii. bending, *Computers & Structures*, **51**(6), 625-641, 1994.

[10.10] T. Wierzbicki and W. Abramowicz, Deep plastic collapse of thin-walled structures, In *Structural Failure*, T. Wierzbicki and N. Jones Eds., John Wiley, New York, 1989.

Part III

NONLINEAR EXPLICIT FINITE ELEMENT METHODS

T. Belytschko
Northwestern University, Evanston, IL, USA

11. GOVERNING EQUATIONS AND WEAK FORM

11.1 Introduction

Finite element discretizations for Lagrangian meshes are described. The notes are based on reference 1 and more details can be found on this book. In Lagrangian meshes, the nodes and elements move with the material. Boundaries and interfaces remain coincident with element edges, so that their treatment is simplified. Quadrature points also move with the material, so constitutive equations are always evaluated at the same material points, which is advantageous for history dependent materials. For these reasons, Lagrangian meshes are widely used for solid mechanics.

The formulations described herein apply to large deformations and nonlinear materials, i.e. they consider both geometric and material nonlinearities. They are only limited by the element's capabilities to deal with large distortions. The limited distortions most elements can sustain without degradation in performance or failure is an important factor in nonlinear analysis with Lagrangian meshes.

Finite element discretizations with Lagrangian meshes are commonly classified as updated Lagrangian formulations and total Lagrangian formulations. Both formulations use Lagrangian descriptions, i.e. the dependent variables are functions of the material (Lagrangian) coordinates and time. In the updated Lagrangian formulation, the derivatives are with respect to the spatial (Eulerian) coordinates; the weak form involves integrals over the deformed (or current) configuration. In the total Lagrangian formulation, the weak form involves integrals over the initial (reference) configuration and derivatives are taken with respect to the material coordinates.

11.2 Governing equations

We consider a body which occupies a domain Ω with a boundary Γ. The governing equations for the mechanical behavior of a continuous body are:
1. conservation of mass (or matter)
2. conservation of linear momentum and angular momentum
3. conservation of energy, often called the first law of thermodynamics
4. constitutive equations
5. strain-displacement equations

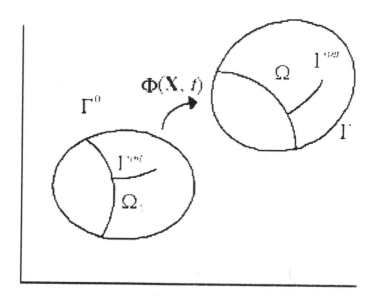

Figure 11.1 Deformed and undeformed body showing a set of admissible lines of interwoven discontinuities Γ^{int} and the notation.

Keep in mind that the first three are basic physical principles, which are exact equations. The last equation is a mathematical definition. It is in the fourth, the constitutive equation, in which the major source for error resides.

We will first develop the updated Lagrangian formulation. The conservation equations have been developed in Chapter 3, of Ref. 1, and are given in both tensor form and indicial form in Box 11.1. As can be seen, the dependent variables in the conservation equations are written in terms of material coordinates but are expressed in terms of what are classically Eulerian variables, such as the Cauchy stress and the rate-of-deformation.

We next give a count of the number of equations and unknowns. The conservation of mass and conservation of energy equations are scalar equations. The equation for the conservation of linear momentum, or momentum equation for short, is a tensor equation which consists of n_{SD} partial differential equations, where n_{SD} is the number of space dimensions. The constitutive equation relates the stress to the strain or strain-rate measure. Both the strain measure and the stress are symmetric tensors, so this provides n_σ equations where

$$n_\sigma \equiv n_{SD}\left(n_{SD}+1\right)/2 \qquad\qquad (11.1)$$

In addition, we have the n_σ equations which express the rate-of-deformation **D** in terms of the velocities or displacements. Thus we have a total of $2n_\sigma+n_{SD}+1$ equations and unknowns. For example, in two-dimensional problems without energy transfer, $n_{SD}=2$, so we have nine partial differential equations in nine unknowns: the two momentum equations, the three constitutive equations, the three equations relating **D** to the velocity and the mass conservation equation. The unknowns are the three stress components (symmetry of the stress follows from angular momemtum conservation), the three components of **D**, the two velocity components, and the density ρ, for a total of 9 unknowns. Additional unknown stresses (plane strain) and strains (plane stress) are evaluated using the plane strain and plane stress conditions, respectively. In three dimensions ($n_{SD}=3$, $n_\sigma=6$), we have 16 equations in 16 unknowns.

When a process is neither adiabatic nor isothermal, the energy equation must be appended to the system. This adds one equation and n_{SD} unknowns, the heat flux vector q_i. The heat flux vector can be determined from a single scalar, the temperature, by Fourier's law for heat conduction so only one unknown is added.

The dependent variables are the velocity $v(\mathbf{X},t)$, the Cauchy stress $\sigma(\mathbf{X},t)$, the rate-of-deformation $\mathbf{D}(\mathbf{X},t)$ and the density $\rho(\mathbf{X},t)$. The expression of all functions in terms of material coordinates is intrinsic in any treatment by a Lagrangian mesh. In principle, the functions can be expressed in terms of the spatial coordinates at any time t by using the inverse of the map $\mathbf{x}=\phi(\mathbf{X},t)$. However, inverting this map is quite difficult. In the formulation, we shall see that it is only necessary to obtain derivatives with respect to the spatial coordinates. This is accomplished by implicit differentiation, so the map corresponding to the motion is never explicitly inverted.

In Lagrangian meshes, the mass conservation equation is used in its integrated form (B11.1.1) rather than as a partial differential equation. This eliminates the need to treat the continuity equation. Although the continuity equation can be used to obtain the density in a Lagrangian mesh, it is simpler and more accurate to use the integrated form (B11.1.1)

The constitutive equation (Eq. B11.1.5), when expressed in rate form in terms of a rate of Cauchy stress, requires a frame invariant rate. For this purpose, any of the frame-invariant rates, such as the Jaumann or the Truesdell rate, can be used. It is not necessary for the constitutive equation in the updated Lagrangian formulation to be expressed in terms of the Cauchy stress or its frame invariant rate. It is also possible to use constitutive equations expressed in terms of the PK2 stress and then to convert the PK2 stress to a Cauchy stress using standard transformations prior to computing the internal forces.

The rate-of-deformation is used as the measure of strain rate in Eq. (B11.1.5). However, other measures of strain or strain-rate can also be used in an updated Lagrangian formulation. For example, the Green strain can be used in updated Lagrangian formulations. As indicated in Ref. 1, simple hypoelastic laws in terms of the rate-of-deformation can cause difficulties in the simulation of cyclic loading because its integral is not path independent. However, for many simulations, such as the single application of a large load, the errors due to the path-dependence of the integral of the rate-of-deformation are insignificant compared to other sources of error, such as inaccuracies and uncertainties in the material data and material model.

The boundary conditions are summarized in Eq. (B11.1.7). In two-dimensional problems, each component of the traction or velocity must be prescribed on the entire boundary; however the same component of the traction and velocity cannot be prescribed at the same point of the boundary, as indicated by Eq. (B11.1.8). Traction and velocity components can also be specified in local coordinate systems, which differ from the global system. An identical rule holds: the same components of traction and velocity cannot be prescribed at the same point of the boundary. A velocity boundary condition is equivalent to a displacement boundary condition: if a displacement is specified as a function of time, then the prescribed velocity can be obtained by time differentiation; if a velocity is specified, then the displacement can be obtained by time integration. Thus a velocity boundary condition will sometimes be called a displacement boundary condition, or vice versa.

The initial conditions can be applied either to the velocities and the stresses or to the displacements and velocities. The first set of initial conditions are more suitable for most engineering problems, since it is usually difficult to determine the initial displacement of a body. On the other hand, initial stresses, often known as residual stresses, can sometimes be measured or estimated by equilibrium solutions. For example, it is almost impossible to determine the displacements of a steel part after it has been formed from an ingot. On the other hand, good estimates of the residual stress field in the engineering component can often be made. Similarly, in a buried tunnel, the notion of initial displacements of the soil or rock enclosing the tunnel is quite meaningless, whereas the initial stress field can be estimated by equilibrium analysis. Therefore, initial conditions in terms of the stresses are more useful.

We have also included the interior continuity conditions on the stresses in Box 11.1 as Eq. (B11.1.11). In this equation, superscripts A and B refer to the stresses and normal on two sides of the discontinuity. These continuity conditions must be met by the tractions wherever stationary discontinuities in certain stress and strain components are possible, such as at material interfaces. They must hold for bodies in equilibrium and in transient problems. In transient problems, moving discontinuities are also possible; however, moving discontinuities are treated in Lagrangian meshes by smearing them over several elements. Thus the moving discontinuity conditions need not be explicitly stated.

11.3 Weak form: principle of virtual power

In this section, the principle of virtual power is developed for the updated Lagrangian formulation. The principle of virtual power is the weak form for the momentum equation, the traction boundary conditions and the interior traction continuity conditions. These three are collectively called *generalized momentum balance*. The relationship of the principle of virtual power to the momentum equations will be described in two parts:

1. The principle of virtual power (weak form) will be developed from the generalized momentum balance (strong form), i.e. strong form to weak form.
2. The principle of virtual power (weak form) will be shown to imply the generalized momentum balance (strong form), i.e. weak form to strong form.

We first define the spaces for the test functions and trial functions. We will consider the minimum smoothness required for the functions to be defined in the sense of distributions, i.e. we allow Dirac delta functions to be derivatives of functions. Thus, the derivatives will not be defined according to classical definitions of derivatives; instead, we will admit derivatives of piecewise continuous functions, where the derivatives include Dirac delta functions.

The space of test functions is defined by:

$$\delta v_j(\mathbf{X}) \in U_0 \qquad U_0 = \left\{ \delta v_i \middle| \delta v_i \in C^0(\mathbf{X}), \delta v_i = 0 \text{ on } \Gamma_{v_i} \right\} \qquad (11.2)$$

This selection of the space for the test functions $\delta \mathbf{v}$ is dictated by what will ensue in the development of the weak form; with this construction, the integral over the kinematic boundary vanishes and the only boundary integral in the weak form is over the traction boundary. The test functions $\delta \mathbf{v}$ are sometimes called the virtual velocities.

The velocity trial functions live in the space given by

$$v_i(\mathbf{X},t) \in U \qquad U = \left\{ v_i \middle| v_i \in C^0(\mathbf{X}), v_i = \overline{v}_i \text{ on } \Gamma_{v_i} \right\} \qquad (11.3)$$

The space of velocities in U is often called kinematically admissible or compatible; they satisfy the continuity conditions required for compatibility and the velocity boundary conditions. Note that the space of test functions is identical to the space of trial functions except that the virtual velocities vanish wherever the trial velocities are prescribed. We have selected a specific class of test and trial spaces that are applicable to finite elements; the weak form holds also for more general spaces, which is the space of functions with square integrable derivatives, called a Hilbert space.

BOX 11.1
Governing Equations for Updated Lagrangian Formulation

conservation of mass
$$\rho(\mathbf{X})J(\mathbf{X}) = \rho_0(\mathbf{X})J_0(\mathbf{X}) = \rho_0(\mathbf{X}) \qquad \text{(B11.1.1)}$$

conservation of linear momentum
$$\nabla \cdot \boldsymbol{\sigma} + \rho \mathbf{b} = \rho \dot{\mathbf{v}} \equiv \rho \frac{D\mathbf{v}}{Dt} \quad \text{or} \quad \frac{\partial \sigma_{ji}}{\partial x_j} + \rho b_i = \rho \dot{v}_i \equiv \rho \frac{Dv_i}{Dt} \qquad \text{(B11.1.2)}$$

conservation of angular momentum: $\boldsymbol{\sigma} = \boldsymbol{\sigma}^T \qquad \text{or} \qquad \sigma_{ij} = \sigma_{ji} \qquad \text{(B11.1.3)}$

conservation of energy:
$$\rho \dot{w}^{\text{int}} = \mathbf{D} : \boldsymbol{\sigma} - \nabla \cdot \mathbf{q} + \rho s \quad \text{or} \quad \rho \dot{w}^{\text{int}} = D_{ij}\sigma_{ji} - \frac{\partial q_i}{\partial x_i} + \rho s \qquad \text{(B11.1.4)}$$

constitutive equation: $\boldsymbol{\sigma}^\nabla = \mathbf{S}_t^{\sigma D}(\mathbf{D}, \boldsymbol{\sigma}, \text{etc.}) \qquad \text{(B11.1.5)}$

rate-of-deformation: $\mathbf{D} = sym(\nabla \mathbf{v}) \qquad D_{ij} = \frac{1}{2}\left(\frac{\partial v_i}{\partial x_j} + \frac{\partial v_j}{\partial x_i}\right) \qquad \text{(B11.1.6)}$

boundary conditions
$$n_j \sigma_{ji} = \overline{t}_i \quad \text{on} \quad \Gamma_{t_i} \qquad v_i = \overline{v}_i \quad \text{on} \quad \Gamma_{v_i} \qquad \text{(B11.1.7)}$$
$$\Gamma_{t_i} \cap \Gamma_{v_i} = 0 \qquad \Gamma_{t_i} \cup \Gamma_{v_i} = \Gamma \qquad i = 1 \text{ to } n_{SD} \qquad \text{(B11.1.8)}$$

initial conditions
$$\mathbf{v}(\mathbf{x},0) = \mathbf{v}_0(\mathbf{x}) \qquad \boldsymbol{\sigma}(\mathbf{x},0) = \boldsymbol{\sigma}_0(\mathbf{x}) \qquad \text{(B11.1.9)}$$
or
$$\mathbf{v}(\mathbf{x},0) = \mathbf{v}_0(\mathbf{x}) \qquad \mathbf{u}(\mathbf{x},0) = \mathbf{u}_0(\mathbf{x}) \qquad \text{(B11.1.10)}$$

interior continuity conditions (stationary)
$$\text{on } \Gamma_{\text{int}}: \quad \langle \mathbf{n} \cdot \boldsymbol{\sigma} \rangle = 0 \quad \text{or} \quad \langle n_i \sigma_{ij} \rangle \equiv n_i^A \sigma_{ij}^A + n_i^B \sigma_{ij}^B = 0 \qquad \text{(B11.1.11)}$$

Since the displacement $u_i(\mathbf{X},t)$ is the time integral of the velocity, the displacement field can also be considered to be the trial function. We shall see that the constitutive equation can be expressed in terms of the displacements or velocities. Whether the displacements or velocities are considered the trial functions is a matter of taste.

The weak form is given as follows: for $v_i(\mathbf{X},t) \in U$

$$\int_\Omega \frac{\partial(\delta v_i)}{\partial x_j} \sigma_{ji} d\Omega - \int_\Omega \delta v_i \rho\, b_i d\Omega - \sum_{j=1}^{n_{SD}} \int_{\Gamma_{ij}} \delta v_j\, t_j\, d\Gamma +$$

$$\int_\Omega \delta v_i \rho\, \dot{v}_i\, d\Omega = 0 \quad \forall \delta v_i \in U_0 \tag{11.4}$$

The above is the weak form of the momentum equation, the traction boundary conditions and the interior continuity conditions. It is known as the principle of virtual power, and it is closely related to the principle of virtual work. This weak form is given in terms of powers with physical meanings in Box 11.2 along the equivalent strong form.

We will next ascribe a physical name to each of the terms in the virtual power equation. This will be useful in systematizing the development of finite element equations. The nodal forces in the finite element discretization will be identified according to the same physical names.

To identify the first integrand in (11.4), note that it can be written as

$$\frac{\partial(\delta v_i)}{\partial x_j} \sigma_{ji} = \delta L_{ij} \sigma_{ji} = \left(\delta D_{ij} + \delta W_{ij}\right) \sigma_{ji} = \delta D_{ij} \sigma_{ji} = \delta \mathbf{D} : \boldsymbol{\sigma} \tag{11.5}$$

Here we have used the decomposition of the velocity gradient into its symmetric and skew symmetric parts and that $\delta W_{ij} \sigma_{ij} = 0$ since δW_{ij} is skew symmetric while σ_{ij} is symmetric. Comparison with (B11.1.4) then indicates that we can interpret $\delta D_{ij} \sigma_{ij}$ as the rate of virtual internal work, or the *virtual internal power*, per unit volume. Observe that $\dot{w}^{\text{int}}$ in (B11.1.4) is power per unit mass, so $\rho \dot{w}^{\text{int}} = \mathbf{D} : \boldsymbol{\sigma}$ is the power per unit volume. The total virtual internal power δP^{int} is defined by the integral of $\delta D_{ij} \sigma_{ij}$ over the domain, i.e.

$$\delta P^{\text{int}} = \int_\Omega \delta D_{ij} \sigma_{ij} d\Omega = \int_\Omega \frac{\partial(\delta v_i)}{\partial x_j} \sigma_{ij} d\Omega \equiv \int_\Omega \delta L_{ij} \sigma_{ij} d\Omega = \int_\Omega \delta \mathbf{D} : \boldsymbol{\sigma}\, d\Omega \tag{11.6}$$

where the third and fourth terms have been added to remind us that they are equivalent to the second term because of the symmetry of the Cauchy stress tensor.

The second and third terms in (11.4) are the *virtual external power:*

$$\delta \mathrm{P}^{ext} = \int_\Omega \delta v_i \rho b_i d\Omega + \sum_{j=1}^{n_{SD}} \int_{\Gamma_{tj}} \delta v_j \overline{t_j} d\Gamma$$

$$= \int_\Omega \delta \mathbf{v} \cdot \rho \mathbf{b} d\Omega + \sum_{j=1}^{n_{SD}} \int_{\Gamma_{tj}} \delta v_j \mathbf{e}_j \cdot \overline{\mathbf{t}} d\Gamma \tag{11.7}$$

This name is selected because the virtual external power arises from the external body forces $\mathbf{b}(\mathbf{x}, t)$ and prescribed tractions $\overline{\mathbf{t}}(x,t)$.

The last term in (11.4) is the *virtual inertial power*

$$\delta \mathrm{P}^{inert} = \int_\Omega \delta v_i \rho \dot{v}_i d\Omega \tag{11.8}$$

which is the power corresponding to the inertial force. The inertial force can be considered a body force in the d'Alembert sense.

Inserting Eqs. (11.6-11.8) into (11.4), we can write the principle of virtual power as

$$\delta P = \delta P^{int} - \delta P^{ext} + \delta P^{inert} = 0 \quad \forall \delta v_i \in U_0 \tag{11.9}$$

which is the weak form for the momentum equation. The physical meanings help in remembering the weak form and in the derivation of the finite element equations. The weak form is summarized in Box 11.2.

REFERENCES

[11.1]   Belytschko, T., Liu, W.K. and Moran, B., *Nonlinear Finite Elements for Continua and Structures*, Wiley, Chichester, U.K., 2000.

BOX 11.2
Weak Form in Updated Lagrangian Formulation:
Principle of Virtual Power

If σ_{ij} is a smooth function of the displacements and velocities and $v_i \in U$, then
if

$$\delta\,\mathrm{P}^{int} - \delta\,\mathrm{P}^{ext} + \delta\,\mathrm{P}^{inert} = 0 \quad \forall \delta v_i \in U_0 \qquad (\mathrm{B}11.2.1)$$

then

$$\frac{\partial\,\sigma_{ji}}{\partial\,x_j} + \rho b_i = \rho \dot{v}_i \quad \text{in } \Omega \qquad (\mathrm{B}11.2.2)$$

$$n_j \sigma_{ji} = \bar{t}_i \quad \text{on } \Gamma_{ti} \qquad (\mathrm{B}11.2.3)$$

$$\left\langle n_j \sigma_{ji} \right\rangle = 0 \quad \text{on } \Gamma_{int} \qquad (\mathrm{B}11.2.4)$$

where

$$\delta\,\mathrm{P}^{int} = \int_\Omega \delta\mathbf{D} : \boldsymbol{\sigma} \; d\Omega = \int_\Omega \delta D_{ij} \sigma_{ij} d\Omega = \int_\Omega \frac{\partial(\delta v_i)}{\partial x_j} \sigma_{ij} d\Omega \qquad (\mathrm{B}11.2.5)$$

$$\delta\,\mathrm{P}^{ext} = \int_\Omega \delta\mathbf{v} \cdot \rho \mathbf{b} d\Omega + \sum_{j=1}^{n_{SD}} \int_{\Gamma_{tj}} \left(\delta\mathbf{v} \cdot \mathbf{e}_j\right) \bar{\mathbf{t}} \cdot \mathbf{e}_j \; d\Gamma$$

$$= \int_\Omega \delta v_i \, \rho b_i \, d\Omega + \sum_{j=1}^{n_{SD}} \int_{\Gamma_{tj}} \delta v_j \bar{t}_j d\Gamma \qquad (\mathrm{B}11.2.6)$$

$$\delta\,\mathrm{P}^{inert} = \int_\Omega \delta\mathbf{v} \cdot \rho \dot{\mathbf{v}} d\Omega = \int_\Omega \delta v_i \rho \dot{v}_i d\Omega \qquad (\mathrm{B}11.2.7)$$

12. FINITE ELEMENT DISCRETIZATION

12.1 Finite element approximation

In this section, the finite element equations for the updated Lagrangian formulation are developed by means of the principle of virtual power. For this purpose the current domain Ω is subdivided into elements Ω_e so that the union of the elements comprises the total domain, $\Omega = \bigcup \Omega_e$. The nodal coordinates in the current configuration are denoted by $x_{iI}, I = 1$ to n_N. Lower case subscripts are used for components, upper case subscripts for nodal values. In two dimensions, $x_{iI} = [x_I, y_I]$, in three dimensions $x_{iI} = [x_I, y_I, z_I]$. The nodal coordinates in the undeformed configuration are X_{iI}.

In the finite element method, the motion $x(X, t)$ is approximated by

$$x_i(X,t) = N_I(X)x_{iI}(t) \quad \text{or} \quad x(X,t) = N_I(X)x_I(t) \qquad (12.1)$$

where $N_I(X)$ are the interpolation (shape) functions and x_I is the position vector of node I. Summation over repeated indices is implied; in the case of lower case indices, the sum is over the number of space dimensions, while for upper case indices the sum is over the number of nodes. The nodes in the sum depend on the entity considered: when the total domain is considered, the sum is over all nodes in the domain, whereas when an element is considered, the sum is over the nodes of the element.

Writing (12.1) at a node with initial position X_J we have

$$x(X_J, t) = x_I(t)N_I(X_J) = x_I(t)\delta_{IJ} = x_J(t) \qquad (12.2)$$

where we have used the interpolation property of the shape functions in the third term. Interpreting this equation, we see that node J always corresponds to the same material point $\mathbf{X}_J$: in a Lagrangian mesh, nodes remain coincident with material points.

We define the nodal displacements by

$$u_{iI}(t) = x_{iI}(t) - X_{iI} \quad \text{or} \quad \mathbf{u}_I(t) = \mathbf{x}_I(t) - \mathbf{X}_I \tag{12.3a}$$

The displacement field is

$$u_i(\mathbf{X},t) = x_i(\mathbf{X},t) - X_i = u_{iI}(t)N_I(\mathbf{X}) \quad \text{or} \quad \mathbf{u}(\mathbf{X},t) = \mathbf{u}_I(t)N_I(\mathbf{X}) \tag{12.3b}$$

which follows from (12.1), and (12.2).

The velocities are obtained by taking the material time derivative of the displacements, giving

$$v_i(\mathbf{X},t) = \frac{\partial u_i(\mathbf{X},t)}{\partial t} = \dot{u}_{iI}(t)N_I(\mathbf{X}) = v_{iI}(t)N_I(\mathbf{X})$$

$$\text{or} \quad \mathbf{v}(\mathbf{X},t) = \dot{\mathbf{u}}_I(t)N_I(\mathbf{X}) \tag{12.4}$$

where we have written out the derivative of the displacement on the left hand side to stress that the velocity is a material time derivative of the displacement, i.e., the partial derivative with respect to time with the material coordinate fixed. Note the velocities are given by the same shape function since the shape functions are constant in time. The superposed dot on the nodal displacements is an ordinary derivative, since the nodal displacements are only functions of time.

The accelerations are similarly given by the material time derivative of the velocities

$$\ddot{u}_i(\mathbf{X},t) = \ddot{u}_{iI}(t)N_I(\mathbf{X}) \quad \text{or} \quad \ddot{\mathbf{u}}(\mathbf{X},t) = \ddot{\mathbf{u}}_I(t)N_I(\mathbf{X}) \tag{12.5}$$

It is emphasized that the shape functions are expressed in terms of the material coordinates in the updated Lagrangian formulation even though we will use the weak form in the current configuration. It is crucial to express the shape functions in terms of material coordinates when a Lagrangian mesh is used because we want the time dependence in the finite element approximation of the motion to reside entirely in the nodal variables.

The velocity gradient is obtained by substituting Eq. (12.5) and finding its gradient, which yields

$$L_{ij} = v_{i,j} = v_{iI} \frac{\partial N_I}{\partial x_j} = v_{iI} N_{I,j} \quad \text{or} \quad \mathbf{L}^T = \mathbf{v}_I \nabla N_I = \mathbf{v}_I N_{I,\mathbf{x}} \qquad (12.6a)$$

and the rate-of-deformation is given by

$$D_{ij} = \tfrac{1}{2}\left(L_{ij} + L_{ji}\right) = \tfrac{1}{2}\left(v_{iI} N_{I,j} + v_{jI} N_{I,i}\right) \qquad (12.6b)$$

In the construction of the finite element approximation to the motion, (12.1), we have ignored the velocity boundary conditions, i.e. the velocities given by (12.4) are not in the space defined by (4.3.2). We will first develop the equations for an unconstrained body with no velocity boundary conditions, and then modify the discrete equations to account for the velocity boundary conditions.

In (12.1), all components of the motion are approximated by the same shape functions. This construction of the motion facilitates the representation of rigid body rotation, which is an essential requirement for convergence.

The test function, or variation, is not a function of time, so we approximate the test function as

$$\delta v_i(\mathbf{X}) = \delta v_{iI} N_I(\mathbf{X}) \quad \text{or} \quad \delta \mathbf{v}(\mathbf{X}) = \delta \mathbf{v}_I N_I(\mathbf{X}) \qquad (12.7)$$

where δv_{iI} are the virtual nodal velocities.

As a first step in the construction of the discrete finite element equations, the test function is substituted into the principle of virtual power giving

$$\delta v_{iI} \int_\Omega \frac{\partial N_I}{\partial x_j} \sigma_{ji} d\Omega - \delta v_{iI} \int_\Omega N_I \rho b_i d\Omega - \sum_{i=1}^{n_{SD}} \delta v_{iI} \int_{\Gamma_{t_i}} N_I \bar{t}_i d\Gamma$$

$$+ \delta v_{iI} \int_\Omega N_I \rho \dot{v}_i d\Omega = 0 \qquad (12.8a)$$

The stresses in (12.8a) are functions of the trial velocities and trial displacements. From the definition of the test space, (11.2), the virtual velocities must vanish wherever the velocities are prescribed, i.e. $\delta v_i = 0$ on Γ_{v_i}

and therefore only the virtual nodal velocities for nodes not on Γ_{v_i} are arbitrary, as indicated above. Using the arbitrariness of the virtual nodal velocities everywhere except on Γ_{v_i}, it then follows that the weak form of the momentum equation is

$$\int_\Omega \frac{\partial N_I}{\partial x_j} \sigma_{ji} d\Omega - \int_\Omega N_I \rho b_i d\Omega - \sum_{j=1}^{n_{SD}} \int_{\Gamma_{t_j}} N_I \bar{t}_i d\Gamma$$

$$+ \int_\Omega N_I \rho \dot{v}_i d\Omega = 0 \quad \forall I, i \notin \Gamma_{v_i} \tag{12.8b}$$

However, the above form is difficult to remember. For purposes of convenience and for a better physical interpretation, it is worthwhile to ascribe physical names to each of the terms in the above equation.

12.2   Internal and external nodal forces

We define the nodal forces corresponding to each term in the virtual power equation. This helps in remembering the equation and also provides a systematic procedure which is found in most finite element software. The internal nodal forces are defined by

$$\delta P^{int} = \delta v_{iI} f_{iI}^{int} = \int_\Omega \frac{\partial(\delta v_i)}{\partial x_j} \sigma_{ji} d\Omega = \delta v_{iI} \int_\Omega \frac{\partial N_I}{\partial x_j} \sigma_{ji} d\Omega \tag{12.9}$$

where the third term is the definition of internal virtual power as given in Eqs. (B11.2.5) and (12.7) has been used in the last term. From the above it can be seen that the internal nodal forces are given by

$$f_{iI}^{int} = \int_\Omega \frac{\partial N_I}{\partial x_j} \sigma_{ji} d\Omega \tag{12.10}$$

These nodal forces are called internal because they represent the *stresses in the body*. These expressions apply to both a complete mesh and to any element or group of elements. Note that this expression involves derivatives of the shape functions with respect to spatial coordinates and integration over the current configuration. Equation (12.10) is a key equation in nonlinear finite element methods for updated Lagrangian meshes; it applies also to Eulerian and ALE meshes.

The external nodal forces are defined similarly in terms of the virtual external power

$$\delta P^{ext} = \delta v_{iI} f_{iI}^{ext} = \int_{\Omega} \delta v_i \rho b_i d\Omega + \sum_{i=1}^{n_{SD}} \int_{\Gamma_{t_i}} \delta v_i \bar{t}_i d\Gamma$$

$$= \delta v_{iI} \int_{\Omega} N_I \rho b_i d\Omega + \sum_{i=1}^{n_{SD}} \delta v_{iI} \int_{\Gamma_{t_i}} N_I \bar{t}_i d\Gamma \tag{12.11}$$

so the external nodal forces are given by

$$f_{iI}^{ext} = \int_{\Omega} N_I \rho b_i d\Omega + \int_{\Gamma_{t_i}} N_I \bar{t}_i d\Gamma \quad \text{or} \quad \mathbf{f}_I^{ext} = \int_{\Omega} N_I \rho \mathbf{b} \, d\Omega + \int_{\Gamma_{t_i}} N_I \mathbf{e}_i \cdot \bar{\mathbf{t}} d\Gamma \tag{12.12}$$

12.3   Mass matrix and inertial forces

The inertial nodal forces are defined by

$$\delta P^{inert} = \delta v_{iI} f_{iI}^{inert} = \int_{\Omega} \delta v_i \rho \dot{v}_i d\Omega = \delta v_{iI} \int_{\Omega} N_I \rho \dot{v}_i \, d\Omega \tag{12.13}$$

so

$$f_{iI}^{inert} = \int_{\Omega} \rho N_I \dot{v}_i d\Omega \quad \text{or} \quad \mathbf{f}_I^{inert} = \int_{\Omega} \rho N_I \dot{\mathbf{v}} d\Omega \tag{12.14}$$

Using the expression (12.5) for the accelerations in the above gives

$$f_{iI}^{inert} = \int_{\Omega} \rho N_I N_J d\Omega \, \dot{v}_{iJ} \tag{12.15}$$

It is convenient to define these nodal forces as a product of a mass matrix and the nodal accelerations. Defining the mass matrix by

$$M_{ijIJ} = \delta_{ij} \int_{\Omega} \rho N_I N_J d\Omega \tag{12.16}$$

it follows from (12.15) and (12.16) that the inertial forces are given by

$$f_{iI}^{inert} = M_{ijIJ} \dot{v}_{jJ} \quad \text{or} \quad \mathbf{f}_I^{inert} = \mathbf{M}_{IJ} \dot{\mathbf{v}}_J \tag{12.17}$$

12.4   Discrete equations

With the definitions of the internal, external and inertial nodal forces, Eqs. (12.9), (12.11) and (12.16), we can concisely write the discrete approximation to the weak form (12.8a) as

$$\delta v_{il}\left(f_{il}^{\text{int}} - f_{il}^{\text{ext}} + M_{ijlJ}\dot{v}_{jJ}\right) = 0 \quad \text{for} \ \forall \delta v_{il} \notin \Gamma_{v_i} \tag{12.18a}$$

We can also write the above as

$$\delta \mathbf{v}^T\left(\mathbf{f}^{\text{int}} - \mathbf{f}^{\text{ext}} + \mathbf{M}\mathbf{a}\right) = 0 \tag{12.18b}$$

where $\mathbf{v}$, $\mathbf{a}$ and $\mathbf{f}$ are column matrices storing the unconstrained virtual velocities, accelerations and nodal forces and $\mathbf{M}$ is the mass matrix related to the unconstrained degrees of freedom.   Invoking the arbitrariness of the unconstrained virtual nodal velocities in (12.18a) and (12.18b), respectively, gives

$$M_{ijlJ}\dot{v}_{jJ} + f_{il}^{\text{int}} = f_{il}^{\text{ext}} \quad \forall I, i \notin \Gamma_{v_i} \quad \text{or} \quad \mathbf{M}\mathbf{a} + \mathbf{f}^{\text{int}} = \mathbf{f}^{\text{ext}} \tag{12.19a,b}$$

The above are the *discrete momentum equations* or *the equations of motion*; they are also called the *semidiscrete momentum equations* since they have not been discretized in time. The implicit sums are over all components and all nodes of the mesh; any prescribed velocity component that appears in the above is not an unknown. Equation (12.19b) can also be written in the form of Newton's second law

$$\mathbf{f} = \mathbf{M}\mathbf{a} \text{ where } \mathbf{f} = \mathbf{f}^{\text{ext}} - \mathbf{f}^{\text{int}} \tag{12.19c}$$

The semidiscrete momentum equations are a system of n_{DOF} ordinary differential equations in the nodal velocities, where n_{DOF} is the number of nodal velocity components which are unconstrained; n_{DOF} is often called the number of degrees of freedom in the system. To complete the system of equations, we append the constitutive equations at the element quadrature points and the expression for the rate-of-deformation in terms of the nodal velocities. Let the n_Q quadrature points in the mesh be denoted by

$$\mathbf{x}_Q(t) = N_I\left(\mathbf{X}_Q\right)\mathbf{x}_I(t) \tag{12.20}$$

Note that the quadrature points are coincident with material points. Let n_σ be the number of independent components of the stress tensor: in a two dimensional plane stress problem, $n_\sigma = 3$, since the stress tensor $\boldsymbol{\sigma}$ is symmetric; in three-dimensional problems, $n_\sigma = 6$.

The semidiscrete equations for the finite element approximation then consist of the following ordinary differential equations in time:

$$M_{ijIJ}\dot{v}_{jJ} + f_{il}^{\text{int}} = f_{il}^{\text{ext}} \quad \text{for} \quad (I,i) \notin \Gamma_{v_i} \tag{12.21}$$

$$\sigma_{ij}^{\nabla}(\mathbf{X}_Q) = S_{ij}\left(D_{kl}(\mathbf{X}_Q), \text{etc}\right) \quad \forall \mathbf{X}_Q \tag{12.22}$$

$$\text{where } D_{ij}(\mathbf{X}_Q) = \tfrac{1}{2}(L_{ij} + L_{ji}) \quad \text{and} \quad L_{ij} = N_{I,j}(\mathbf{X}_Q)v_{il} \tag{12.23}$$

This is a standard initial value problem, consisting of first-order ordinary differential equations in the velocities $v_{il}(t)$ and the stresses $\sigma_{ij}(\mathbf{X}_Q,t)$. If we substitute (12.23) into (12.22) to eliminate the rate-of-deformation from the equations, the total number of unknowns is $n_{DOF} + n_\sigma n_Q$. This system of ordinary differential equations can be integrated in time by any of the methods for integrating ordinary differential equations, such as Runge-Kutta methods or the central difference method; this is discussed in Chapter 13.

The nodal velocities on prescribed velocity boundaries, v_{il}, $(I,i) \in \Gamma_{v_i}$, are obtained from the boundary conditions, Eq. (B11.1.7b). The initial conditions (B11.1.9) are applied at the nodes and quadrature points

$$v_{il}(0) = v_{il}^0 \tag{12.24}$$

$$\sigma_{ij}(\mathbf{X}_Q,0) = \sigma_{ij}^0(\mathbf{X}_Q) \tag{12.25}$$

where v_{il}^0 and σ_{ij}^0 are initial data at the nodes and quadrature points. If data for the initial conditions are given at a different set of points, the values at the nodes and quadrature points can be estimated by least square fits, as in Section 2.4.5 in Ref. 1.

For an equilibrium problem, the accelerations vanish and the governing equations are

$$f_{il}^{\text{int}} = f_{il}^{\text{ext}} \quad \text{for} \quad (I,i) \notin \Gamma_{v_i} \quad \text{or} \quad \mathbf{f}^{\text{int}} = \mathbf{f}^{\text{ext}} \tag{12.26}$$

along with (12.22) and (12.23). The above are called the discrete *equilibrium equations*. If the constitutive equations are rate-independent, then the discrete equilibrium equations are a set of nonlinear algebraic equations in the stresses and nodal displacements. For rate-dependent materials, any rate terms must be discretized in time to obtain a set of nonlinear algebraic equations.

12.5   Element coordinates

Finite elements are usually developed with shape functions expressed in terms of parent element coordinates, which we will often call element coordinates for brevity. Examples of element coordinates are triangular coordinates and isoparametric coordinates. We will next describe the use of shape functions expressed in terms of element coordinates. As part of this description, we will show that the element coordinates can be considered an alternative set of material coordinates in a Lagrangian mesh. Therefore, expressing the shape functions in terms of element coordinates in a Lagrangian mesh is intrinsically equivalent to expressing them in terms of material coordinates. We denote the parent element coordinates by ξ_i^e, or $\boldsymbol{\xi}^e$ in tensor notation, and the parent domain by Δ; the superscript e will only be carried in the beginning of this description. The shape of the parent domain depends on the type of element and the dimension of the problem; it may be a biunit square, a triangle, or a cube, for example.

When a Lagrangian element is treated in terms of element coordinates, we are concerned with three domains that correspond to an element:

1. the parent element domain Δ;
2. the current element domain $\Omega^e = \Omega^e(t)$;
3. the initial (reference) element domain Ω_0^e

The following maps are pertinent:

1. parent domain to current configuration: $\mathbf{x} = \mathbf{x}\left(\boldsymbol{\xi}^e, t\right)$
2. parent domain to initial configuration: $\mathbf{X} = \mathbf{X}\left(\boldsymbol{\xi}^e\right)$
3. initial configuration to the current configuration, i.e. the motion
    $\mathbf{x} = \mathbf{x}(\mathbf{X}, t) \equiv \boldsymbol{\phi}(\mathbf{X}, t)$

The map $\mathbf{X} = \mathbf{X}(\xi^e)$ corresponds to $\mathbf{x} = \mathbf{x}(\xi^e,0)$. These maps are illustrated in Fig. 4.1 for a triangular element where a space-time plot of a two-dimensional triangular element is shown.

The motion in each element is described by a composition of these maps

$$\mathbf{x} = \mathbf{x}(\mathbf{X},t) = \mathbf{x}\big(\xi^e(\mathbf{X}),t\big) \quad \mathbf{x}(\mathbf{X},t) = \mathbf{x}\big(\xi^e,t\big) \circ \xi^e(\mathbf{X}) \quad \text{in } \Omega_e \qquad (12.27)$$

where $\xi^e(\mathbf{X}) = \mathbf{X}^{-1}(\xi^e)$. For the motion to be well defined and smooth, the inverse map $\mathbf{X}^{-1}(\xi^e)$ must exist and the function $\mathbf{x} = \mathbf{x}(\xi^e,t)$ must be sufficiently smooth and meet certain conditions of regularity so that $\mathbf{x}^{-1}(\xi^e,t)$ exists; these conditions are given in Section 12.8. The inverse map $\mathbf{x}^{-1}(\xi^e,t)$ is usually not constructed because in most cases it cannot be obtained explicitly, so instead the derivatives with respect to the spatial coordinates are obtained in terms of the derivatives with respect to the parent coordinates by implicit differentiation.

The motion is approximated by

$$x_i(\xi,t) = x_{iI}(t)N_I(\xi) \quad \text{or} \quad \mathbf{x}(\xi,t) = \mathbf{x}_I(t)N_I(\xi) \qquad (12.28)$$

where we have dropped the superscript e on the element coordinates. As can be seen in the above, the shape functions $N_I(\xi)$ are only functions of the parent element coordinates; the time dependence of the motion resides entirely in the nodal coordinates. The above represents a time dependent mapping between the parent domain and the current configuration of the element.

Writing this map at time $t=0$ we obtain

$$X_i(\xi) = x_i(\xi,0) = x_{iI}(0)N_I(\xi) = X_{iI}N_I(\xi) \quad \text{or} \quad \mathbf{X}(\xi) = \mathbf{X}_I N_I(\xi) \qquad (12.29)$$

It can be seen from (12.29) that the map between the material coordinates and the element coordinates is time invariant in a Lagrangian element. If this map is one-to-one and onto, then *the element coordinates can in fact be considered surrogate material coordinates in a Lagrangian mesh*, since each material point in an element then has a unique element coordinate label. To establish a unique correspondence between element coordinates and the material coordinates in Ω_0, the element number must be part of the label. The use of the initial coordinates $\mathbf{X}$ as material coordinates in fact originates mainly in analysis; in finite element methods, the use of element coordinates as material labels is more natural.

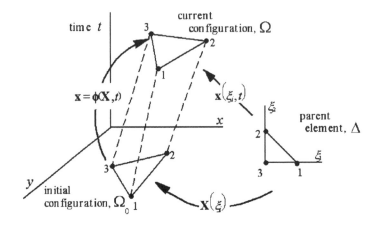

Figure 12.1 Initial and current configurations of an element and their relationships to the parent element.

As before, since the element coordinates are time invariant, we can express the displacements, velocities and accelerations in terms of the same shape functions

$$u_i(\xi,t)=u_{iI}(t)N_I(\xi) \quad \text{or} \quad \mathbf{u}(\xi,t)=\mathbf{u}_I(t)N_I(\xi) \tag{12.30}$$

$$\dot{u}_i(\xi,t)=v_i(\xi,t)=v_{iI}(t)N_I(\xi) \quad \text{or} \quad \dot{\mathbf{u}}(\xi,t)=\mathbf{v}(\xi,t)=\mathbf{v}_I(t)N_I(\xi) \tag{12.31}$$

$$\dot{v}(\xi,t)=\dot{v}_{iI}(t)N_I(\xi) \quad \text{or} \quad \dot{\mathbf{v}}(\xi,t)=\dot{\mathbf{v}}_I(t)N_I(\xi) \tag{12.32}$$

where we have obtained (12.31) by taking material time derivative of (12.30) and we have obtained (12.32) by taking the material time derivative of (12.31). The time dependence, as before, resides entirely in the nodal variables, since the element coordinates are independent of time.

REFERENCES

[12.1]   Belytschko, T., Liu, W.K. and Moran, B. *Nonlinear Finite Elements for Continua and Structures*, Wiley, Chichester, U.K., 2000.

Box 12.1
Discrete Equations and Internal Nodal Force Algorithm
for the Updated Lagrangian Formulation

Equations of Motion (discrete momentum equation)

$$M_{ijIJ}\dot{v}_{jJ} + f_{iI}^{int} = f_{iI}^{ext} \quad \text{for} \quad (I,i) \notin \Gamma_{v_i} \tag{B12.1.1}$$

Internal Nodal Forces

$$f_{iI}^{int} = \int_\Omega B_{Ij}\sigma_{ji}d\Omega = \int_\Omega \frac{\partial N_I}{\partial x_j}\sigma_{ji}d\Omega \quad \text{or} \quad \left(\mathbf{f}_I^{int}\right)^T = \int_\Omega \mathbf{B}_I^T \boldsymbol{\sigma}\, d\Omega \tag{B12.1.2}$$

$$\mathbf{f}_I^{int} = \int_\Omega \mathbf{B}_I^T \{\boldsymbol{\sigma}\}d\Omega \quad \text{in Voigt notation}$$

External Nodal Forces

$$f_{iI}^{ext} = \int_\Omega N_I \rho b_i d\Omega + \int_{\Gamma_{t_i}} N_I \bar{t}_i d\Gamma \quad \text{or} \quad \mathbf{f}_I^{ext} = \int_\Omega N_I \rho \mathbf{b}\, d\Omega + \int_{\Gamma_{t_i}} N_I \mathbf{e}_i \cdot \bar{\mathbf{t}}\, d\Gamma \tag{B12.1.3}$$

Mass Matrix (total Lagrangian)

$$M_{ijIJ} = \delta_{ij}\int_{\Omega_0}\rho_0 N_I N_J d\Omega_0 = \delta_{ij}\int_\Delta \rho_0 N_I N_J J_\xi^0 d\Delta \tag{B12.1.4}$$

$$\mathbf{M}_{IJ} = \mathbf{I}\tilde{M}_{IJ} = \mathbf{I}\int_{\Omega_0}\rho_0 N_I N_J\, d\Omega_0 \tag{B12.1.5}$$

Internal nodal force computation for element

1. $\mathbf{f}^{int} = 0$

2. for all quadrature points ξ_Q

     i. compute $\left[B_{Ij}\right] = \left[\partial N_I(\xi_Q)/\partial x_j\right]$ for all I

     ii. $\mathbf{L} = \left[L_{ij}\right] = \left[v_{iI}B_{Ij}\right] = \mathbf{v}_I \mathbf{B}_I^T; \quad L_{ij} = \dfrac{\partial N_I}{\partial x_j}v_{iI}$

     iii. $\mathbf{D} = \frac{1}{2}\left(\mathbf{L}^T + \mathbf{L}\right)$

     iv. if needed compute $\mathbf{F}$ and $\mathbf{E}$ by procedures in Box 4.7

v. compute the Cauchy stress $\boldsymbol{\sigma}$ or the PK2 stress $\mathbf{S}$ or by
   constitutive equation

vi. if $\mathbf{S}$ computed, compute $\boldsymbol{\sigma}$ by $\boldsymbol{\sigma} = J^{-1}\mathbf{FSF}^{T}$

vii. $\mathbf{f}_{I}^{int} \leftarrow \mathbf{f}_{I}^{int} + \mathbf{B}_{I}^{T}\boldsymbol{\sigma}J_{\xi}\overline{w}_{Q}$ for all I

end loop

$\overline{w}_{Q}$ = quadrature weights

13. EXPLICIT SOLUTION METHODS

13.1   Central difference method.

The central difference method is among the most popular of the explicit methods in computational mechanics and physics.   The central difference method is developed from central difference formulas for the velocity and acceleration. We consider here its application to Lagrangian meshes with rate-independent materials. Geometric and material nonlinearities are included, and in fact have little effect on the time integration algorithm.

For the purpose of developing the integrator the notation described next will be used. Let the time of the simulation $0 \le t \le t_E$ be subdivided into time steps Δt^n, $n=1$ to n_{TS} where n_{TS} is the number of time steps and t_E is the end of the simulation; Δt^n is a time increment. The superscript; indicates the time step: t^n and $\mathbf{d}^n \equiv \mathbf{d}(t^n)$ are the time and displacement, respectively, at time step n.

We consider here an algorithm with a variable time step. This is necessary in most practical calculations since the stable time step changes as the mesh deforms and the wave speed changes dues to the stress. For this purpose, we define the time increments by

$$\Delta t^{n+\frac{1}{2}} = t^{n+1} - t^n, \qquad t^{n+\frac{1}{2}} = \tfrac{1}{2}\left(t^{n+1} + t^n\right) \qquad \Delta t^n = t^{n+\frac{1}{2}} - t^{n-\frac{1}{2}} \quad (13.1)$$

The central difference formula for the velocity is

$$\dot{\mathbf{d}}^{n+\frac{1}{2}} \equiv \mathbf{v}^{n+\frac{1}{2}} = \frac{\mathbf{d}^{n+1} - \mathbf{d}^n}{t^{n+1} - t^n} = \frac{1}{\Delta t^{n+\frac{1}{2}}}\left(\mathbf{d}^{n+1} - \mathbf{d}^n\right) \qquad (13.2a)$$

where the definition of $\Delta t^{n+\frac{1}{2}}$ from (13.1) has been used in the last step. This difference formula can be converted to an integration formula by rearranging the terms as follows:

$$\mathbf{d}^{n+1} = \mathbf{d}^n + \Delta t^{n+\frac{1}{2}} \mathbf{v}^{n+\frac{1}{2}}$$

(13.2b)

The acceleration and the corresponding integration formula are

$$\ddot{\mathbf{d}}^n \equiv \mathbf{a}^n = \left(\frac{\mathbf{v}^{n+\frac{1}{2}} - \mathbf{v}^{n-\frac{1}{2}}}{t^{n+\frac{1}{2}} - t^{n-\frac{1}{2}}} \right) \qquad \mathbf{v}^{n+\frac{1}{2}} = \mathbf{v}^{n-\frac{1}{2}} + \Delta t^n \mathbf{a}^n$$

(13.3a)

As can be seen from the above, the velocities are defined at the midpoints of the time intervals, which are called half-steps or midpoint steps. By substituting (13.2a) and its counterpart for the previous time step into (13.3a), the acceleration can be expressed directly in terms of the displacements

$$\ddot{\mathbf{d}}^n \equiv \mathbf{a}^n = \frac{\Delta t^{n-\frac{1}{2}} \left(\mathbf{d}^{n+1} - \mathbf{d}^n \right) - \Delta t^{n+\frac{1}{2}} \left(\mathbf{d}^n - \mathbf{d}^{n-1} \right)}{\Delta t^{n-\frac{1}{2}} \Delta t^n \Delta t^{n+\frac{1}{2}}}$$

(13.3b)

For the case of equal time steps the above reduces to

$$\ddot{\mathbf{d}}^n \equiv \mathbf{a}^n = \frac{\left(\mathbf{d}^{n+1} - 2\mathbf{d}^n + \mathbf{d}^{n-1} \right)}{\left(\Delta t^n \right)^2}$$

(13.3c)

This is the well known central difference formula for the second derivative of a function.

We now consider the time integration of the equations of motion, (12.19), which at time step n are given by

$$\mathbf{M}\mathbf{a}^n = \mathbf{f}^n = \mathbf{f}^{ext}(\mathbf{d}^n, t^n) - \mathbf{f}^{int}(\mathbf{d}^n, t^n)$$

(13.4a)

$$\text{subject to } g_I(\mathbf{d}^n) = 0, \ I = 1 \text{ to } n_c$$

(13.4b)

The equations of motion are ordinary differential equations of second order in time. They are often called semidiscrete, since they have been discretized in space but not in time. Equation (13.4b) is a generalized representation of the n_c displacement boundary conditions and other constraints on the model.

These constraints are linear or nonlinear algebraic functions of the nodal displacements. If the constraint involves integral or differential relationships, it can be put in the above form by using difference equations or a numerical approximation of the integral. The mass matrix is constant for a Lagrangian mesh.

The internal and external nodal forces are functions of the nodal displacements and the time. The external loads are usually prescribed as functions of time; they may also be functions of the nodal displacements because they may depend on the configuration of the structure, as when pressure forces are applied to surfaces which undergo large deformations. The dependence of the internal nodal forces on displacements is obvious: the nodal internal forces depend on the stresses, which depend on the strain and strain rates by the constitutive equations, which in turn depend on the displacement and their derivatives. The internal nodal forces can also depend directly on time, e.g. when the temperature is prescribed as a function of time, then the stresses and hence the internal nodal forces are directly functions of time.

The equations for updating the nodal velocities and displacements are obtained as follows. Substituting (13.4a) into (13.3b) gives

$$\mathbf{v}^{n+\frac{1}{2}} = \mathbf{v}^{n-\frac{1}{2}} + \Delta t^n \mathbf{M}^{-1} \mathbf{f}^n \tag{13.5}$$

At any time step n, the displacements $\mathbf{d}^n$ is known. The nodal forces $\mathbf{f}^n$ can be determined by sequentially evaluating the strain-displacement equations, the constitutive equation expressed in terms of $\mathbf{D}^{n-1/2}$ or $\mathbf{E}^n$ and the nodal internal forces. Thus the entire right hand side of (13.5) can be evaluated, and (13.5) can be used to obtain $\mathbf{v}^{n+1/2}$. The displacements $\mathbf{d}^{n+1}$ can be then be determined by (13.2b).

The update of the nodal velocities and nodal displacements can be accomplished without solving any system equations provided that the mass matrix $\mathbf{M}$ is diagonal. This is the salient characteristic of an explicit method: *in an explicit method, the time integration of the discrete momentum equations for a does not require the solution of any equations.* The avoidance of any solution of equations of course hinges critically on the use of a diagonal mass matrix.

In numerical analysis, integration methods are classified according to the structure of the time difference equation. The difference equations for first and second derivatives, respectively, can be written in the general expressions

$$\sum_{k=0}^{n_S}\left(\alpha_k \mathbf{d}^{n_S-k} - \Delta t\, \beta_k\, \dot{\mathbf{d}}^k\right)=0 \quad \sum_{k=0}^{n_S}\left(\overline{\alpha}_k\, \mathbf{d}^{n_S-k} - \Delta t^2\, \overline{\beta}_k\, \ddot{\mathbf{d}}^k\right)=0 \quad (13.6)$$

where n_S is the number of steps in the difference equation; the time steps are constant. The difference formula for the first or second derivatives is called explicit if $\beta_0 = 0$ or $\overline{\beta}_0 = 0$, respectively. Thus a difference formula is called explicit if the equation for the function at time step n_S only involves the derivatives at previous time steps. In the central difference formula for the second derivative, (13.3c), $\overline{\beta}_0 = 0$, $\overline{\beta}_1 = 1$, $\overline{\beta}_2 = 0$, so it is explicit. Difference equations which are explicit according to this classification generally lead to solution schemes which require no solution of equations. In most cases there is no benefit in using explicit schemes, which involve the solution of equations, so such explicit schemes are rare. There are exceptions. For example, the consistent mass is sometimes used with the central difference method in wave propagation problems, and as can be seen from (13.5), the update then involves an inverse, hence the solution of equations.

13.2   Implementation.

Box 13.1 gives a flow chart for explicit time integration of a finite element model with rate-independent materials. The damping is modeled by a linear viscous force $\mathbf{f}^{damp} = \mathbf{C}^{damp}\mathbf{v}$, so that the total force in (13.5) is $\mathbf{f} - \mathbf{C}^{damp}\mathbf{v}$. The implementation of the velocity update is broken into two substeps (13.3a) by

$$\mathbf{v}^n = \mathbf{v}^{n-\frac{1}{2}} + \left(t^n - t^{n-\frac{1}{2}}\right)\mathbf{a}^n \quad \mathbf{v}^{n+\frac{1}{2}} = \mathbf{v}^n + \left(t^{n+\frac{1}{2}} - t^n\right)\mathbf{a}^n$$

$$(13.6b)$$

This enables energy balance to be checked at integer time steps.

The cardinal dependent variables in this flowchart are the velocities and the Cauchy stresses. Initial conditions must be given for the velocities, the Cauchy stresses, and all material state variables. The initial displacements are assumed to vanish (initial displacements are meaningless in nonlinear analysis except for a hyperelastic material since the stress generally depends on the history of deformation).

Box 13.1

Flowchart for Explicit Time Integration

1. Initial conditions and initialization:

   set $\mathbf{v}^0, \boldsymbol{\sigma}^0$, and initial values of other material state variables;

   $\mathbf{d}^0 = \mathbf{0}$, $n = 0$, $t = 0$; compute $\mathbf{M}$

2. $getforce\left(\mathbf{f}^n, \Delta t^{n+\frac{1}{2}} \right)$

3. compute accelerations $\mathbf{a}^n = \mathbf{M}^{-1}\left(\mathbf{f}^n - \mathbf{C}^{damp}\mathbf{v}^{n-\frac{1}{2}} \right)$

4. time update: $t^{n+1} = t^n + \Delta t^{n+\frac{1}{2}}$, $t^{n+\frac{1}{2}} = \frac{1}{2}\left(t^n + t^{n+1} \right)$

5. first partial update nodal velocities: $\mathbf{v}^{n+\frac{1}{2}} = \mathbf{v}^n + \left(t^{n+\frac{1}{2}} - t^n \right)\mathbf{a}^n$

6. enforce velocity boundary conditions:

   if node I on Γ_{v_i}: $v_{iI}^{n+\frac{1}{2}} = \bar{v}_i\left(\mathbf{x}_I, t^{n+\frac{1}{2}} \right)$

6b. check energy balance at time step n, see (13.12-16)

7. update nodal displacements: $\mathbf{d}^{n+1} = \mathbf{d}^n + \Delta t^n \mathbf{v}^{n+\frac{1}{2}}$

8. $getforce\left(\mathbf{f}^{n+1}, \Delta t^{n+\frac{1}{2}} \right)$

9. compute accelerations $\mathbf{a}^{n+1} = \mathbf{M}^{-1}\left(\mathbf{f}^{n+1} - \mathbf{C}^{damp}\mathbf{v}^{n+\frac{1}{2}} \right)$

10. second partial update nodal velocities:

   $\mathbf{v}^{n+1} = \mathbf{v}^{n+\frac{1}{2}} + \left(t^{n+1} - t^{n+\frac{1}{2}} \right)\mathbf{a}^{n+1}$

11. update counter and time: $n \leftarrow n+1$

12. output, if simulation not complete, go to 5

Subroutine *getforce* $\left(\mathbf{f}^n, \Delta t \right)$

0. initialization: $\mathbf{f}^n = 0$, $\Delta t_{crit} = \infty$

1. compute global external nodal forces $\mathbf{f}_{ext}^n$

2. loop over elements e

     i. GATHER element nodal displacements and velocities

     ii. $\mathbf{f}_{int}^{e,n} = \mathbf{0}$

     iii. loop over quadrature points ξ_Q

         1. if $n=0$, go to 4

         2. compute measures of deformation: $\mathbf{D}^{n-\frac{1}{2}}\left(\xi_Q \right), \mathbf{F}^n\left(\xi_Q \right), \mathbf{E}^n\left(\xi_Q \right)$

         3. compute stress $\sigma^n\left(\xi_Q \right)$ by constitutive equation

         4. $\mathbf{f}_e^{int,n} \leftarrow \mathbf{f}_e^{int,n} + \mathbf{B}^T \sigma^n \overline{w}_Q J \big|_{\xi_Q}$

     END quadrature point loop

     iv. compute external nodal forces on element, $\mathbf{f}_e^{ext,n}$

     v. $\mathbf{f}_e^n = \mathbf{f}_e^{ext,n} - \mathbf{f}_e^{int,n}$

     vi. compute Δt_{crit}^e, if $\Delta t_{crit}^e < \Delta t_{crit}$ then $\Delta t_{crit} = \Delta t_{crit}^e$

     vii. SCATTER $\mathbf{f}_e^n$ to global $\mathbf{f}^n$

3. END loop over elements

4. $\Delta t = \alpha \Delta t_{crit}$

The main part of the procedure is the calculation of the nodal forces, which is done in *getforce*. The major steps in this subroutine are:

1. extract the nodal displacements of the element from the global array of nodal displacements by the *gather operation*;
2. the strain measures are computed at each quadrature point of the element;
3. the stresses are computed by the constitutive equation at each quadrature point;
4. evaluate the internal nodal forces by integrating the product of the **B** matrix and the Cauchy stress over the domain of the element;
5. the nodal forces of the element are *scattered* into the global array.

In the first time step, the strain measures and the stresses are not computed. Instead, as shown in the flowchart, the internal nodal forces are computed directly from the initial stresses, $\sigma(\mathbf{X},0)$.

In the flowchart, the matrix form of the internal force computation, in which the stress tensor is stored as a square matrix, is shown. To change to Voigt form, replace $\mathbf{B}$ by $\mathbf{B}^T$ and the square matrix of stresses by the column matrix $\{\sigma\}$.

Most essential boundary conditions are easily handled in explicit methods. For example, if the velocities or displacements are prescribed as functions of time along any boundary, then the velocity/displacement boundary conditions can be enforced by setting the nodal velocities according to the data:

$$v_{iI}^{n+\frac{1}{2}} = \bar{v}_i\left(\mathbf{x}_I, t^{n+\frac{1}{2}}\right)$$

(13.7)

If the data is not available on the nodes, they can be obtained by the least square procedure. When the boundary conditions are posed in terms of displacements, imposition of (13.7) involves a numerical differentiation of the prescribed displacements to obtain velocities; the velocities are then obtained from the displacements. This round-about procedure can be avoided by setting the prescribed boundary displacements after step 8.

The velocity boundary conditions can also be enforced in local coordinate systems. In that case, the equations of motion at these nodes must be expressed in the local coordinate components; the nodal forces must be transformed to in the local coordinate components before assembly. The orientation of the local coordinate system may vary with time but the time integration formulas must then be modified to account for the rotation of the coordinate system.

When essential boundary conditions are given as linear or nonlinear algebraic equations relating the displacements (13.4b), the implementation is more complicated. The penalty or Lagrange multiplier methods are commonly used.

Any damping in the system lags by a half time step: see step 3 in Box 13.1. This also holds for any rate-dependent terms in the constitutive equation evaluation in step 3 of *getforce*.   The time lag is unavoidable if the implementation is to be fully explicit, i.e. not require the solution of any equations. This decreases the stable time step for the method.

As can be seen from the flowchart, an explicit method is easily implemented. Furthermore, explicit time integration is very robust, by which we mean that the explicit procedure seldom aborts due to failure of the numerical algorithm. The salient disadvantage of explicit integration, the price you pay for the simplicity of the method and its avoidance of the solution of equations, is *the conditional stability of explicit methods*.   If the time step exceeds a critical value Δt_{crit}, some modes will grow unboundedly and the solution will be useless.

The *critical time step* is also called the *stable time step*. A stable time step for a mesh of constant strain elements is given by

$$\Delta t = \alpha \Delta t_{crit} \qquad \Delta t_{crit} = \frac{2}{\omega_{max}} \leq \min_{e,I} \frac{1}{\omega_I^e} = \min_e \frac{\ell_e}{c_e} \tag{13.11}$$

where ω_{max} is the maximum frequency of the linearized system, l^e is a characteristic length of element e, c^e the current wavespeed in element e and α is a reduction factor that accounts for the destabilizing effects of nonlinearities; a good choice for α is $0.8 \leq \alpha \leq 0.98$ [13.1].

The above is called the Courant condition in finite difference methods after one of its discoverers; the result was first published by Courant, Friedrichs and Lewy [13.2]. The ratio of the time step to the critical time step, α, is called the Courant number.   From (13.11) it can be seen that the critical time step decreases with mesh refinement and increasing stiffness of the material. The cost of an explicit simulation is independent of the frequency range that is of interest and depends only on the size of the model and the number of time steps.

An interesting question for elastic-plastic materials is whether the slower wavespeed in the plastic response (see Chapter 5 of Ref. 1) enables one to increase the time step. Based on our experience, the answer appears to be negative.   An elastic-plastic material can unload at any moment, and in numerical solutions unanticipated unloading often occurs due to numerical noise. During elastic unloading, the critical time step depends on the elastic wavespeed, and a larger time step results in instability.

The mesh time step is obtained from element time steps. For each element, an element time step is calculated and the minimum element time step is chosen as the mesh time step. The theoretical justification for setting the critical time step on an element basis is given in Chapter 6 of Ref. 1.

13.3 Energy balance

The above stability conditions, emanate from an analysis of the stability of the integrator for the linear equations of motion. At this time, there are no stability theorems that cover the range of nonlinear phenomena encountered in engineering problems, such as contact-impact, tearing, etc.

It is possible for instabilities to develop even when (13.11) is observed. In contrast to linear problems, where an instability leads to an exponential growth of the solution and cannot be overlooked, unstable solutions of nonlinear problems are sometimes not readily discernible. For example Belytschko [13.3] describes a numerical phenomenon called an *arrested instability*. The scenario is as follows. An instability is triggered by nonlinearities, such as geometric stiffening, while the material is elastic. This instability causes local exponential growth of the solution, which in turn leads to plastic behavior. The plastic response softens the structure and decreases the wavespeed, so the integrator regains stability.

Such arrested instabilities can lead to a large overprediction of displacements, but they are not detectable by perusing the results. However, they can easily be detected by an energy balance check. Any instability results in the spurious generation of energy, which leads to a violation of the conservation of energy. Therefore, whether stability was maintained during a nonlinear computation can be established by checking the energy balance.

In low order methods like the central difference method, the energy is usually integrated in time by a method of similar order, such as the trapezoidal rule. The internal and external energies are integrated as follows:

$$
\begin{aligned}
W_{int}^{n+1} &= W_{int}^{n} + \frac{\Delta t^{n+1/2}}{2}\left(\mathbf{v}^{n+\frac{1}{2}}\right)^{T}\left(\mathbf{f}_{int}^{n} + \mathbf{f}_{int}^{n+1}\right) \\
&= W_{int}^{n} + \tfrac{1}{2}\Delta\mathbf{d}^{T}\left(\mathbf{f}_{int}^{n} + \mathbf{f}_{int}^{n+1}\right)
\end{aligned}
\tag{13.12}
$$

$$W_{ext}^{n+1} = W_{ext}^n + \frac{\Delta t^{n+1/2}}{2} \left(\mathbf{v}^{n+1/2}\right)^T \left(\mathbf{f}_{ext}^n + \mathbf{f}_{ext}^{n+1}\right)$$
$$= W_{ext}^n + \tfrac{1}{2}\Delta\mathbf{d}^T \left(\mathbf{f}_{ext}^n + \mathbf{f}_{ext}^{n+1}\right) \qquad (13.13)$$

The kinetic energy is given by

$$W_{kin}^n = \frac{1}{2}\left(\mathbf{v}^n\right)^T \mathbf{M}\mathbf{v}^n \qquad (13.14)$$

Note that integer time steps are used for the velocities, which is why the half step formulas are used twice in Box 13.2.

The internal energies can also be computed on the element or quadrature point level by

$$W_{int}^{n+1} = W_{int}^n + \tfrac{1}{2}\sum_e \Delta\mathbf{d}_e^T \left(\mathbf{f}_{e,int}^n + \mathbf{f}_{e,int}^{n+1}\right)$$
$$= W_{int}^n + \frac{\Delta t^{n+\frac{1}{2}}}{2}\sum_e \sum_{n_Q} \bar{w}_Q \mathbf{D}_Q^{n+\frac{1}{2}} : \left(\mathbf{s}_Q^n + \mathbf{s}_Q^{n+1}\right) J_{\xi Q} \qquad (13.15)$$

(see Section 4.5.4 for quadrature notation). Energy conservation requires that

$$\left|W_{kin} + W_{int} - W_{ext}\right| \le \varepsilon \, \max(W_{ext}, W_{int}, W_{kin}) \qquad (13.16)$$

where ε is a small tolerance, generally on the order of 10^{-2}.

If the system is very large, on the order of 10^5 nodes or larger, the energy balance should be performed on subdomains of the model. The internal forces from adjacent subdomains are then treated as external forces for each subdomain.

13.4 Accuracy

The central difference method is second order in time, i.e. the truncation error is of order Δt^2 in the displacements. We will see that the spatial error in the displacements in the L_2 norm for linear complete elements is of order h^2, where h is the element size. Although there are some technical differences between these two measures of error, the outcome is similar. Since the time step and the element size must be of the same order to meet the stability condition, (13.11), the time integration error and the spatial error are of the same order for central difference time integration. However, for materials with rapidly varying stiffness, such as viscoplastic materials, the accuracy of the central difference method is sometimes inadequate. In those cases, we suggest Runge-Kutta methods. The Runge-Kutta method need not be used on all equations: it can be applied to the constitutive equation while integrating the equations of motion by the central difference method.

13.5   Mass scaling, subcycling and dynamic relaxation

When a model contains a few very small or stiff elements, the efficiency of explicit integration is compromised severely, since the time step of the entire mesh is set by these very stiff elements. Several techniques are available for circumventing this difficulty:

1. mass scaling: the masses of stiffer elements are set so that the time step is not decreased by these elements;
2. subcycling: a smaller time step is used for the stiffer elements,

Mass scaling should be used for problems where high frequency effects are not important. For example, in sheet-metal forming, which is essentially a static process, it causes no difficulties. On the other hand, if high frequency response is important, mass scaling is not recommended.

Subcycling was introduced by Belytschko, Yen and Mullen [13.4]. In this technique the model is split into subdomains and each is integrated with its own stable time step. The crucial issue in subcycling is the treatment of the interface between subdomains. The initial methods used linear interpolation. These can be shown to be stable for first order systems, Belytschko, Smolinski, and Liu [13.5]. But, as shown by Daniel [13.6], in second order systems, linear interpolation leads to narrow bands of instability. These can be eliminated either by the addition of artificial viscosity, or by other subcycling methods, which are more complex. Stable subcycling methods for second order systems are given by Smolinski, Sleith and Belytschko [13.7] and Daniel [13.6].

Dynamic relaxation is often used in explicit codes to obtain static solutions. The basic idea is to apply the load very slowly in the dynamic system equations and add enough damping so that oscillations are minimized. In path-dependent materials, dynamic relaxation often yields poor solutions.

Furthermore, it is very slow. Newton methods combined with effective iterative solvers, such as preconditioned conjugate gradient or multigrid methods, are much faster and more accurate.

REFERENCES

[13.1]　Belytschko, T., Liu, W.K. and Moran, B. *Nonlinear Finite Elements for Continua and Structures*, Wiley, Chichester, U.K., 2000.

[13.2]　Courant, R., Friedrichs, K.O. and Lewy, H., Über die partiellen differenz enschleingen der mathematischen physik, *Math. Ann.*, **100**, 32, 1928.

[13.3]　Mullen, R. and Belytschko, T., An analysis of an unconditionally stable explicit method, *Computers and Structures*, **16**, 691-696, 1983.

[13.4]　Belytschko, T., Yen, H.J. and Mullen, R., Mixed methods in time integration, *Comp. Methods in Appl. Mech. and Eng.*, **17/18**, 259-275, 1979.

[13.5]　Belytschko, T., Smolinski, P. and Lin, W.K., Stability of multi-time step partitioned integrators for first-order finite element systems, *Comp. Methods in Appl. and Eng.*, **49**, 281-297, 1985.

[13.6]　Daniel, W.J.T, Analysis and implementation of a new constant acceleration subcycling algorithm, *Int. J. Num. Meth. for Engng.*, **40**, 2481-2855, 1997.

[13.7]　Smolinski, P., Sleith, S. and Belytschko, T., Explicit-subcycling for linear structural dynamics, *Comp. Mechanics*, **18**, 236-244, 1996.

14. CONTACT MECHANICS

14.1   Contact interface equations

14.1.1. Notation and Preliminaries.

Contact-impact algorithms in general purpose software can treat the interaction of many bodies, but for purposes of simplicity, we limit ourselves to two bodies as illustrated in Fig. 14.1. The treatment of multi-body contact is identical: the interaction of any pair of bodies is exactly like the two body problem. We have denoted the configurations of the two bodies by Ω^A and Ω^B and denote the union of the two bodies by Ω. The boundaries of the bodies are denoted by Γ^A and Γ^B. Although the two bodies are interchangeable with respect to their mechanics, it is sometimes useful to express the equations in term of one of the bodies, which is called the master; body A is designated as the master, body B as the slave. When we wish to distinguish field variables that are associated with a particular body, we append a superscript A or B; when neither of these superscripts appears, the field variable applies to the union of the two bodies. Thus the velocity field $\mathbf{v}(\mathbf{X},t)$ refers to the velocity field in both bodies, whereas $\mathbf{v}^A(\mathbf{X},t)$ refers to the velocity in body A.

The contact interface consists of the intersection of the surfaces of the two bodies and is denoted by Γ_c.

$$\Gamma_c = \Gamma^A \cap \Gamma^B \tag{14.1}$$

This contact interface consists of the two physical surfaces of the two bodies which are in contact, but since they are theoretically coincident we often refer to a single interface Γ_c. In numerical solutions, the two surfaces will usually not be coincident. Moreover, although the two bodies may be in contact on several disjoint interfaces, we designate their union by Γ_c. The contact interface is a function of time, and its determination is an important part of the solution of the contact-impact problem.

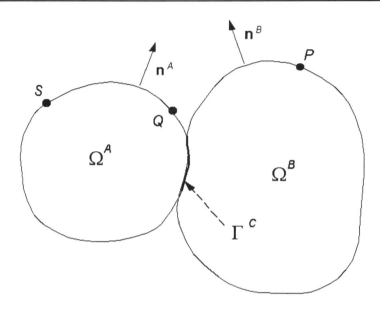

Figure 14.1 Model problem for contact-impact showing notation.

In constructing the equations on the contact interface, it is convenient to express vectors in terms of local components. A local coordinate system is set up at each point of the master contact surface as shown in Fig. 14.2. At each point, we construct unit vectors tangent to the surface of the master body $\hat{\mathbf{e}}_1^A \equiv \hat{\mathbf{e}}_x^A$ and $\hat{\mathbf{e}}_2^A \equiv \hat{\mathbf{e}}_y^A$. The procedure for obtaining these unit vectors is identical to that used in shell elements. The normal for body A is given by

$$\mathbf{n}^A = \hat{\mathbf{e}}_1^A \times \hat{\mathbf{e}}_2^A \tag{14.2}$$

On the contact surface, the normals of the two bodies are in opposite directions:

$$\mathbf{n}^A = -\mathbf{n}^B \tag{14.3}$$

The velocity fields can be expressed in the local coordinates of the contact surface by

$$\mathbf{v}^A = v_N^A \mathbf{n}^A + \hat{v}_\alpha^A \hat{\mathbf{e}}_\alpha^A = v_N^A \mathbf{n}^A + \mathbf{v}_T^A \tag{14.4a}$$

$$\mathbf{v}^B = v_N^B \mathbf{n}^A + \hat{v}_\alpha^B \hat{\mathbf{e}}_\alpha^A = v_N^A \mathbf{n}^B + \mathbf{v}_T^A \tag{14.4b}$$

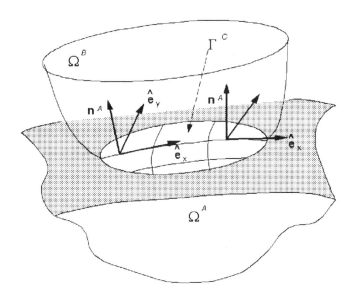

Figure 14.2 Contact interface showing local unit vectors referred to master surface A.

where the range of Greek subscripts is 2 in three dimensional problems. When the problem is two dimensional, the contact surface becomes a line, so we have a single unit vector $\hat{\mathbf{e}}_1 \equiv \hat{\mathbf{e}}_x$ tangent to this line; the range of the Greek subscripts in (14.4) is then one and the tangential component is a scalar. As can be seen in the above, the components are expressed in terms of the local coordinate system of the master surface. The normal velocities are given by

$$v_N^A = \mathbf{v}^A \cdot \mathbf{n}^A \quad v_N^B = \mathbf{v}^B \cdot \mathbf{n}^A \tag{14.5}$$

which can easily be seen by taking the dot product of the expressions in (14.4) with $\mathbf{n}^A$ and using the fact that the normal is orthogonal to the unit vectors tangent to the plane $\hat{\mathbf{e}}_i^A$.

The bodies are governed by the standard field equations given in Box 11.1: conservation of mass, momentum and energy, a strain measure, and the constitutive equations. Contact adds the following conditions: the bodies can not interpenetrate and the tractions must satisfy momentum conservation on the interface. Furthermore, the normal traction across the contact interface cannot be tensile. We classify the requirements on the displacements and velocities as kinematic conditions and the requirements on the tractions as kinetic conditions.

14.1.2. *Traction Conditions.*

The tractions must observe the balance of momemtum across the contact interface. Since the interface has no mass, this requires that the sum of the tractions on the two bodies vanish

$$\mathbf{t}^A + \mathbf{t}^B = \mathbf{0} \tag{14.6a}$$

The tractions on the surfaces of the two bodies are defined by Cauchy's law

$$\mathbf{t}^A = \boldsymbol{\sigma}^A \cdot \mathbf{n}^A \quad or \quad t_i^A = \sigma_{ij}^A n_j^A \qquad \mathbf{t}^B = \boldsymbol{\sigma}^B \cdot \mathbf{n}^B \quad or \quad t_i^B = \sigma_{ij}^B n_j^B \tag{14.6b}$$

The normal tractions are defined by

$$t_N^A = \mathbf{t}^A \cdot \mathbf{n}^A \quad or \quad t_N^A = t_j^A n_j^A \qquad t_N^B = \mathbf{t}^B \cdot \mathbf{n}^A \quad or \quad t_N^B = t_j^B n_j^A \tag{14.6c}$$

Note that the normal components refer to the master body. The normal component of momentum balance can be obtained by taking a dot product of (14.6a) with the normal vector $\mathbf{n}^A$, which gives

$$t_N^A + t_N^B = 0 \tag{14.6d}$$

We do not consider any adhesion between the contact surfaces in the normal direction, so the normal tractions cannot be tensile. We will subsequently often use the phrase that the normal tractions must be compressive, although the normal tractions can also vanish. The condition that the normal tractions cannot be tensile can be stated as

$$t_N \equiv t_N^A(\mathbf{x}, t) = -t_N^B(\mathbf{x}, t) \le 0 \tag{14.6e}$$

The condition that the normal tractions be compressive requires t_N^B to be positive since t_N^B is the projection of the traction on body B onto the unit normal of A, which points into body B.

The tangential tractions are defined by

$$\mathbf{t}_T^A = \mathbf{t}^A - t_N^A \mathbf{n}^A \qquad \mathbf{t}_T^B = \mathbf{t}^B - t_N^B \mathbf{n}^A \tag{14.7a}$$

so the tangential tractions are the total tractions projected on the master contact surface. Momentum balance requires that

$$\mathbf{t}_T^A + \mathbf{t}_T^B = \mathbf{0} \tag{14.7b}$$

The above equation can be obtained by substituting (14.7a) into (14.6a) and using (14.6d).

When a frictionless model of contact is used, the tangential tractions vanish:

$$\mathbf{t}_T^A = \mathbf{t}_T^B = 0 \tag{14.7c}$$

We have used the phrase "frictionless model of contact" to emphasize that it is not implied that friction is absent, but rather that friction is neglected in the model because it is deemed unimportant. Subsequently we shall just say frictionless contact, but it should be understood that friction never vanishes in reality.

Although one of the bodies has been chosen as the master body in developing the preceding contact interface equations, these equations are symmetrical with respect to the bodies when the two contact surfaces are coincident and Eq. (14.3) is observed. Thus it does not matter which body is chosen as the master body. However, when the two surfaces are not coincident, as in most numerical solutions, then the choice of the master body changes the equations somewhat.

14.2   Friction models

14.2.1 Classification

The models used for the computation of the tangential tractions are collectively called friction models. There are basically three types of friction models:

1. Coulomb friction models, which are based on the classical theories of friction commonly taught in undergraduate mechanics and physics courses;
2. Interface constitutive equations, which approximate the behavior of the tangential forces by equations similar to constitutive equations used for materials;
3. Asperity-lubricant models, which model the behavior of the physical characteristics of the interface, often on a microscale.

The demarcations between these classes are not sharp; some models adopt features of more than one of the above classes, but the above roughly describes the current state of affairs.

14.2.2 Coulomb Friction.

Coulomb friction models originate from modeling of frictional forces between rigid bodies. In the application of classical Coulomb friction models to continua, they are applied at each point of the contact interface. A direct translation of the Coulomb friction law from rigid body mechanics to a continuum law gives

if A and B are in contact at $\mathbf{x}$, *and*

$$a)\ if\ \left\|\mathbf{t}_T(\mathbf{x},t)\right\| < -\mu_F t_N(\mathbf{x},t),\ then\ \boldsymbol{\gamma}_T(\mathbf{x},t) = 0 \qquad (14.8)$$

$$b)\ if\ \left\|\mathbf{t}_T(\mathbf{x},t)\right\| = -\mu_F t_N(\mathbf{x},t),\ \mathbf{t}_T(\mathbf{x},t)$$
$$= -\alpha(\mathbf{x},t)\mathbf{t}_T(\mathbf{x},t),\ \ \alpha \geq 0 \qquad (14.9)$$

where μF is the coefficient of friction and α is a variable which is determined by the solution of the complete problem. The condition that the two bodies are in contact at a point implies that the normal traction $t_N \leq 0$, so the RHS of the two expressions, $-\mu_F t_N$, is always positive. Condition (a) is known as the stick condition: when the tangential traction at a point is less than the critical value, no relative tangential motion is permitted, i.e. *the two bodies stick.* Condition (b) corresponds to frictional sliding, and the second part of that equation expresses the fact that the tangential frictional traction must be opposite to the direction of the relative tangential velocity.

Classical Coulomb friction closely resembles a rigid-plastic constitutive equation. If the tangential velocity $\boldsymbol{\gamma}_T$ is interpreted as a strain and the tangential traction components are interpreted as stresses, the first relation in (14.8) can be interpreted as a yield function. According to (14.8), when the yield criterion is not met, the tangential velocity vanishes. Once the yield function is satisfied, the tangential velocity is in the direction of the tangential traction but its magnitude is unspecified. These attributes of the response parallel the rigid plastic model.

The stick condition of Coulomb friction is its most troublesome characteristic, since it induces discontinuities into the time history of the relative tangential velocity. When the motion of a point changes from slip to stick, the relative tangential velocity γ_T discontinuously jumps to zero. Thus the tangential velocities are not smooth, but exhibit the same discontinous character as the normal velocities at the time of impact. Furthermore, the inequalities in Coulomb friction result in weak forms which are inequalities. Therefore, Coulomb friction is difficult to handle in numerical solutions and we consider it only for some special cases.

14.2.3 Interface Constitutive Equations

A different approach to defining interface laws has been pioneered by Michalowski and Mroz [14.1] and Curnier [14.2]. This approach is motivated by the theory of plasticity and the analogy between Coulomb friction and elasto-plasticity we alluded to above. Constitutive models of interface behavior are motivated by the fact that microscopic examination of even the smoothest surfaces reveals surface roughness due to asperities, such as shown in Fig. 14.3. Even when the surfaces appear smooth, friction results from the interaction of these asperities during sliding. Sliding initially causes elastic deformations of these asperities, so a true stick condition cannot occur in nature, i.e. the stick condition is an idealization of observed behavior. Elastic deformation of the asperities is followed by "grinding" down of the asperities as the sliding proceeds. The elastic deformations of the asperities are reversible, whereas the grinding down is irreversible, so ascribing an elastic character to the initial sliding and a plastic character to subsequent sliding is natural.

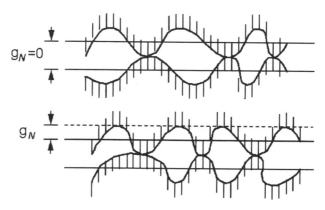

Figure 14.3   Asperities on contacting surface.

As an example of an interface constitutive law we describe an adaptation of Curnier's [14.2] plasticity theory for friction. This model contains all of the ingredients of a plasticity theory for continua: a decomposition of deformation into reversible and irreversible components, a yield function and a flow law. In this description of the Curnier model we have replaced displacements by velocities, which appears appropriate for problems involving arbitrary time histories and large relative sliding.

14.3   Weak forms

14.3.1 Notation and Preliminaries

The weak form of the momentum equation and the contact interface conditions will be developed for a Lagrangian mesh. This development is also applicable to an ALE mesh when the contact surface is treated as Lagrangian. For simplicity, we start with frictionless contact and defer the treatment of tangential tractions to the last part of this Section. We restrict the following developments to the case where all traction or velocity components are prescribed on a traction or dispacement boundary, respectively.

The contact surface is neither a traction nor a displacement boundary. The total boundary of body A is given by

$$\Gamma^A = \Gamma_t^A \cup \Gamma_u^A \cup \Gamma_C \tag{14.10a}$$

$$\Gamma_t^A \cap \Gamma_u^A = 0 \quad \Gamma_t^A \cap \Gamma_C = 0 \quad \Gamma_u^A \cap \Gamma_C = 0 \tag{14.10b}$$

Similar relations hold for body B.

The trial solutions are in the space of *kinematically admissible velocities*, and as in Chapter 11 we choose the velocities to be the cardinal dependent variable; the displacements can be obtained by time integration. The trial solution is $v(X,t) \in U$ where the space of trial functions is defined by

$$U = \left\{ v(X,t) \mid v \in C^0(\Omega^A), v \in C^0(\Omega^B), v = \overline{v} \, on \, \Gamma_u \right\} \tag{14.11}$$

This space of trial functions is similar to that for the single body problem, but the velocities are separately approximated in the two bodies; the velocity fields in U are not continuous across the contact interface. The admissible velocity fields are here given as C^0, i.e. in H^1, but for purposes of convergence analysis in linear elastostatics the displacements should be in $H^{1/2}$, see Kikuchi and Oden [14.3]. This is the same function space that is used in fracture mechanics to handle the singular stresses at the crack tip. In contact problems, singularities occur at the edge. However, unlike in fracture mechanics, the singularities in contact problems do not appear to be of any engineering significance, since the roughness of surfaces appears to eliminate the appearance of even near singular behavior in the stresses.

The space of test functions is defined by

$$U_0 = U \cap \{\delta v(\Omega) \mid \delta v = 0 \text{ on } \Gamma_u\} \qquad (14.12)$$

which parallels the definition in Section 11.3.

14.3.2 Lagrange Multiplier Weak Form

A common approach to imposing the contact constraints is by means of Lagrange multipliers. We follow the description given by Belytschko and Neal [14.4]. Let the Lagrange multiplier trial function be $\lambda(\zeta,t)$ and the corresponding test functions be $\delta\lambda(\zeta,t)$, which reside in the following spaces

$$\lambda(\zeta,t) \in J^+, \quad J^+ = \{\lambda(\zeta,t) \mid \lambda \in C^{-1}, \lambda \geq 0 \text{ on } \Gamma^c\} \qquad (14.13)$$

$$\delta\lambda(\zeta) \in J^-, \quad J^- = \{\delta\lambda(\zeta) \mid \delta\lambda \in C^{-1}, \delta\lambda \leq 0 \text{ on } \Gamma^c\} \qquad (14.14)$$

The weak form is:

$$\delta P_L(v, \delta v, \lambda, \delta\lambda) \equiv \delta P + \delta G_L \geq 0 \quad \forall \delta v \in U_0, \forall \delta\lambda \in J^- \qquad (14.15)$$

$$\text{where } \delta G_L = \int_{\Gamma^c} \delta(\lambda \gamma_N) d\Gamma \qquad (14.16)$$

In the above, δP is defined in Table B11.2 and $v \in U$, $\lambda \in J^+$. This weak form is equivalent to the momentum equation, the traction boundary conditions, the interior continuity conditions (generalized momentum balance) and the following contact interface conditions: impenetrability, momentum balance on normal tractions (14.6d) and the frictionless condition (14.7c). The requirement that the normal interface traction be compressive is among the constraints on the trial space for the Lagrange multipliers. Note that the above *weak form is an inequality.*

The above is a standard way of appending a constraint to a weak form by means of a Lagrange multipliers: compare to the Hu-Washizu variational principle. The only difference from the Hu-Washizu form is that the constraint is an inequality.

The equivalence of the weak form to the momentum equation, the traction boundary `conditions and the contact conditions is shown by a procedure that parallels that in Chapter 5 of Ref. 5. Recall that δP is given in Box 11.2 as

$$\delta P = \int_{\Omega} \left[\delta v_{i,j}\sigma_{ji} - \delta v_i(\rho b_i - \rho \dot{v}_i) \right] d\Omega - \int_{\Gamma_t} \delta v_i \bar{t}_i d\Gamma \tag{14.17}$$

where we have used commas to denote derivatives with respect to the spatial variables and a superposed dot to denote the material time derivative. All integrals in the above apply to the union of both bodies, i.e. $\Omega = \Omega^A \cup \Omega^B$, $\Gamma_t = \Gamma_t^A \cup \Gamma_t^B$. The first step is to integrate the internal virtual power by parts and apply Gauss's theorem:

$$\int_{\Omega} \left(\delta v_i \sigma_{ji} \right)_{,j} d\Omega = \int_{\Gamma_t} \delta v_i \sigma_{ji} n_j d\Gamma + \int_{\Gamma^c} \left(\delta v_i^A t_i^A + \delta v_i^B t_i^B \right) d\Gamma \tag{14.18}$$

We have used the fact that the integral on the displacement boundary Γ_u vanishes because $\delta v_i = 0$ on Γ_u and Cauchy's law has been applied to obtain the expressions in the last integral. The first integral on the RHS of the above applies to both bodies. An integral over the contact surface appears on each body when Gauss's theorem is applied we wish to express the result as a single integral over the contact surface, so the identity of the body is indicated by the superscripts A and B on the filed variables.

The integrand of the second integral on the RHS of the above is now broken up into components normal and tangential to the contact surface. In indicial notation this gives

$$\delta v_i^A t_i^A = \delta v_N^A t_N^A + \delta v_\alpha^A t_\alpha^A \tag{14.19}$$

where the range of alpha is 1 for two dimesional problems and 2 for three dimensional problems. A similar relationship can be written for body B. The above is clearer to some people in vector notation, where the above is

$$\delta \mathbf{v}^A \cdot \mathbf{t}^A = \left(\delta v_N^A \mathbf{n}^A + \delta \mathbf{v}_T^A \right) \cdot \left(t_N^A \mathbf{n}^A + \mathbf{t}_T^A \right) = \delta v_N^A t_N^A + \delta \mathbf{v}_T^A \cdot \mathbf{t}_T^A \tag{14.20}$$

The last term is obtained by noting that $\mathbf{n}$ is normal to the tangent vectors $\mathbf{t}_N$ and $\mathbf{v}_N$. The last term in the above is an alternative expression for $\hat{t}_\alpha \hat{v}_\alpha$.

Substituting (14.18) and (14.19) into (14.17) gives

$$\delta P = \int_{\Gamma^C} \delta v_i(\rho v_i - b_i - \sigma_{ij,j})d\Omega + \int_{\Gamma_t} \delta v_i(\sigma_{ji}n_j - \bar{t}_i)d\Gamma$$
$$+ \int_{\Gamma^c} (\delta v_N^A t_N^A + \delta v_N^B t_N^B + \delta \hat{v}_\alpha^A \hat{t}_\alpha^A + \delta \hat{v}_\alpha^B \hat{t}_\alpha^B)d\Gamma \tag{14.21}$$

Now consider (14.16):

$$\delta G_L = \int_{\Gamma^C} \delta(\lambda \gamma_N)d\Gamma = \int_{\Gamma^C} (\delta\lambda \gamma_N + \delta\gamma_N \lambda)d\Gamma \tag{14.22}$$

Substituting (14.2) into the above gives

$$\delta G_L = \int_{\Gamma^c} (\delta\lambda\gamma_N + \lambda(\delta v_N^A - \delta v_N^B))d\Gamma \tag{14.23}$$

Combining (14.21) and (14.23) yields

$$0 \le \delta P_L = \int_\Omega \delta v_i \left(\sigma_{ji,j} - \rho b_i - \rho \dot{v}_i\right)d\Omega + \int_{\Gamma_t} \delta v_i(\sigma_{ji}n_j - \bar{t}_i)d\Gamma$$
$$+ \int_{\Gamma^c} [\delta v_N^A(t_N^A + \lambda) + \delta v_N^B(t_N^B - \lambda) + (\delta\hat{v}_\alpha^A \hat{t}_\alpha^A + \delta\hat{v}_\alpha^B \hat{t}_\alpha^B) + \delta\lambda\gamma_N]d\Gamma \tag{14.24}$$

Extracting the strong form from the weak inequality is similar to the procedure described in Section 11.2. Whenever the test function is unconstrained, there is then no restriction on the sign of the term which multiplies the test function and the term must vanish by the density theorem. Thus it follows from the first two integrals of the above that

$$\sigma_{ji,j} - \rho b_i = \rho \dot{v}_i \quad in\,\Omega \tag{14.25}$$

$$\sigma_{ji}n_j = \bar{t}_i \quad on\,\Gamma_t \tag{14.26}$$

i.e. that the momentum equation and the natural boundary conditions are satisfied in bodies A and B. In all terms of the integrand on the contact surface except the last, the test function is also unconstrained, and we obtain the equalities

$$\hat{t}_\alpha^A = 0 \ \text{and} \ \ \hat{t}_\alpha^B = 0 \ on\Gamma^c, \quad or \ \mathbf{t}_T^A = \mathbf{t}_T^B = 0 \ on\Gamma^c \tag{14.27}$$

$$\lambda = -t_N^A \ \text{and} \ \lambda = t_N^B \ on \ \Gamma^c \tag{14.28}$$

where the inequalitites in (14.28) follow because the trial functions for λ are constrained to be positive, see (14.13): it follows from (14.28) that the normal traction on the contact interface is compressive. By eliminating λ from (14.28) we obtain the momentum balance condition on the normal tractions

$$t_N^A + t_N^B = 0 \ \ on \ \Gamma^c \tag{14.29a}$$

In the last term of the integrand of (14.23), the variation $\delta\lambda$ is constrained to be negative. Therefore, its coefficient γ_N does not necessarily vanish. However it can be deduced that the coefficient of γ_N must be nonpositive, i.e. that the weak *inequality* implies $\gamma_N \leq 0 \ on \ \Gamma^c$, which is the interpenetration inequality (14.2).

Equations (14.25-27), (14.29) and (14.10) constitute the strong form corresponding to the weak form (14.15). This set includes the momentum equation, the interior continuity conditions, and the traction (natural) boundary conditions on both bodies. On the contact surface, the strong form includes momentum balance of the normal tractions and the inequality on the interpenetration rate. The compressive character of the normal tractions follows from the restriction on the Lagrange multiplier field (14.13).

14.3.3 Interpenetration-Dependent Penalty

The above form of the penalty method often performs poorly since it may allow excessive interpenetration. The normal penalty traction is applied only when the relative velocities lead to continued interpenetration. As soon as the relative velocities of contiguous points of the two surfaces become equal or negative, the normal traction vanishes. Substantial interpenetration may consequently persist in the solution. Therefore, it is recommended that the normal traction also be a function of the interpenetration. For this purpose, we define the following relation for the normal traction:

$$p = p(g_N, \gamma_N) H(g_N) \tag{14.30}$$

where g_N is the interpenetration. The weak form is then given by

$$\delta G_p = \int_{\Gamma_c} \delta \gamma_N p \, d\Gamma \tag{14.31}$$

The same procedure as before then gives

$$t_N^A + p = 0 \quad \text{and} \quad t_N^B - p = 0 \quad \text{on } \Gamma_c \tag{14.32}$$

Combining the two above equations gives

$$t_N^A = -t_N^B = -p(g_N, \gamma_N) H(g_N) \tag{14.33}$$

Thus the tractions are always compressive and satisfy momentum balance. The tractions are functions of the interpenetration and rate of interpenetration. An example of a penalty function for normal traction is

$$p = (\beta_1 g_N + \beta_2 \gamma_N) H(\beta_1 g_N + \beta_2 \gamma_N) \tag{14.34a}$$

where β_1, β_2 are penalty parameters. The step function in this expression avoids tensile normal tractions across the interface.

REFERENCES

[14.1]   Michalowski, R. and Mroz, Z., Associated and non-associated sliding rules in contact friction problems, *Arch. Mech.*, **30**, 259-276, 1978.

[14.2]   Curnier, A., A theory of friction, *Int. J. Solids Structures*, **30**, 637-647, 1984.

[14.3]   Kikuchi, N. and Oden, J.T., *Contact Problems In Elasticity: A Study Of Variational Inequalities And Finite Element Methods*, SIAM, Philadelphia, Pennsylvania, 1988.

[14.4]   Belytschko, T. and Neal, M., Contact-impact by the pinball algorithm with penalty and Lagrange methods, *Int. J. of Num. Methods for Eng.*, **31**, 547-572, 1991.

[14.5]   Belytschko, T., Liu, W.K. and Moran, B., *Nonlinear Finite Elements for Continua and Structures*, Wiley, Chichester, U.K., 2000.

Box 14.1
Weak Forms

$\delta P_C = \delta P + \delta G + \delta G_T$ **NOTE:** $\gamma \equiv \gamma_N$

Tangential tractions: $\delta G = \int_{\Gamma_c} \delta \gamma_T \cdot \lambda_T \, d\Gamma \equiv \int_{\Gamma_c} \delta \hat{\gamma}_\alpha \hat{\lambda}_\alpha \, d\Gamma$

Lagrangian: $\delta G = \delta G_L = \int_{\Gamma_c} \delta(\lambda \gamma) \, d\Gamma, \quad \delta P_C \geq 0$

Penalty: $\delta G = \delta G_P = \int_{\Gamma_c} \frac{1}{2} \beta \delta(\gamma^2) \, d\Gamma, \quad \delta P_C = 0$

Augmented Lagrangian: $\delta G = \delta G_{AL} = \int_{\Gamma_c} \delta(\lambda \gamma + \frac{\alpha}{2} \gamma^2) \, d\Gamma, \quad \delta P_C \geq 0$

Perturbed Lagrangian: $\delta G_N = \delta G_{PL} = \int_{\Gamma_c} \delta(\lambda \gamma - \frac{1}{2\beta} \lambda^2) \, d\Gamma, \quad \delta P_C = 0$

15. FINITE ELEMENTS IN CONTACT-IMPACT

15.1   Finite element discretization

15.1.1 Overview

The finite element equations for the various treatments of contact-impact are developed. The weak statements for all of the approaches to the contact-impact problem, (penalty, Lagrange multiplier, etc.) involve a sum of the standard virtual power and a contribution from the contact interface. The standard virtual power is discretized exactly as in the absence of contact, so we will use the results developed in Chapter 12. This chapter concentrates on the discretization of the various contact interface weak forms. The development closely follows Ref. 1. Ref 2 is useful for more background.

The developments that follow here are applicable to Lagrangian meshes with both updated and total Lagrangian formulations. However in total Lagrangian formulations, the contact interface conditions must be imposed in terms of the tractions on the undeformed surface areas, t^0. These are more difficult to handle.

The following discretizations are also applicable to ALE formulations as long as the nodes on the contact surface are Lagrangian. *They are not directly applicable to Eulerian formulations* since we assume that we have at our disposal a referential coordinate that describes the contact surface. Such a coordinate system cannot easily be defined in an Eulerian mesh. In a Lagrangian mesh, the contact surface corresponds to a subset of the boundary of the mesh.

We will first develop the FEM discretization for the Lagrangian multiplier method in indicial notation. Indicial notation enables us to go through some subtle steps, which will subsequently be skipped in the matrix derivations; anyone who wishes to replicate these steps for other formulations can rederive these in indicial notation.

15.1.2 Lagrange Multiplier Method

The velocity $v(\mathbf{X},t)$ is approximated by C^0 interpolants in each body as in the single body problem; as can be seen from (11.3), the velocities of the two bodies need not be continuous across the contact interface; the interpenetration condition will emanate from the discretization of the weak form. The velocity field can also be expressed in terms of the element coordinates ζ on the contact surface. As in Chapter 11, we note that the approximation of the velocity field also defines the approximation of the displacement field, we could have chosen either as the primal variable.

The finite element approximation for the velocity field is expressed in terms of the material coordinates since we are dealing with a Lagrangian mesh. To clarify certain issues, we will initially discard the summation convention on repeated nodal indices and indicate sums explicitly. The velocity field is

$$v_i^A(\mathbf{X},t) = \sum_{I \in \Omega^A} N_I(\mathbf{X})v_{iI}^A(t) \qquad v_i^B(\mathbf{X},t) = \sum_{I \in \Omega^B} N_I(\mathbf{X})v_{iI}^B(t) \qquad (15.1a)$$

If the node numbers of bodies A and B are distinct, then the two velocity fields can be written as a single expression

$$v_i(\mathbf{X},t) = N_I(\mathbf{X})v_{iI}(t) \qquad v_i(\xi,t) = N_I(\xi)v_{iI}(t) \qquad (15.1b,c)$$

where an implicit summation is implied on the repeated nodal indices over all nodes of both bodies. Since the two bodies do not share any nodes, the velocity field is discontinuous across the contact interface. We have also written the velocity field as a function of element coordinates ξ, since as pointed out in Chapter 11 the two sets of coordinates are one-to-one.

The Lagrange multiplier field $\lambda(\zeta,t)$, is a function of time and the projection of ξ onto the contact surface, denoted by S. As can be seen from (4.5) and (4.6), the Lagrange multiplier need only be a C^{-1} field. The shape functions for the Lagrange multiplier field frequently differ from those used for the velocities, so a different symbol is used for its approximation:

$$\lambda(\zeta,t) = \sum_{I \in \Gamma_c} \Lambda_I(\zeta)\lambda_I(t) \equiv \Lambda_I(\zeta)\lambda_I(t) \quad \lambda(\zeta,t) \geq 0 \qquad (15.2)$$

where $\Lambda_I(\zeta)$ are C^{-1} shape functions. These shape functions can be element-specific. In small displacement problems, where the nodes of the two bodies may be contiguous, the projection of the body mesh onto the contact surface may be used for the Lagrange multiplier mesh. In large displacement problems, a separate mesh should usually be constructed for the Lagrange multipliers; this is discussed later.

The test functions are given by

$$\delta v_i(\mathbf{X}) = N_I(\mathbf{X})\delta v_{Ii}, \quad \delta\lambda(\zeta) = \Lambda_I(\zeta)\delta\lambda_I \quad \delta\lambda(\zeta) \leq 0 \qquad (15.3)$$

where the implicit sums are defined in Chapter 5 of Ref. 1.

To develop the semidiscrete equations, the above approximations for the velocity and Lagrange multiplier fields and the test functions are substituted into the weak form in Box 14.1. The nodal forces emerging from δP are identical to those developed in Chapter 11, so they will not be rederived; the results are given in Table B11.1. From (B11.1.) it follows that

$$\delta P = \delta v_{il} r_{il} = \delta v_{il}\left(f_{il}^{int} - f_{il}^{ext} + M_{iljJ}\dot{v}_{jJ}\right)$$
$$= \delta\dot{\mathbf{d}}^T\mathbf{r} \equiv \delta\dot{\mathbf{d}}^T\left(\mathbf{f}^{int} - \mathbf{f}^{ext} + \mathbf{M}\ddot{\mathbf{d}}\right) \qquad (15.4)$$

The interpenetration rate can be expressed in terms of the nodal velocities:

$$\gamma_N = \sum_{I \in \Gamma^c \cap \Gamma^A} N_I v_{il}^A n_i^A + \sum_{I \in \Gamma^c \cap \Gamma^B} N_I v_{il}^B n_i^B \qquad (15.5)$$

where the first sum, as indicated, is over the nodes of body A which are on the contact interface, and the second sum is over the nodes of body B which are on the contact interface. The normal components are defined by

$$v_{NI} = v_{Nil}^A n_i^A \text{ if } I \text{ in } A, \quad v_{NI} = v_{il}^B n_i^A \text{ if } I \text{ in } B \qquad (15.6)$$

so that

$$\gamma_N = \hat{N}_I v_{NI} \quad \text{where} \quad \hat{N}_I = N_I n_i^A \text{ on } A, \quad \hat{N}_I = N_I n_i^B \text{ on } B \qquad (15.7)$$

Then using the approximations (15.1-4) it follows that

$$\int_{\Gamma^c} \delta(\lambda \gamma_N) d\Gamma = \delta v_{NI} \hat{G}_{IJ}^T \lambda_J + \delta \lambda_I \hat{G}_{IJ} v_{NJ} \quad \text{where} \quad \hat{G}_{IJ} = \int_{\Gamma^c} \Lambda_I N_J d\Gamma \quad (15.8)$$

superposed hat has been placed on $\hat{G}_{IJ}$ to indicate that it pertains to the velocities in the local coordinate system of the contact interface. Combining (15.4), and (15.8) we can write the *discrete weak form* as

$$\sum_{I \in \Omega - \Gamma_c} \delta v_{iI} r_{iI} + \sum_{I \in \Gamma_c} \left(\delta v_{NI} \hat{G}_{IJ}^T \lambda_J + \delta \lambda_I \hat{G}_{IJ} v_{NJ} + \delta v_{NI} r_{NI} + \delta v_{\alpha I} r_{\alpha I} \right) \geq 0 \quad (15.9)$$

where the implicit sum on the index J holds, but the sums on the index I are explicitly specified; the summation over components applies as always. The residual $\mathbf{r}$ is defined in (15.4) and (13.1) and we have expressed the residual on the contact nodes in terms of normal and tangential components. The discrete weak form, like the weak form, is an inequality.

The governing equations must be extracted carefully because of the inequalities and the different sums. The equations for nodes, which are not on the contact interface, can be directly extracted from the first sum since these nodal velocities are arbitrary and do not appear in the second sum, which yields

$$r_{iI} = 0 \quad \text{or} \quad M_{IJ} \dot{v}_{jJ} = f_{iI}^{ext} - f_{iI}^{int} \quad \text{for } I \in \Omega - \Gamma_c \text{ or } \Gamma_u \quad (15.10)$$

These are the standard equations of motion. To obtain the equations on the contact interface, what remains of the first sum after extracting (15.10) is considered. Since the test nodal velocities are unconstrained, the weak inequality yields the following equalities:

$$\hat{r}_{\alpha I} = 0 \quad \text{or} \quad M_{IJ} \dot{v}_{\alpha J} = \hat{f}_{\alpha I}^{ext} - \hat{f}_{\alpha I}^{int} \quad \text{for } I \in \Gamma_c \quad (15.11)$$

$$r_{NI} + \hat{G}_{IJ}^T \lambda_J = 0 \quad \text{or} \quad M_{IJ} \dot{v}_{NJ} + f_{NI}^{ext} - f_{NI}^{int} + \hat{G}_{IJ}^T \lambda_J = 0 \quad \text{for } I \in \Gamma_c \quad (15.12)$$

To extract the equations associated with the Lagrange multipliers, we note that the test (or variation of) Lagrange multipliers must be negative. Therefore the weak inequality (15.9) implies an *inequality*

$$\hat{G}_{IJ} v_{NJ} \leq 0 \quad (15.13)$$

In addition, we have from (14.14) the requirement that the test function for the Lagrange multiplier field must be positive $\lambda(S,t) \geq 0$. This inequality is difficult to enforce. For elements with piecewise linear displacements along the edges, this condition can be enforced only at the nodes by $\lambda_I \geq 0$, since minima of $\lambda(S)$. For higher order approximations, the condition must be checked more exhaustively.

The above, in conjunction with the strain-displacement equations and the constitutive equation, comprise the complete system of equations for the semidiscrete model of contact-impact. The discrete equations consist of the equations of motion and the contact interface conditions. The equations of motion for nodes not on the contact interface are unchanged from the unconstrained case. On the contact interface, additional forces $\hat{G}_{IJ}\lambda_J$, which represent the normal contact tractions, appear. In addition, the impenetrability constraint in weak form (15.12) is imposed. Like the equations without contact, the semidiscrete equations are ordinary differential equations, but the variables are subject to inequality constraints. These inequality constraints substantially complicate the time integration, since the smoothness, which is implicitly assumed by most time integration procedures, is lost.

For the subsequent developments, it is convenient to define

$$\gamma_N = \Phi_{il}(\zeta)\, v_{il}(t) \quad \text{or} \quad \gamma_N = \Phi \mathbf{v} \tag{15.14}$$

where

$$\Phi_{il}(\zeta) = \begin{cases} N_I(\xi)\, n_i^A & \text{if } I \text{ on } A \\ N_I(\xi)\, n_i^B & \text{if } I \text{ on } B \end{cases} \tag{15.15}$$

The term associated with normal contact in the discrete weak form (15.9) is then given by

$$\delta G_L = \int_{\Gamma^c} \delta(\lambda_I \Lambda_I \Phi_{jJ} v_{jJ})\, d\Gamma = \lambda^T \mathbf{G} \mathbf{v} \tag{15.16a}$$

where

$$G_{IjJ} = \int_{\Gamma_c} \Lambda_I \Phi_{jJ}\, d\Gamma \qquad \mathbf{G} = \int_{\Gamma_c} \Lambda^T \Phi\, d\Gamma \tag{15.16b}$$

where jJ has been converted to a single index by the Voigt column matrix rule to form the matrix expression on the left, see Ref.1, appendix 1.

Once we have grasped the subtleties of the extracting the discrete equations, it is convenient to use following matrix notation for the discrete weak form:

$$\delta \mathbf{v}^T \left(\mathbf{f}^{int} - \mathbf{f}^{ext} + \mathbf{M}\dot{\mathbf{v}} \right) + \delta \left(\mathbf{v}^T \mathbf{G}^T \boldsymbol{\lambda} \right) = 0 \quad \forall \delta v_{il} \notin \Gamma_u \quad \forall \delta \lambda_I \leq 0 \qquad (15.17)$$

The matrix forms of the equations of motion and the interpenetration condition are

$$\mathbf{M}\dot{\mathbf{v}} + \mathbf{f}^{int} - \mathbf{f}^{ext} + \mathbf{G}^T \boldsymbol{\lambda} = 0 \quad \forall \delta v_{il} \notin \Gamma_u \quad \forall \delta \lambda_I \leq 0 \qquad (15.18a)$$

$$\mathbf{G}\mathbf{v} \leq 0 \qquad (15.18b)$$

15.1.2.1   Lagrange Multiplier Mesh.

The construction of the interpolation for the Lagrange multipliers poses some difficulties. In general, the nodes of the two contacting bodies are not coincident. Therefore it is necessary to develop a scheme to deal with noncontiguous nodes. In two dimensions this is not too difficult. One possibility is indicated in Fig. 15.1, where the nodes for the Lagrange multiplier field are chosen to be the nodes of the master contact surface. This simple scheme is not effective when the elements of body B is much smaller than the elements of body A, since the coarse mesh for the Lagrange multipliers will then lead to interpenetration. An alternative is to place Lagrange multiplier nodes wherever a node appears in either body A or B, as shown in Fig. 15.1. However, the latter scheme can not be extended to three-dimensional problems, since the elements will then be very oddly shaped, which can lead to ill-conditioning of the equations. In general three-dimensional applications, either the finer mesh of the two should be chosen as the multiplier mesh. In complex engineering problems, a separate mesh must be constructed for the multipliers since it is difficult to select a body mesh that is ideal. In any case, the use of a separate multiplier mesh then entails the establishment of the coincidence the nodes of the body meshes and the multiplier mesh. For large meshes this can be time consuming unless sophisticated search techniques are used.

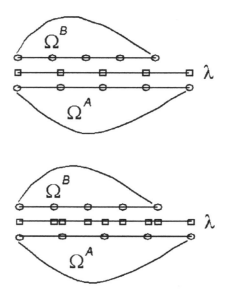

Figure 15.1 Nodal arrangements for two contacting bodies with noncontiguous nodes showing (a) a Lagrange multiplier mesh based on the master body and (b) a Lagrange multiplier mesh with nodes wherever they occur in either body.

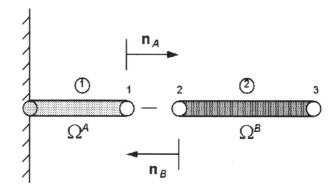

Figure 15.2 One dimensional example of contact of two 2-node elements

15.1.2.2   Example: One-Dimensional Contact-Impact.

Consider the two rods shown in Fig. 15.2.   The cross-sectional area A=1. The contact interface consists of the nodes 1 and 2. The unit normals, as shown in Fig. 15.2, are $n_x^A = 1, n_x^B = -1$.   The contact interface in one-dimensional problems is rather odd since it consists of a single point.

The velocity fields in the two elements are

$$v(\xi,t) = \mathbf{N}(\xi)\dot{\mathbf{d}}(t) = \begin{bmatrix} \xi^A, & 1-\xi^B, & \xi^B \end{bmatrix}\dot{\mathbf{d}} \tag{E15.1}$$

where $\dot{\mathbf{d}}^T = \begin{bmatrix} v_1 & v_2 & v_3 \end{bmatrix}$. The $\mathbf{G}$ matrix is given by (15.15) and (15.16); in a one-dimensional problem, the integral is replaced by a single function value, with the function evaluated at the contact point:

$$\mathbf{G}^T = \begin{bmatrix} \xi^A \cdot n^A, & (1-\xi^B)n^B, & \xi^B \end{bmatrix}\Big|_{\xi^A=1, \xi^B=0} = \begin{bmatrix} (1)\,(1), & 1(-1), & 0 \end{bmatrix} = \begin{bmatrix} 1, & -1, & 0 \end{bmatrix}$$

The impenetrability condition in rate form, (15.18b), is given by

$$\mathbf{G}^T\dot{\mathbf{d}} \le 0 \;\; or \;\; \begin{bmatrix} 1 & -1 & 0 \end{bmatrix}\dot{\mathbf{d}} = v_1 - v_2 \le 0 \tag{E15.2}$$

The above can easily be obtained by inspection: when the two nodes are in contact, the velocity of node 1 must be less or equal than the velocity of node 2 to preclude overlap. If they are equal, they remain in contact, whereas when the inequality holds, they release. These conditions are not sufficient to check for initial contact, which should be checked in terms of the nodal displacements: $x_1 - x_2 \ge 0$ indicates contact has occurred during the previous time step.

Lagrange Multiplier Method: Since there is only one point of contact, only a single Lagrange multiplier is needed to enforce the contact constraint. The equations of motion are

$$\begin{bmatrix} M_{11} & M_{12} & M_{13} \\ M_{21} & M_{22} & M_{23} \\ M_{31} & M_{32} & M_{33} \end{bmatrix}\begin{Bmatrix} \ddot{d}_1 \\ \ddot{d}_2 \\ \ddot{d}_3 \end{Bmatrix} - \begin{Bmatrix} f_1 \\ f_2 \\ f_3 \end{Bmatrix} + \begin{Bmatrix} 1 \\ -1 \\ 0 \end{Bmatrix}\lambda_1 = 0 \;\; and \;\; \lambda_1 \ge 0 \tag{E15.3}$$

The last terms on the RHS in the above are the nodal forces resulting from contact between nodes 1 and 2. These forces on the nodes are equal and opposite and vanish when the Lagrange multiplier vanishes. The equations of motion are identical to the equations for an unconstrained finite element mesh except at the nodes, which are in contact. The equations for a diagonal mass matrix are

$$M_1 a_1 - f_1 + \lambda_1 = 0, \quad M_2 a_2 - f_2 - \lambda_1 = 0, \quad M_3 a_3 - f_3 = 0 \qquad \text{(E15.4)}$$

where $a_I = \ddot{d}_I$. The equations for small-displacement elastostatics are obtained by combining the **G** matrix, with the assembled linear stiffness giving

$$
\begin{bmatrix}
k_1 & 0 & 0 & 1 \\
0 & k_2 & -k_2 & -1 \\
0 & -k_2 & k_2 & 0 \\
1 & -1 & 0 & 0
\end{bmatrix}
\begin{Bmatrix}
d_1 \\
d_2 \\
d_3 \\
\lambda_1
\end{Bmatrix}
=
\begin{Bmatrix}
f_1 \\
f_2 \\
f_3 \\
0
\end{Bmatrix}^{ext}
\begin{matrix} \\ \\ \\ \geq \end{matrix}
\qquad \text{(E15.5)}
$$

where K_e is the stiffness of element e. The assembled stiffness matrix is singular in the absence of contact, i.e. the upper left hand 3×3 matrix. But with the addition of the contact interface conditions, the complete 4×4 matrix becomes regular.

Penalty Method. To write the equation for the penalty method, we will use the penalty law $p = \beta g = \beta (x_1 - x_2)\,\text{H}(g) = \beta(X_1 - X_2 + u_1 - u_2)\,\text{H}(g)$. Then

$$
\mathbf{f}^c = \int_{\Gamma_c} \boldsymbol{\phi}^T p\, d\Gamma =
\begin{Bmatrix}
1 \\
-1 \\
0
\end{Bmatrix}
\beta g
\qquad \text{(E15.6)}
$$

The above integral consists of the integrand evaluated at the interface point since Γ_c is a point. For a diagonal mass the equations are

$$M_1 a_1 - f_1 + \beta g = 0, \quad M_2 a_2 - f_2 - \beta g = 0, \quad M_3 a_3 - f_3 = 0 \qquad \text{(E15.7)}$$

These equations are identical to that for the Lagrange multiplier method, except that the penalty force replaces the Lagrange multiplier.

To construct the small displacement, elastostatic equations for the penalty method, we first evaluate $\mathbf{P}_c$:

$$
\mathbf{P}_c = \int_{\Gamma_c} \bar{\beta} \boldsymbol{\Phi}^T \boldsymbol{\Phi}\, d\Gamma = \bar{\beta}
\begin{bmatrix}
+1 \\
-1 \\
0
\end{bmatrix}
\begin{bmatrix} +1 & -1 & 0 \end{bmatrix}
= \bar{\beta}
\begin{bmatrix}
+1 & -1 & 0 \\
-1 & +1 & 0 \\
0 & 0 & 0
\end{bmatrix}
\qquad \text{(E15.8)}
$$

where $\bar{\beta} = \beta_1 \mathrm{H}(g)$. Adding $\mathbf{P}_c$ to the linear stifness yields

$$
\begin{bmatrix}
k_1 + \bar{\beta} & -\bar{\beta} & \\
-\bar{\beta} & k_2 + \bar{\beta} & -k_2 \\
& -k_2 & k_2
\end{bmatrix}
\begin{Bmatrix}
d_1 \\
d_2 \\
d_3
\end{Bmatrix}
=
\begin{Bmatrix}
f_1 \\
f_2 \\
f_3
\end{Bmatrix}^{ext}
\tag{E15.9}
$$

It can be seen from the above that the penalty method adds a spring with a stiffness $\bar{\beta}$ between nodes 1 and 2. The above equation is nonlinear since $\bar{\beta}$ is a nonlinear function of $g = u_1 - u_2$.

15.2   Explicit methods

In this Section we describe the procedures for treating contact impact with explicit time integration. Explicit time integration is well suited to contact-impact problems because the small time steps imposed by numerical stability are well suited to the discontinuities in contact-impact. The large time steps made possible by unconditionally stable implicit methods are not effective for discontinuous response. Furthermore, contact-impact introduces discontinuities in the Jacobian, which impedes the convergence of Newton methods.

Another advantage of explicit algorithms is that the bodies can first be integrated completely independently, as if they were not in contact. This uncoupled update correctly indicates which parts of the body are in contact. The contact conditions are imposed after the two bodies have been updated in an uncoupled manner; no iterations are needed to establish the contact interface. An explicit algorithm with contact-impact is almost identical to the algorithm described in Chapter 13 except that the bodies are checked for interpenetration. In each time step, the displacements and velocities of those nodes, which have penetrated into another body, are modified to reflect momentum balance and impenetrability on the interface.

15.2.1 Contact in one dimension.

The one-dimensional example is shown in Fig. 15.2. We first consider the premise that uncoupled updates of bodies A and B followed by modifications of the interpenetrating nodes for contact-impact lead to consistent solutions. For the two points R and S, which correspond to nodes 1 and 2, respectively, of bodies A and B, there are four possibilities during a contact-impact problem

1. R and S are not in contact and do not contact during the time step;
2. R and S are not in contact but impact during the time step;
3. R and S are in contact and remain in contact;
4. R and S are in contact and separate during a time step, known as release.

For case 3, the statement "remain in contact" does not imply that if two points must remain contiguous, because relative tangential motion, or sliding, which separates contiguous points is always possible. When two bodies remain in contact, they are assumed not to separate.

All of these possibilities can be correctly accounted for by integrating the two bodies independently, as if they were not in contact, and subsequently adjusting the velocities and the displacements. The possibilities which need to be explained are cases 2, 3 and 4.

We use the equations of motion for the nodes 1 and 2; although the problem shown in Fig. 15.2 is somewhat different, the equations for the contact nodes are unchanged. We will show that when the velocities from the uncoupled update predict initial or continuing contact, then the Lagrange multiplier $\lambda \geq 0$. The accelerations of nodes 1 and 2 when the two bodies are updated as uncoupled are

$$M_1\bar{a}_1 - f_1 = 0 \ , \quad M_2\bar{a}_2 - f_2 = 0 \ , \quad \bar{v}_1 = v_1^- + \Delta t \bar{a}_1 \quad \bar{v}_2 = v_2^- + \Delta t \bar{a}_2 \qquad (15.19)$$

where $(\bullet)^+ \equiv (\bullet)^{n+1/2}, (\bullet)^- \equiv (\bullet)^{n-1/2}$ and bars have been superposed on *trial accelerations and velocities*; and all unmarked variables are at time step n. The trial acclerations are computed with the uncoupled equations of motion. The correct velocites are given by the central difference update of (15.3 a,b)

$$M_1 v_1^+ - M_1 v_1^- - \Delta t f_1 + \Delta t \lambda = 0, \quad M_2 v_2^+ - M_2 v_2^- - \Delta t f_2 - \Delta t \lambda = 0 \qquad (15.20a,b)$$

When contact occurs during the time step, the velocities must be equal at the end of the time step, $v_1^+ = v_2^+$. Eliminating λ from the above equations using $v_1^+ = v_2^+$ gives

$$v_1^+ = v_2^+ = \frac{M_2 v_1^- + M_2 v_1^- + \Delta t(f_1 + f_2)}{M_1 + M_2} \qquad (15.21)$$

The velocities are updated by the above when impact occurs or when the nodes were in contact in the previous time step. The above can be recognized as conservation of momentum for plastic impact of rigid bodies; more will be said on this later.

We will now show that when the nodal velocities which interpenetrate are modified by (15.21), then the Lagrange multiplier will be positive, i.e. the interface force will be compressive. In other words, (15.21), when applied to all nodes which interpenetrate during the time step, yields the correct sign on the Lagrange multipliers. This corresponds to showing that

$$\text{if } \overline{v}_1^+ \geq \overline{v}_2^+, \text{ then } \lambda \geq 0. \qquad (15.22)$$

Multiplying (15.20a) by M_2 and (15.20b) by M_1 and subtracting gives

$$M_1 M_2 \left(v_1^- - v_2^-\right) + \Delta t (M_2 f_1 - M_1 f_2) = \lambda \Delta t (M_1 + M_2) \qquad (15.23)$$

Substituting the expressions for f_1 and f_2 from (15.19) into the above and rearranging gives

$$\frac{\Delta t(M_1 + M_2)}{M_1 M_2} \lambda = \left(v_1^- - v_2^-\right) + \Delta t(\overline{a}_1 - \overline{a}_2) = \overline{v}_1^+ - \overline{v}_2^+ \qquad (15.24)$$

where the last equality is obtained by using the central difference formulas for the uncoupled integration of the two bodies: $\overline{v}_I^+ = v_I^- + \Delta t a_I$. The coefficient of λ is positive, so the sign of the RHS gives the sign of λ. Thus (15.22) has been demonstrated.

To examine this finding in more detail, we now consider the three cases listed above (case 1 is trivial since it requires no modification of the nodal velocities since there is no contact):

case 2 (no contact /contacts during Δt): then $\bar{v}_1^+ > \bar{v}_2^+$ and $\lambda \geq 0$ by (15.24)

case 3 (in contact/remains in contact): then $\bar{v}_1^+ > \bar{v}_2^+$ and $\lambda \geq 0$ by (15.24)

case 4 (in contact/release during Δt): then $\bar{v}_1^+ < \bar{v}_2^+$ and $\lambda < 0$ by (15.24)

Thus the velocities obtained by uncoupled integration correctly predict the sign of the Lagrange multiplier λ.

Two other interesting properties of explicit integration that can be learned from this example are:

1.   initial contact, i.e. impact cannot occur in the same time step as release;
2.   energy is dissipated during impact;

The first statement rests on the fact that the Lagrange multiplier at time step n is computed so that the velocities at time step $n+1/2$ match. Hence there is no mechanism in an explicit method for forcing release during the time step in which impact occurs. This property is consistent with the mechanics of wave propagation. In the mechanics of impacting bodies, release is caused by rarefaction waves, which develop when the compressive waves due to impact reflect from a free surface and reach the point of contact. When the magnitude of this rarefaction is sufficient to cause tension across the contact interface, release occurs. Therefore the minimum time required for release subsequent to impact is two traversals of the distance to the nearest free surface (unless a rarefaction wave from some other event reaches the contact surface). The stable time step, you may recall, allows the any wave generated by impact to move at most to the node nearest to the contact nodes. Therefore, in explicit time integration, there is insufficient time in a stable time step for the waves to traverse twice the distance to the nearest free surface.

The second statement can be explained by (15.21) which shows that the post-impact velocities are obtained by the plastic impact conditions. These always dissipate energy. The dissipation decreases with the refinement of the mesh. In the continuous impact problem, i.e. the solution to the PDES, no energy is dissipated because the condition of equal velocities after impact is limited to the impact surfaces. A surface is a set of measure zero, so a change of energy over the surface has no effect on the total energy. (For one-dimensional problems the impact surface is a point, which is also a set of measure zero.) In a discrete model, the impacting nodes represent the material layer of thickness $h/2$ adjacent to the contact surface. Therefore, the dissipation in a discrete model is always finite. It should be stressed that these arguments do not apply to multi-body models with beams, shells, or rods, where the stiffness through the thickness direction is not modeled. The release and impact conditions are then more complex.

15.2.2 Penalty method

The discrete equation at the impacting nodes for the two body problem can be taken as:

$$M_1 a_1 - f_1 + f_1^c = 0 \qquad M_2 a_2 - f_2 - f_2^c = 0 \qquad (15.25)$$

where the contact forces f_I^c replaces the Lagrange multiplier in (15.20). When the nodes are initially coincident, then $x_1 = x_2$ and the interface normal traction can be written as

$$f^c = p = \beta_1 g + \beta_2 \dot{g} = \beta_1 (u_1 - u_2) H(g) + \beta_2 (v_1 - v_2) H(g) \qquad (15.26)$$

The unitary condition is now violated since the normal traction is positive while the interpenetration rate is positive, so its product no longer vanishes. The post-impact velocities depend on the penalty parameters. The velocities of the two nodes are not equal since the penalty method only enforces the impenetrability constraint approximately. As the penalty parameter is increased, the condition of impenetrability is observed more closely. However in the solution of dynamic problems by explicit methods the penalty parameter cannot be arbitrarily large, since the stable time step is inversely proportional to the penalty parameter.

REFERENCES

[15.1] Belytschko, T., Liu, W.K. and Moran, B., *Nonlinear Finite Elements for Continua and Structures*, Wiley, Chichester, U.K., 2000.

[15.2] Wriggers, P., Finite element algorithms for contact problems, *Arch. Comp. Meth. Eng.*, **2**(4), 1-49, 1995.

Part IV

MULTIBODY DYNAMICS TOOLS FOR STRUCTURAL AND BIOMECHANICS CRASHWORTHINESS

J.A.C. Ambrósio

IDMEC, Instituto Superior Técnico, Lisbon, Portugal

NOTATION

a, **b**,	Column vectors (boldface lower-case characters)
A, **B**,	Matrices (boldface upper-case characters)
$(...)'$	Components of (...) in body-fixed coordinates
$(...)$	Components of (...) in inertial frame coordinates
$\boldsymbol{\lambda}$	Vector of Lagrange multipliers
$\xi\eta\zeta_i$	Body-fixed Cartesian coordinate system
$\boldsymbol{\Phi}$	Vector of kinematic constraints
D	Jacobian matrix of constraints
q	Vector of Cartesian coordinates

16. RIGID MULTIBODY SYSTEMS: THE PLASTIC HINGE APPROACH

by
J.A.C. AMBRÓSIO, M. SEABRA PEREIRA[1] and J.F.A. MILHO[1]

16.1   Introduction

Multibody systems are generally complex arrangements of structural and mechanical subsystems with different design purposes and mechanical behavior. Depending on the type of applications, operating speeds, external or internal loading of the components, the multibody system may experience small or large deformations that lead to a change of the system performance. This is well known in aerospace mechanisms where slender elements are present. Other cases include vehicle components under extreme conditions or machine rods operating at high speeds. The structures, on the other hand, may behave as multibody systems due to their large rotations or because they develop well defined mechanisms of deformation, as in crashworthiness applications.

Based on rigid multibody dynamics the system deformations can be described using lumped deformations modeled by spring-damper elements. Due to its simplicity, this approach has found its way into multibody systems where the components are made of slender elements such as beams. Methodologies to describe flexibility effects such as the finite segment approach [16.1] or the plastic hinge technique [16.2-5] represent these methods. The advantage of

[1] Instituto de Engenharia Mecânica, Instituto Superior Técnico, Av. Rovisco Pais, 1049-001 Lisboa, Portugal

these procedures lies in the simple mechanics associated with them and the small number of parameters required to describe the structural behavior that characterizes the lumped stiffness. This fact makes the lumped stiffness methods well suited to the optimal design of flexible multibody systems in vehicle dynamics and crashworthiness [16.6-7]. Both the finite segment approach and the plastic hinge technique are unable to describe the coupling resulting from loading in different directions, such as beam-columns, or to capture other nonlinear flexible effects. These methods assume that the mechanisms of deformation are known beforehand. However, when used with care, these methods constitute the basis for very efficient and reliable design tools suitable to the early phases of the design process. Another advantage of the plastic hinge approach is that it enables the analyst to include in the multibody model the structural or biomechanical model structural information obtained through the use of experimental procedures, finite element method, macro element method or any other.

Many novel methodologies for the description of rigid multibody systems have been proposed in the past two decades. In this work, Cartesian coordinates are used to describe the rigid multibody dynamics, not because of their efficiency but because they are well understood. The objective is to describe the local nonlinear deformations that appear in the rigid multibody system as a result of contact or impact between the system components and external or internal obstacles and to incorporate the elastodynamics in the description of flexible multibody systems. Other formulations for rigid multibody systems have relative advantages and drawbacks but they will not be discussed here, as they are summarized in recent books [16.8-13].

16.2   Rigid Multibody System Equations of Motion

16.2.1 General multibody system

A multibody system is a collection of rigid and flexible bodies joined together by kinematic joints and force elements as depicted in Figure 16.1. The presence of the kinematic joints restricts the relative motion between adjacent bodies reducing the number of degrees of freedom of the system.

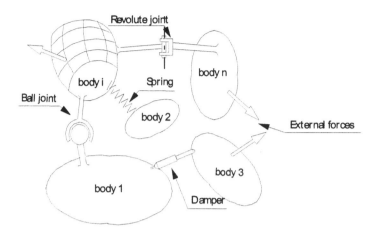

Figure 16.1 Schematic representation of a multibody system

Depending on the choice of coordinates used to describe the system, the number of degrees-of-freedom may not coincide with the number of coordinates. This is the case of the Cartesian coordinates [16.8], adopted throughout this part, for which 6 coordinates are used to represent the position and orientation of each rigid body. The use of such coordinates gives raise to a system of differential-algebraic-equations (DAEs) that need to be solved and integrated forward in time. Due to the existence of kinematic constraints, subsets of dependent and independent coordinates can be identified. The existence of dependent coordinates must be accounted for during the solution procedure.

16.2.2 Rigid body equation of motion

Reference Cartesian coordinates identify the position of each body through the position of a point and a set of parameters to define its angular orientation. For the ith body of the system, represented in Figure 16.2, vector $\mathbf{q}_i$ contains the translational coordinates $\mathbf{r}_i$, and a set of rotational coordinates $\mathbf{p}_i$. A vector of velocities for a rigid body i is defined as $\mathbf{v}_i$ containing a 3-vector of translational velocities $\dot{\mathbf{r}}_i$ and a 3-vector of angular velocities $\boldsymbol{\omega}_i^{\,2}$. The vector of accelerations for the body, denoted by $\dot{\mathbf{v}}_i$, is the time derivative of $\mathbf{v}_i$.

---

[2] Note that the velocity vector $\mathbf{v}_i$ is not necessarily the time derivative of vector $\mathbf{q}_i$. In fact, for what follows, the orientation of the rigid body is represented by Euler parameters (vector $\mathbf{p}_i$ has 4 coordinates) while its angular velocity is not represented by the time derivative of the Euler parameters, $\dot{\mathbf{p}}_i$, but by the angular velocity $\boldsymbol{\omega}_i$ (3 component vector)[16.8].

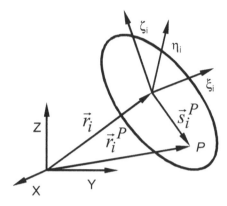

Figure 16.2 Position and orientation of a single rigid body using a body fixed coordinate
frame attached to the body center of mass

Assuming that the body fixed coordinate frame is fixed to the body center
of mass the translation equations of motion for a single free rigid body are

$$m_i \ddot{\mathbf{r}}_i = \mathbf{f}_i \tag{16.1a}$$

and the rotational equations of motion are

$$\mathbf{J}'_i \dot{\boldsymbol{\omega}}'_i = \mathbf{n}'_i - \tilde{\boldsymbol{\omega}}'_i \mathbf{J}'_i \boldsymbol{\omega}'_i \tag{16.1b}$$

the inertia tensor $\mathbf{J}'_i$ is constant and, if the body fixed frame is aligned with the
principal inertia directions of the rigid body, is diagonal. In equations (16.1) $\mathbf{f}_i$
represents the resultant of all forces applied on the rigid body and $\mathbf{n}'_i$ is the
moment obtained by summing all external applied moments and moments-of-
force resulting from the transport of the external applied forces to the body
center of mass.

Equations (16.1) are re-written in a matrix form as

$$\begin{bmatrix} m\mathbf{I} & \\ & \mathbf{J}' \end{bmatrix}_i \begin{bmatrix} \ddot{\mathbf{r}} \\ \dot{\boldsymbol{\omega}}' \end{bmatrix}_i = \begin{bmatrix} \mathbf{f} \\ \mathbf{n}' - \tilde{\boldsymbol{\omega}}' \mathbf{J}' \boldsymbol{\omega}' \end{bmatrix}_i \equiv \mathbf{M}_i \mathbf{v}_i = \mathbf{g}_i \tag{16.2}$$

For a system of nb unconstrained bodies there are $6 \times nb$ equations of motion.
By expressing the external applied forces $\mathbf{f}_i$ as functions of crushable structural
elements connecting the rigid bodies, it would be possible to use a system of
this type to model certain types of structural and vehicle impact scenarios. The
models proposed by Kamal [16.14], for instance, use this type of description.

16.2.3 Kinematic constraints

The existence of kinematic constraints, restricting the relative mobility of adjacent rigid bodies, is what differentiates a multibody system from a system of many free bodies. Most of the practically used kinematic constraints can be built by setting algebraic relations between vectors defined in the rigid bodies. For what follows, the vector represented in figure 16.3 are used.

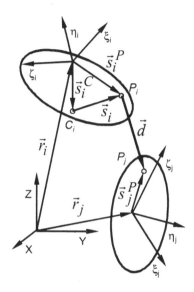

Figure 16.3 System of two rigid bodies where vectors with constant and varying magnitudes are represented.

The condition for two vectors of constant magnitude to remain perpendicular all times is realized by the single constraint equation, corresponding to set their dot product to zero,

$$\Phi^{(n1,1)} \equiv \quad \mathbf{s}_i^T \mathbf{s}_j = \mathbf{s}_i'^T \mathbf{A}_i^T \mathbf{A}_j \mathbf{s}_j' = 0 \tag{16.4}$$

where $\mathbf{A}_i$ and $\mathbf{A}_j$ are the transformation matrices from body fixed coordinates to the inertial frame coordinates, for bodies i and j respectively. The condition of perpendicularity, now using vectors with varying and constant magnitudes, is

$$\Phi^{(n2,1)} \equiv \quad \mathbf{s}_i^T \mathbf{d} = \mathbf{s}_i'^T \mathbf{A}_i^T (\mathbf{r}_j + \mathbf{A}_j \mathbf{s}_j'^P - \mathbf{r}_i - \mathbf{A}_i \mathbf{s}_i'^P) = 0 \tag{16.5}$$

For two vectors to remain parallel two equations, resulting from setting the vector product between them to zero, are required. For two vectors with constant magnitude this condition is given by

$$\Phi^{(p1,2)} \equiv \quad \tilde{\mathbf{s}}_i \mathbf{s}_j = \mathbf{A}_i \tilde{\mathbf{s}}_i' \; \mathbf{A}_i^T \mathbf{A}_j \mathbf{s}_j' = \mathbf{0} \tag{16.6}$$

In the same manner, if one of the vectors has varying magnitude the constraint is

$$\Phi^{(p2,2)} \equiv \quad \tilde{\mathbf{s}}_i \mathbf{d} = \mathbf{A}_i \tilde{\mathbf{s}}_i' \; \mathbf{A}_i^T (\mathbf{r}_j + \mathbf{A}_j \mathbf{s}_j'^P - \mathbf{r}_i - \mathbf{A}_i \mathbf{s}_i'^P) = \mathbf{0} \tag{16.7}$$

Note that out of the three equations obtained from the vector product, both in equations (16.6) and (16.7), only two are independent.

16.2.3.1   Spherical joint

The spherical joint, depicted by figure 16.4, is realized by setting the coordinates of points P_i and P_j to be coincident all times. The three constraint equations that represent this condition are

$$\Phi^{(s,3)} \equiv \quad \mathbf{r}_i + \mathbf{A}_i \mathbf{s}_i'^P - \mathbf{r}_j - \mathbf{A}_j \mathbf{s}_j'^P = \mathbf{0} \tag{16.8}$$

Therefore, the number of relative degrees-of-freedom between two bodies connected by a spherical joint is three.

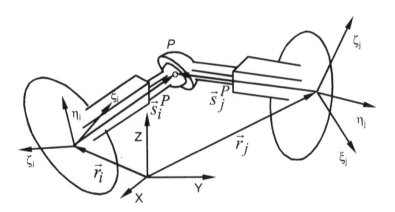

Figure 16.4 Spherical joint.

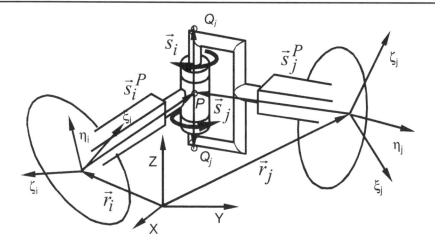

Figure 16.5 Revolute joint.

16.2.3.2   Revolute joint

The revolute joint, depicted by figure 16.5, is obtained by adding to the spherical joint constraints two extra constraints, which correspond to the requirement for vectors $\vec{s}_i$ and $\vec{s}_j$ to remain parallel given by equation (16.6). The full set of constraints for a revolute joint is

$$\Phi^{(s,3)} \equiv \mathbf{r}_i + \mathbf{A}_i\mathbf{s}_i'^P - \mathbf{r}_j - \mathbf{A}_j\mathbf{s}_j'^P = \mathbf{0}$$
$$\Phi^{(p1,2)} \equiv \tilde{\mathbf{s}}_i\mathbf{s}_j = \mathbf{A}_i\tilde{\mathbf{s}}_i'\,\mathbf{A}_i^T\mathbf{A}_j\mathbf{s}_j' = \mathbf{0}$$

$$(16.9)^3$$

Therefore, the number of relative degrees-of-freedom between two bodies connected by a revolute joint is one.

16.2.3.3   Universal joint

The revolute joint, depicted by figure 16.6, is obtained by adding to the spherical joint constraints an extra constraint corresponding to the requirement that vectors $\vec{s}_i$ and $\vec{s}_j$ remain perpendicular, which is given by equation (16.4). The full set of constraints for a universal joint is

---

[3] An alternative formulation is realized by the spherical joint and by two dot products between vector $\vec{s}_i$ and two vectors fixed to body j and perpendicular to $\vec{s}_j$.[16.8]

$$\Phi^{(s,3)} \equiv \mathbf{r}_i + \mathbf{A}_i \mathbf{s}_i'^P - \mathbf{r}_j - \mathbf{A}_j \mathbf{s}_j'^P = 0$$

$$\Phi^{(n1,1)} \equiv \mathbf{s}_i^T \mathbf{s}_j = \mathbf{A}_i^T \mathbf{s}_i'^T \mathbf{A}_j \mathbf{s}_j' = 0 \tag{16.10}$$

Finally there are two relative degrees-of-freedom between two bodies connected by a universal joint.

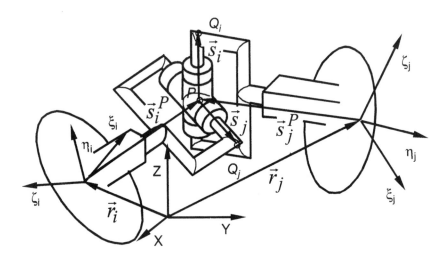

Figure 16.6 Universal joint.

16.2.3.4 Kinematic equations

For a multibody system containing *nb* bodies, the vectors of coordinates, velocities, and accelerations are $\mathbf{q}$, $\mathbf{v}$ and $\dot{\mathbf{v}}$ that contain the elements of $\mathbf{q}_i$, $\mathbf{v}_i$ and $\dot{\mathbf{v}}_i$, respectively, for *i*=1,...,*nb*. Let the kinematic constraints representing all joints between rigid bodies of the multibody system, i.e., equations (16.4) through (16.10), be grouped in the *mr* independent equations as

$$\Phi(\mathbf{q},t) = 0 \tag{16.11}$$

The first and second time derivatives of the constraints yield the kinematic velocity and acceleration equations, given respectively by

$$\dot{\Phi}(\mathbf{q},t) = 0 \quad \equiv \quad \mathbf{D}\,\mathbf{v} = \mathbf{v} \tag{16.12}$$

$$\ddot{\Phi}(\mathbf{q},\dot{\mathbf{q}},t) = 0 \quad \equiv \quad \mathbf{D}\,\dot{\mathbf{v}} = \gamma \tag{16.13}$$

where $\mathbf{D}$ is the Jacobian matrix of the constraints, $\mathbf{v}$ is the partial derivative of the constraint equations with respect to time and γ groups all terms of the acceleration equations that depend exclusively on the position, velocity and time (explicitly).

Many other types of constraint equations, representing kinematic joints between system components, can be devised. In particular, it is of great importance to many of multibody models used in crashworthiness the translational joints (both prismatic and cylindrical joints). The interested reader can find in references [16.8-10] their treatment.

16.2.4 Applied forces

Through the models for the external applied forces it is possible to represent most of the loads applied over the rigid bodies. In what follows, by external loads it is meant all loading internal or external to the system, excluding the joint reaction forces resulting from the kinematic joints. Contact forces, seat belt loads, springs and dampers representing suspension elements or crushable parts of the structural components are all described mathematically as applied forces. Even kinematic joints for which no constraint equations are set can be described as external forces applied over a particular rigid body. That is the case of a wide number of biomechanical models where the biomechanical joints are described by surface contact instead of mechanical joints associated to the kinematic constraints.

16.2.4.1   External forces

Let an external force $\mathbf{f}_i$ be applied on point P_i of rigid body i, as pictured in figure 16.7. This force is added to the vector of forces of body i, which appears in equation (16.1a). Generally, the external force is not applied in the body center of mass. Therefore, its moment about the origin must be calculated and added to the moment resultant vector $\mathbf{n}_i$ appearing in equation (16.1b). This moment is given by

$$\mathbf{n}_i' = \tilde{\mathbf{s}}_i^P \mathbf{f}_i \qquad (16.14)$$

When modeling external applied loads, such as those resulting from contact or seat belts, the position of its application point is generally know, and used in the evaluation of the force itself. Consequently, the evaluation of the moment represented by equation (16.14) is obtained inexpensively.

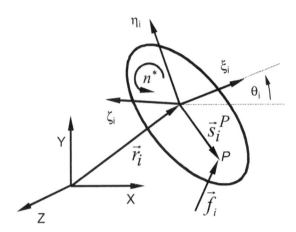

Figure 16.7 External force applied on a rigid body.

16.2.4.2   Springs and dampers

The translational spring-damper-actuator element, represented in figure 16.8, is of particular importance in modeling vehicle suspension elements. Also, the element is the basis of the plastic hinge approach and the finite segment method, used to represent the structural deformations of system components. These force elements may have a linear or a nonlinear behavior, which has its implications on their computational implementation.

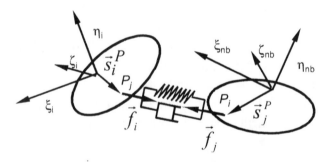

Figure 16.8 Translational spring-damper-actuator element.

For what follows let it be assumed that the spring, damper and actuator are parallel elements attached to points P_i and P_j of bodies i and j respectively. The forces of the element on bodies applied on bodies i and j are given by

$$f_i^{(sda)} = (f_i^{(s)} + f_i^{(d)} + f_i^{(a)})\ \mathbf{u} \tag{16.15a}$$

$$\mathbf{f}_j^{(sda)} = -\mathbf{f}_i^{(sda)} \tag{16.15b}$$

where $f^{(s)}$, $f^{(d)}$ and $f^{(a)}$ are the spring, damper and actuator force magnitudes respectively. $\mathbf{u}$ is a unit vector aligned with the element force vector

$$\mathbf{u} = \frac{\mathbf{l}}{l}\ ; \quad \mathbf{l} = \mathbf{r}_j + \mathbf{A}_j\,\mathbf{s}_j^{\prime P} - \mathbf{r}_i - \mathbf{A}_i\,\mathbf{s}_i^{\prime P} \tag{16.16}$$

the magnitude of vector $\mathbf{l}$ is evaluated as $l = (\mathbf{l}^T\mathbf{l})^{1/2}$.

The magnitude of the spring force, for a linear spring element, is

$$f_i^{(s)} = k(l - l^0) \tag{16.17}$$

where l^0 is the spring undeformed length. For a nonlinear element, the spring force is a nonlinear function of its state of deformation, written as

$$f_i^{(s)} = f(l) \tag{16.18}$$

The translational damper force is a function of the element elongation/shortening velocity, written as

$$f_i^{(d)} = d\,\dot{l}\ ; \quad \dot{l} = \frac{\mathbf{l}^T\dot{\mathbf{l}}}{l}\ ; \quad \dot{\mathbf{l}} = \dot{\mathbf{r}}_j + \omega_j \mathbf{B}_j\,\mathbf{s}_j^{\prime P} - \dot{\mathbf{r}}_i - \omega_j \mathbf{B}_i\,\mathbf{s}_i^{\prime P} \tag{16.19}$$

where $\dot{l}$ represents the time rate of the variation of length of vector $\mathbf{l}$ while $\dot{\mathbf{l}}$ represents its velocity. Once again, if the damper is nonlinear, the damping coefficient is a function of the element variation of length and of its time rate.

The translational actuator element force is written is a general form as

$$f_i^{(a)} = f(\mathbf{q}, \dot{\mathbf{q}}, t) \tag{16.20}$$

Note that generally the actuator forces are function of the state variables and time. This element is used to model hydraulic actuators, controllers, or firing systems, for instance. In the computer implementation, there is no qualitative difference between the actuator element and the nonlinear spring or damper elements.

The rotational spring-damper-actuator elements are also widely used in control and crashworthiness applications of multibody systems. In particular, the models for plastic hinges of beams, due to bending, are realized through the use of this type of elements. In what follows it will be assumed that the rotational spring-damper-actuator element is used together with revolute or universal joints. Therefore, the direction of the moment resulting from this element is the same as the axis of the joint about which it is used. If that is not the case in any particular application foreseen, the direction of the resulting moment must be evaluated before it can be finally applied on the rigid bodies.

For the linear rotational spring acting between bodies i and j, represented in figure 16.9, the moment is found as

$$n^{(r-s)} = \beta \, (\theta - \theta^0) \tag{16.21}$$

where β is the spring torsional stiffness and θ^0 is its undeformed angle. For a nonlinear spring the torsional stiffness is dependent on its state of deformation. The representation of the nonlinear torsional spring, as well as the torsional damper and actuator, is similar to those of the translational elements and will not be repeated here.

16.2.5 Equations of motion for a multibody system

16.2.5.1   Equations of motion

The equations of motion for a single free rigid body are given by equation (16.2). For a system of nb unconstrained bodies equation (16.2) is repeated nb times leading to

$$\mathbf{M}\,\dot{\mathbf{v}} = \mathbf{g} \tag{16.22}$$

For a system of constrained bodies the effect of the kinematic joints can be included in equations (16.22) by adding to their right-hand-side the equivalent joint reaction forces $\mathbf{g}^{(c)}$, leading to

$$\mathbf{M}\,\ddot{\mathbf{q}} = \mathbf{g} + \mathbf{g}^{(c)} \tag{16.23}$$

By using the Lagrange multiplier technique, the joint reaction forces are related to the kinematic constraints by [16.8]

$$\mathbf{g}^{(c)} = -\mathbf{D}^T \boldsymbol{\lambda} \tag{16.24}$$

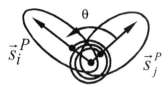

Figure 16.9 Rotational spring element.

where λ is a vector with nc unknown Lagrange multipliers, being nc the number of independent kinematic constraints. Substituting equation (16.24) into equation (16.23) leads to the equations of motion of a multibody system

$$\mathbf{M}\dot{\mathbf{v}} + \mathbf{D}^T \lambda = \mathbf{g} \tag{16.25}$$

Equation (16.25) has $nb+nc$ unknowns (nb accelerations and nc Lagrange multipliers) and can only be solved together with the acceleration equation (16.13). The resulting system of differential-algebraic equations is

$$\begin{bmatrix} \mathbf{M} & \mathbf{D}^T \\ \mathbf{D} & \mathbf{0} \end{bmatrix} \begin{bmatrix} \dot{\mathbf{v}} \\ \lambda \end{bmatrix} = \begin{bmatrix} \mathbf{g} \\ \gamma \end{bmatrix} \tag{16.26}$$

It should be pointed out that the solution of DAE, such as equation (16.26), obtained for this formulation, present some numerical difficulties resulting from the need to ensure that the kinematic constraints are not violated during the integration process. Other types of coordinates leading to a minimum set of equations do not present this problem, when used for open-loop systems.

16.2.5.2   Solution of the equations of motion

The forward dynamic analysis of a multibody system requires that the initial conditions of the system, i.e. the position vector $\mathbf{q}^0$ and the velocity vector $\mathbf{v}$, are given. With the state information and the model data it is possible to build the mass and Jacobian matrices and the force and right-hand-side acceleration vectors. Equation (16.26) is assembled and solved for the unknown accelerations, which are in turn integrated in time together with the velocities. This gives raise to the system position and velocity of the new time step. The process, schematically shown in figure 16.10, proceeds until the system response is obtained for the complete time period of the analysis.

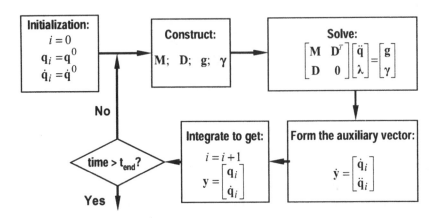

Figure 16.10 Flowchart representing the forward dynamic analysis of a multibody system.

In the process of solving equation (16.26) the Lagrange multipliers are calculated. The joint reaction forces are then obtained through equation (16.24). Note from the computational point of view these reaction forces are obtained as a side result, therefore their computation is inexpensive.

16.2.5.3   Integration methods and constraint stabilization

Equations (16.26) constitute a system of differential and algebraic equations that must be solved together to obtain the system accelerations. For the numerical solution of the system equations of motion a predictor-corrector, variable order algorithm is used [16.15] together with a constraint violation stabilization scheme [16.16] or with an Augmented Lagrangian methodology [16.12]. This algorithm also includes error estimation and step-size control, which is particularly suitable for impact simulations where sudden forces may appear as a result of contact between different parts of the impacting structures.

16.3   Flexible multibody dynamics by lumped deformations

A natural way to describe flexibility effects is to assume specific patterns of deformation for the flexible components, modeling the deformation itself by linear or nonlinear spring elements. In the framework of linear deformations, the finite segment approach uses this idea to model slender multibody components made of beams and bars [16.1]. The plastic hinge approach also uses the same principles but it is applied to models of systems experiencing nonlinear deformations [16.4]. Both formulations are reviewed here.

16.3.1. Finite segment approach to flexible multibody systems

Let the flexible components of the multibody system be made of slender components such as the connecting rods of high-speed machinery or the structural frame of buses and trucks. In this case each slender component can be modeled as a collection of rigid bodies connected by linear springs, as shown in Figure 16.11. These springs, representing the axial, bending and torsion properties of the beams, capture the flexibility of the whole component.

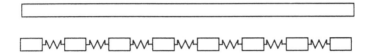

Figure 16.11 Slender component and its finite segment model

Twelve generalized displacements are associated to each finite segment, this is, three translations and three rotations at each end. When the beams deform the reference frames attached to the rigid bodies used for their model rotate and translate with respect to each other. Then the relative displacements between the ends of adjacent segments can be easily calculated.

Forces and moments applied to the rigid bodies can be calculated from the relative end displacements and rotations assuming that each two adjacent bodies are connected by springs and, eventually, by dampers. For this purpose to each rigid body there are deformation elements attached to each end as depicted by Figure 16.12. The characteristics of these springs are related with the material and geometric properties of the system components.

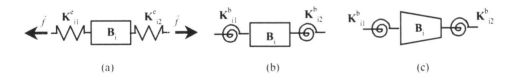

(a)   (b)   (c)

Figure 16.2 Finite segments and their combinations: (a) extensional straight; (b) bending straight; (c) tapered bending

Using the principles of structural analysis the stiffness coefficients for these springs are calculated [16.1]. For instance, the straight extensional segment and the straight bending, respectively represented in figures 16.12a and 16.12b have their stiffness coefficients, respectively, given by:

$$k_i^e = k_j^e = 2\frac{E_i A_i}{l_i} \qquad (16.27a)$$

$$k_i^e = k_j^e = 2\frac{E_i I_i}{l_i} \qquad (16.27b)$$

where E_l is the Young modulus, A_l is the cross-section area, I_i the cross-section moment of inertia and l_l the length of the finite segment.

16.3.2 Plastic hinges in multibody nonlinear deformations

In many impact situations, the individual structural members are overloaded, principally in bending, giving rise to plastic deformations in highly localized regions, called plastic hinges. These deformations, presented in Figure 16.3, develop at points where maximum bending moments occur, load application points, joints or locally weak areas [16.17] and, therefore, for most practical situations, their location is predicted well in advance. Multibody models obtained with this method are relatively simple, which makes the procedure adequate for the early phases of vehicle design. The methodology described herein is known in the automotive, naval and aerospace industries as conceptual modeling [16.3,16.5, 16.18-19].

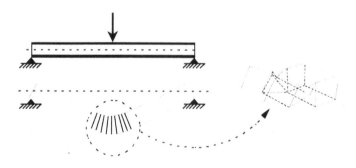

Figure 16.13 Localized deformations on a beam and a plastic hinge

The plastic hinge concept has been developed by using generalized spring elements to represent constitutive characteristics of localized plastic deformation of beams and kinematic joints to control the deformation kinematics [16.4], as illustrated in Figure 16.14. The characteristics of the spring-damper that describes the properties of the plastic hinge are obtained by experimental component testing, finite element nonlinear analysis or simplified analytical methods. For a flexural plastic hinge the spring stiffness is expressed as a function of the change of the relative angle between two adjacent bodies connected by the plastic hinge, as shown in Figure 16.15.

For a bending plastic hinge the revolute joint axis must be perpendicular to the neutral axis of the beam and to the plastic hinge bending plane simultaneously. The relative angle between the adjacent bodies measured in the bending plane is:

$$\theta_{ij} = \theta_i - \theta_j - \theta_{ij}^0 \qquad (16.28)$$

where θ_{ij}^0 is the initial relative angle between the adjacent bodies. Note that for the case of flexible adjacent bodies the relative angular values also include information on the nodal rotational displacements.

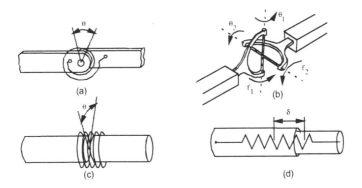

Figure 16.4 Plastic hinge models for different loading conditions: a) one axis bending b) bending with two axis; c) torsion; d) axial

The typical torque-angle constitutive relationship, as in Figure 16.15, is found based on a kinematic folding model [16.20-21] for the case of a steel tubular cross section. This model is modified accounting for elastic-plastic material properties including strain hardening and strain rate sensitivity of some materials. A dynamic correction factor is used to account for the strain rate sensitivity [16.22].

$$P_d / P_s = 1 + 0.07 \ V_0^{0.82} \qquad (16.29)$$

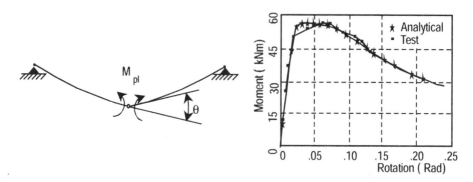

Figure 16.15 Plastic hinge bending moment and its constitutive relationship

Here P_d and P_s are the dynamic and static forces, respectively, and V_0 is the relative velocity between the adjacent bodies. The coefficients appearing in equation (16.29) are dependent on the type of cross section and material.

16.4   Continuous contact force model

A model for the contact force must consider the material and geometric properties of the surfaces, contribute to a stable integration and account for some level of energy dissipation. Based on a Hertzian description of the contact forces between two solids [16.23], Lankarani and Nikravesh [16.24] propose a continuous force contact model that accounts for energy dissipation during impact. The procedure is used for rigid body and finite element nodal contact.

Let the contact force between two bodies or a system component and an external object be a function of the pseudo-penetration δ and pseudo-velocity of penetration $\dot{\delta}$

$$\mathbf{f}_{s,i} = \left(K\delta^n + D\dot{\delta} \right) \mathbf{u} \tag{16.30}$$

where K is the equivalent stiffness, D is a damping coefficient and $\mathbf{u}$ is a unit vector normal to the impacting surfaces. The hysteresis dissipation is introduced in equation (16.30) by $D\dot{\delta}$. The damping coefficient is given by

$$D = \frac{3K\left(1 - e^2\right)}{4\dot{\delta}^{(-)}} \delta^n \tag{16.31}$$

This coefficient is a function of the impact velocity $\dot\delta^{(-)}$, stiffness of the contacting surfaces and restitution coefficient e. For a fully elastic contact $e=1$ while for a fully plastic contact $e=0$. The generalized stiffness coefficient K depends on the geometry material properties of the surfaces in contact. For the contact between a sphere and a flat surface the stiffness is [16.25]

$$K = \frac{0.424\sqrt{r}}{\left(\dfrac{1-v_i^2}{\pi E_i} + \dfrac{1-v_j^2}{\pi E_j}\right)} \tag{16.32}$$

where v_l and E_l are the Poisson's ratio and the Young's modulus associated to each surface and r is the radius of the impacting sphere.

The nonlinear contact force is obtained by substituting equation (16.31) into equation (16.30), leading to

$$\mathbf{f}_{s,i} = K\ \delta^n \left[1 + \frac{3\left(1-e^2\right)}{4}\frac{\dot\delta}{\dot\delta^{(-)}}\right]\mathbf{u} \tag{16.33}$$

This equation is valid for impact conditions, in which the contacting velocities are lower than the propagation speed of elastic waves, i.e., $\dot\delta^{(-)} \le 10^{-5}\sqrt{E/\rho}$.

16.5   Impact of a rotating beam

A rotating beam model based on the multibody methodologies is presented and analyzed and its results are compared with those obtained from experimental testing [16.4,16.21]. The experimental procedure consists in a beam, with a length of 1 m, a rectangular cross section of 50x25 mm and a thickness of 2.65 mm, made of E36 steel, articulated at an end and with a mass of 5.95 Kg in the other end. The beam is accelerated in order for the velocity of the tip to be 11.85 m/s when it collides with a stop, shown in Figure 16.16, located 0.5 m from the beam support. The accelerations of the ballasting mass are measured with accelerometers mounted on it.

The observation of the impacting beam behavior clearly indicates that a plastic hinge develops in the region of contact, as pictured in Figure 16.17. The permanent bending angle observed for the region of the plastic hinge is 22°.

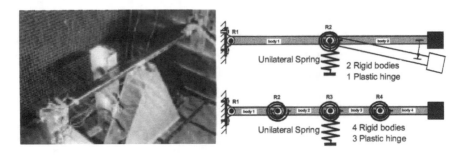

Figure 16.16 Experimental configuration and multibody models for the impacting beam

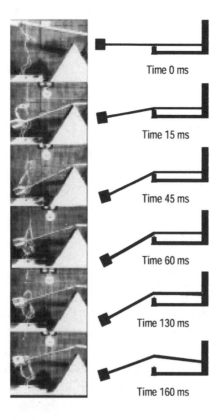

Figure 16.17 Evolution of the bending angle: experiment and simulation

Computer simulations are carried out with different models, shown in Figure 3.6, where the contact between the beam and the rigid block is modeled with the continuous contact force described by equation (16.33). Figure 16.17

illustrates the displacement of the multibody model made of two rigid bodies and a plastic hinge. It is observed that the gross motion predicted is similar to the experimental behavior with a permanent bending angle of 21.25° between the rigid bodies.

The measured and simulated accelerations of the ballasting beam are shown in Figure 16.18. The acceleration levels for the experimental and numerical results are similar, in particular during the contact stage.

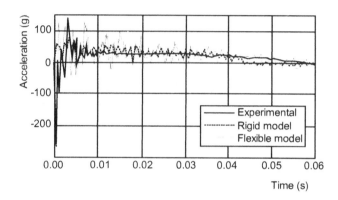

Figure 16.18 Acceleration response of the ballasting mass obtained in the experimental test and from the simulation of two multibody models where the bodies interconnected by plastic-hinges are modeled as rigid and flexible bodies respectively.

16.6   Application of the lumped deformation methods to railway impact

The methodology presented is applied to the simulation of train collision [18.26]. Table 16.1 presents the arrangement of a train set with eight individual car-bodies and it includes their individual length and the mass [16.27].

Table 16.1   Train set pattern

	HE	LE	LE	LE	LE	LE	LE	LE	HE
	1	2	3	4	5	6	7	8	
Length (m)	20	26	26	26	26	26	26	26	
Mass (10^3 Kg)	68	51	34	34	34	34	34	51	

For identification of the regions in the train where energy absorption occurs during the collision, the designations of high-energy (HE) and low-energy (LE) zones are adopted. The high-energy zones are located in the extremities of the train set in the frontal zone of the motor car-body and in the opposing back zone, in the last car-body. The high-energy zones are potential impact extremities between two train sets. The low-energy zones are located in the remaining extremities of the train car-bodies and correspond to regions of contact between cars of the same train set.

The multibody model for each individual car-body is schematically represented in Figure 16.19. Eight rigid bodies, B_1 through B_8 representing the passenger compartment, the bogies chassis, the bogies connecting supports and the deformable end extremities of the car-bodies are used in the car-body model. The relative motion of the multibody system is restricted by three revolute joints R_1, R_2, R_3 and by four translation joints T_1, T_2, T_3 and T_4.

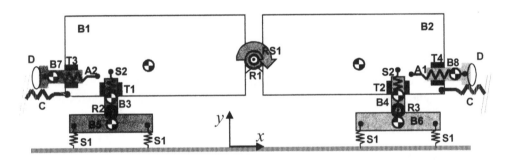

Figure 16.19 Car-body multibody model

Due to modeling assumptions, where the wheels are considered to move only on the rails along the x direction, the mass of the wheels is added to the mass of the bogies in this direction. The initial positions of the bodies along y direction are obtained considering that the static position of the global center of masses of the system of bodies that defines the car-body, is located 1.3 m above the rail. Direction x of the global coordinate system is coincident with the longitudinal direction of the rail, therefore the initial positions of the bodies in this direction are obtained from the car-bodies geometric dimensions [16.27].

In order to represent the passenger compartment bending deformation, a rotational spring RS_1 between two rigid bodies B_1 and B_2 is used in the model. This spring introduces stiffness in the revolute joint R_1. Considering the passenger compartment as a beam, simply supported on the secondary suspensions and assuming the bending stiffness for this type of car-bodies to be $EI = 2.7 \times 10^9$ Nm^2 [16.27] the spring stiffness is obtained

$$K_\theta = \frac{M_{bend} \ell}{4\delta} = 0.569 \times 10^9 \ Nm / rad$$

where l is in the distance between the beam supports, M_{bend} the bending moment and δ the maximum deflection at the beam half length.

16.6.1 Structural deformation elements

The structural behavior of the car-body coupler is represented by a deformation element. This element C is defined in the region between two car-bodies, as shown in Figure 16.19, connecting two half passenger compartments B_2 of one car-body and B_1 of the next car-body. A resisting compressive force of 1000 KN characterizes the coupler while no contact between the car-body ends occurs. The coupler ceases to actuate once the coupling of the anti-climber device occurs, i.e., when contact between the car extremities is detected.

The representation of the structural behavior of the end extremities of each car-body and the inter-vehicle couplers is done by defining deformation elements, represented by nonlinear force elements, which include energy absorption. Their nonlinear characteristics are defined according to the force-deformation curves of the end extremities and couplers as shown in Figure 16.20. Linear segments for loading and unloading represent these elements.

An auxiliary switch variable R is introduced to describe the transition between the elasto-plastic loading and the plastic deformation lines. The loading and unloading lines are parallel to the straight-line portion of the linear elastic line. In the initial step of the force element loading, the switch R is set to be null. Otherwise, the switch R is defined by

$$R = \begin{cases} 1, & (d < d_1) \wedge (d > d^{(t-)}) \wedge R = 0 \\ 0, & (d < d_R) \wedge R = 1 \end{cases} \tag{16.34}$$

where $d = l - l_0$, being l the element current length and l_0 its undeformed length. In equation (16.34) $d^{(t-)}$ is the deformation in the time step which precedes the current one. The coordinates pair (d_R, F_R), corresponding to the point where plastic deformation unloading starts, is defined by

$$(d_R, F_R) = (d^{(t-)}, f^{(t-)}), \qquad R^{(t-)} = 0 \wedge R = 1 \tag{16.35}$$

where $R^{(t-)}$ and $f^{(t-)}$ are respectively the auxiliary variable and the force in the time step previous to the current one. The nonlinear force element is defined as

$$f = \begin{cases} T_0 & \text{if } \left(d > \dfrac{T_0}{K_1}\right) \wedge R = 0 \\[2mm] K_1 d & \text{if } \left(d_1 < d < \dfrac{T_0}{K_1}\right) \wedge R = 0 \\[2mm] \displaystyle\sum_{m=1}^{a} K_m (d_m - d_{m-1}) + \\ \quad + K_{a+1}(d - d_a) & \text{if } (d_{a+1} < d < d_a) \wedge R = 0 \\[2mm] F_R - K_1(d_R - d) & \text{if } \left((d_R - d) > \dfrac{F_R - T_0}{K_1}\right) \wedge R = 1 \\[2mm] T_0 & \text{if } \left((d_R - d) < \dfrac{F_R - T_0}{K_1}\right) \wedge R = 1 \end{cases} \tag{16.36}$$

where K_i is the slope of the linear force segment i. The force is applied in the bodies along the direction of the line connecting the attachment points of the nonlinear force element.

The car-bodies end extremities structural behavior is modeled by the deformation elements A_1 and A_2, which are represented in Figure 16.19. Resisting forces to compression according to their deformation regions, as described in Table 16.2, characterize the end extremities.

Table 16.2   Force and deformation levels of the car-bodies end extremities

Deformation region	High Energy		Low Energy	
	Force (KN)	Deformation (mm)	Force (KN)	Deformation (mm)
Nonstructural	2000	800	2000	250
Structural	3500	1000	3000	500
Passenger compartment	5000		5000	

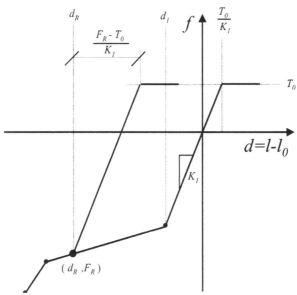

Figure 16.20 Force-deformation curve of the nonlinear force elements

16.6.2 Suspension elements

The car-body suspension system involves four suspension elements S_1 representing the primary suspension and two suspension elements S_2 representing the secondary suspension, as presented in Figure 16.19. In parallel to each suspension spring element a linear damper is assembled in the suspension systems. To represent the spring behavior of the primary and secondary suspension systems of the car-bodies, a suspension element represented by a nonlinear elastic spring is defined. The relation between the spring force f and the length variation d, is presented in Figure 16.21.

The slope K_1 corresponds to the suspension spring stiffness while slope K_2 corresponds to the equivalent stiffness of the end course of the suspensions, ie, the lower and upper bump-stops. The nonlinear elastic spring force is defined as function of its length variation

$$f = \begin{cases} K_1 d_{min} + K_2 (d - d_{min}) & if \quad d < d_{min} \\ K_1 d & if \quad d_{min} \le d \le d_{max} \\ K_1 d_{max} + K_2 (d - d_{max}) & if \quad d > d_{max} \end{cases} \qquad (16.37)$$

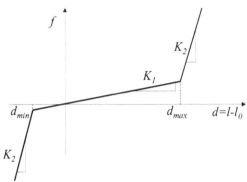

Figure 16.21 Force-length variation of the nonlinear elastic spring

The force is applied in the bodies and has the direction coincident with the line connecting the attachment points of the nonlinear elastic spring.

The parameters of the nonlinear elastic springs of the suspension elements and the dampers are indicated in table 16.3.

Table 16.3   Characteristic of the nonlinear elastic springs and dampers of the suspensions

Suspension system	Stiffness K_1 (N/mm)	Stiffness K_2 (N/mm)	Lower bump-stop (mm)	Upper bump-stop (mm)	Damping (Ns/mm)
Primary	2026	100000	30	35	17
Secondary	1191	100000	30	35	28

16.6.3 Contact forces on the anti-climbers

The anti-climbers are devices designed to ensure the axial alignment of the train during a crash event. Their representation is defined by two rigid bodies, which correspond to the car-bodies end extremities. The continuous contact force model represents the contact force between these bodies, in what normal penetration is concerned. The objective of the anti-climbers is to maintain the alignment between the train carbodies during the crash, therefore no slipping between the carbodies extremities is allowed. This is achieved in the numerical model by imposing stiction between the end extremities.

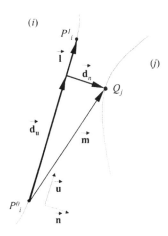

Figure 16.22 Vectors and projections used in the contact description

Let the contact surface in body i be defined by a straight-line segment $P^0{}_iP^l{}_i$ and let a contact point Q_j be defined in body j as shown in Figure 16.22. The contact occurrence between bodies i and j is described by the penetration of point Q_j in the surface $P^0{}_iP^l{}_i$. It is considered that during the coupling of the anti-climber device, the penetration of the point in the surface corresponds to a local deformation. Considering the projections d_u and d_n of vector **m** in the contact surface, along the tangential and normal unit vectors **u** and **n** respectively, the two necessary conditions for contact are expressed as

$$0 < d_u < l \tag{16.38}$$

$$d_n < 0 \tag{16.39}$$

where l represents the magnitude of vector **l**. Once these conditions are meet, it is necessary to identify the contact point C_i of body i, located on the contact surface. As presented in Figure 16.23, the location of this point is determined by the intersection of the segments $P^0{}_iP^l{}_i$ and $Q_j^{(t-)}Q_j$, where $Q_j^{(t-)}$ is the position of point Q_j in the time step previous to the penetration occurrence,. The relative velocity between the contact points $\dot{\delta}$ is defined by the magnitude of the difference of velocities of points C_i and Q_j,

$$\dot{\delta} = \sqrt{\left(\dot{\mathbf{r}}_i^C - \dot{\mathbf{r}}_j^Q\right)^T \left(\dot{\mathbf{r}}_i^C - \dot{\mathbf{r}}_j^Q\right)} \tag{16.40}$$

Similarly, the approaching relative velocity between the contact points $\dot{\delta}^{(-)}$, defined by the magnitude of the difference of velocities of point C_i and Q_j, is given by equation (16.40) for the time step in which contact is initialized. The direction of the application of the contact force, represented by vector **t**, is

$$t = r_i^C - r_j^Q \tag{16.41}$$

therefore the pseudo penetration δ is obtained as

$$\delta = (t^T t)^{1/2} \tag{16.42}$$

and the unit vector t_δ is defined by

$$t_\delta = \frac{1}{\delta} t \tag{16.43}$$

It should be noted that due to the stiction assumption for the contact, used to model the anti-climbers, there is no need to decompose the force vector into a tangential and a normal component with respect to the surface.

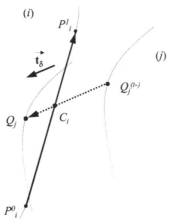

Figure 16.23 Contact point location

16.6.4 Results of the train impact simulations

The simulation scenarios are characterized by a moving train traveling from right to left, which collides with a train located to his left, that is parked with brakes applied. The trains are guided in the same rail and the collision velocities scenarios are 30, 40 and 55 km/h. Of special importance to the anti-climber design are the simulation results for the contact forces and the relative displacements between car-bodies extremities. The contact force results obtained in the analysis are treated using a second order Butterworth filter, with a cut frequency of 40 Hz [16.28].

The vertical relative displacement between the points of the contact surfaces defining the anti-climber devices is described by the distance *g* measured along the contact surface, between points *A* and *B*. These points are initially leveled with the same quota y, as represented in figure 16.24. This displacement is calculated in the time step when contact occurs between the end extremities of the car-bodies, corresponding to the start of penetration of point *B* in the contact surface. The vertical relative displacement is considered positive when the coordinate of point *B* along the *y* direction is higher than point *A*, as in figure 16.24. The values of vertical relative displacement in the interfaces between car-bodies obtained in the different analysis are illustrated in figure 16.25.

The vertical relative displacement in the high-energy interface is negligible given the geometric symmetry of collision, i.e., the tendency for the leading cars to pitch down symmetrically. The vertical relative displacement tends to increase for the interfaces away from the high-energy interface, reaching maximum levels in the colliding train.

The tangential force in the anti-climber device is described as the tangential component of the contact force between the end extremities of the car-bodies. The maximum values of the tangential force are lower than half of the weight of the passenger compartment. The higher levels for the tangential force occur at the interfaces away from the high-energy interface, where the vertical gap reaches higher levels, and are located predominantly in the colliding train, as illustrated in Figure 16.26.

The tangential force at the interfaces, tend to increase both in magnitude and frequency in the final stage of the train impact. This situation is illustrated in Figure 16.27 for 30 Km/h collision velocities.

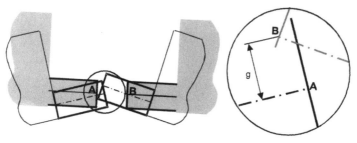

Figure 16.24   Anti-climber device contact geometry

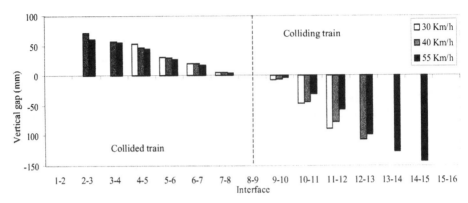

Figure 16.25 Vertical relative displacement along the car-bodies interfaces

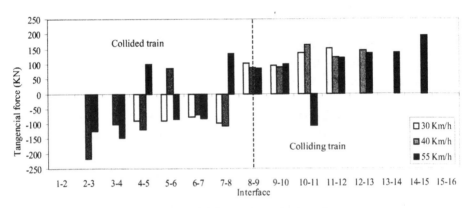

Figure 16.26   Maximum tangential force along the interfaces

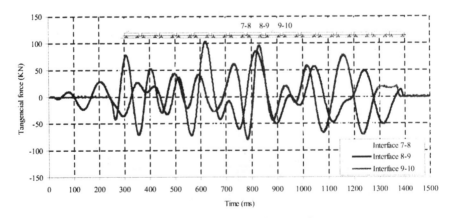

Figure 16.27 Tangential force in train front car-bodies interfaces

REFERENCES

[16.1]   Huston, R.L. and Wang, Y., Flexibility effects in multibody systems, In *Computer Aided Analysis Of Rigid And Flexible Mechanical Systems*, M. Pereira and J. Ambrósio (Eds.) NATO ASI Series E. Vol. 268, Kluwer Academic Publishers, Dordrecht, Netherlands, 351-376, 1994.

[16.2]   Kamal, M. M. Analysis and simulation of vehicle to barrier impact. SAE Paper No. 700414, Society of Automotive Engineers, Warrendale PA, 1970.

[16.3]   Nikravesh, P.E. ,Chung, I.S. and Benedict, R.L., Plastic hinge approach to vehicle simulation using a plastic hinge technique, *J. Comp. Struct.* **16**, 385-400, 1983.

[16.4]   Ambrósio, J.A.C., Pereira, M.S. and Dias, J., Distributed and discrete nonlinear deformations on multibody systems, *Nonlinear Dynamics* **10**(4), 359-379, 1996

[16.5]   Kindervater, C.M., Aircraft and helicopter crashworthiness: design and simulation, In *Crashworthiness Of Transportation Systems: Structural Impact And Occupant Protection*, J.A.C. Ambrósio, M.S. Pereira and F.P Silva (Eds.), NATO ASI Series E. Vol. 332, Kluwer Academic Publishers, Dordrecht, Netherlands, 525-577, 1997

[16.6]   Haug, E.J. and Arora, J.S., *Applied Optimal Design*, John Wiley & Sons, New York, New York, 1979

[16.7]   Dias, J.P. and Pereira, M.S., Design for vehicle crashworthiness using multibody dynamics, *Int. J. of Vehicle Design*, **15**(6), 563-577, 1994

[16.8]   Nikravesh, P.E. Computer aided analysis of mechanical systems, Prentice-Hall, Englewood Cliffs, New Jersey, 1988

[16.9]   Roberson , R.E. and Schwertassek, R., Dynamics *Of Multibody Systems*, Springer-Verlag, Berlin, Germany, 1988

[16.10] Haug, E.J., *Computer-Aided Kinematics And Dynamics Of Mechanical Systems Volume 1: Basic Methods*, Allyn and Bacon, Boston, Massachusetts, 1989

[16.11] Shabana, A., *Dynamics Of Multibody Systems*, John Wiley & Sons, New York, New York, 1989

[16.12] García de Jalón, J. and Bayo, E., *Kinematic And Dynamic Simulation Of Multibody Systems - The Real Time Challenge*, Springer-Verlag, New York, New York, 1993

[16.13] Pereira M.S., Ambrósio J.A.C. (Eds.), *Computer Aided Analysis Of Rigid And Flexible Mechanical Systems*, NATO ASI Series E. Vol. 268, Kluwer Academic Publishers, Dordrecht, Netherlands, 1994

[16.14] Kamal, M.M. and Lin K.H., Collision simulation, In *Modern Automotive Structural Analysis*, M.M. Kamal and J.A. Wolf (Eds.), Van Nostrand Reinhold Comp., New York, New York, 316-355, 1982

[16.15] Shampine, L.F. and Gordon, M.K. *Computer Solution Of Ordinary Differential Equations: The Initial Value Problem*, V.H. Freeman and Co, San Francisco, California, 1975

[16.16] Baumgarte, J., Stabilization of constraints and integrals of motion, *Computer Methods in Applied Mechanics Engineering*, **1**, 1-16, 1972

[16.17] Murray, N.W., The static approach to plastic collapse and energy dissipation in some thin-walled steel structures, In *Structural Crashworthiness*, N. Jones and T. Wierzbicki (Eds.), Butterworths, London, Englend, 44-65, 1983

[16.18] Ambrósio, J.A.C., Pereira, M.S. Multibody dynamic tools for crashworthiness and impact, In *Crashworthiness Of Transportation Systems: Structural Impact And Occupant Protection*, J.A.C. Ambrósio, M.S. Pereira and F.P Silva (Eds.), NATO ASI Series E. Vol. 332, Kluwer Academic Publishers, Dordrecht, Netherlands, 475-521, 1997

[16.19] Matolcsy, M., Crashworthiness of bus structures and rollover protection. In *Crashworthiness Of Transportation Systems: Structural Impact And Occupant Protection*, J.A.C. Ambrósio, M.S. Pereira and F.P Silva (Eds.), NATO ASI Series E. Vol. 332, Kluwer Academic Publishers, Dordrecht, Netherlands, 321-360, 1997

[16.20] Kecman, D., Bending collapse of rectangular and square section tubes, *Int. J. of Mech. Sci*, **25**(9-10), 623-636, 1983

[16.21] Anceau, J., Drazetic, P. and Ravalard, I. *Plastic Hinges Behaviour in Multibody Systems*, Mécanique Matériaux Électricité. n° 444, 1992

[16.22] Winmer, A., Einfluß der belastungsgeschwindigkeit auf das festigkeits- und verformungsverhalten am beispiel von kraftfarhzeugen, *ATZ*, **77**(10), 281-286, 1977

[16.23] Hertz, H., *Gesammelte Werk*, Leipzig, Germany, 1895.

[16.24] Lankarani, H.M. and Nikravesh, P.E., Continuous contact force models for impact analysis in multibody systems, *Nonlinear Dynamics*, **5**, 193-207, 1994

[16.25] Lankarani, H.M., Ma, D. and Menon, R., Impact dynamics of multibody mechanical systems and application to crash responses of aircraft occupant/structure, In *Computer Aided Analysis Of Rigid And Flexible Mechanical Systems*, M. Pereira and J. Ambrósio (Eds.) NATO ASI Series E. Vol. 268, Kluwer Academic Publishers, Dordrecht, Netherlands, 239-265, 1995.

[16.26] Milho, J., Ambrósio, J. and Pereira M., A multibody methodology for the design of anti-climber devices for train crashworthiness simulation, In *Proceedings of the International Crash Conference IJCRASH2000*, London, United Kingdom, September 6-8, 2000.

[16.27] ADT/SOR, *Safetrain Train Crashworthiness for Europe,* Task 4 Technical Document Project, 1998

[16.28] European Rail Research Institute, *Materiel Roulant Voyageurs*, ERRI B 106/RP, Utrecht, 1995.

17. FLEXIBLE MULTIBODY DYNAMICS IN CRASH ANALYSIS

17.1   Introduction

The design requirements of advanced mechanical and structural systems exploit the ease of use of the powerful computational resources available today to create virtual prototyping environments.   These advanced simulation facilities play a fundamental role in the study of systems that undergo large rigid body motion while their components experience material or geometric nonlinear deformations, such as vehicles in impact and crash scenarios. If in one hand the nonlinear finite element method is the most powerful and versatile procedure to describe the flexibility of the system components on the other hand the multibody dynamic formulations are the basis for the most efficient computational techniques that deal with large overall motion. Therefore, the efficiency of the nonlinear finite elements to handle the system deformation can be combined with advantage with the representation of the system components large overall motion using a multibody dynamic approach.

In most of the earlier work dealing with the elastodynamics of mechanical systems the deformations of the system components, assumed elastic and small, is superimposed to their large rigid body motion [17.1-3]. Using reference frames fixed to planar flexible bodies, Song and Haug [17.4] suggest a finite element based methodology, which yields coupled gross rigid body motion and small elastic deformations. The idea behind Song and Haug's approach is further developed and generalized by Shabana and Wehage [17.5-6] that use substructuring and the mode component synthesis to reduce the number of generalized coordinates required to represent the flexible components. However, the use of the mode component synthesis prevents the application of the methodology in crashworthiness scenarios.

The work of researchers in the finite element community, such as that by Belytschko and Hsieh [17.7], Simo and Vu-Quoc [17.8] or Bathe and Bolourchi [17.9] among others, addressing the same type of problems can be easily adapted to the framework of flexible multibody dynamics. Recognizing the problem posed and using some of the approaches well in line with those of the finite element community Cardona and Geradin proposed formulations for the nonlinear flexible bodies using either a geometrically exact model [17.10].

Another approach taken by Ambrósio and Nikravesh [17.11] to model geometrically nonlinear flexible bodies is to relax the need for the structures to exhibit small moderate rotations about the floating frame by using an incremental finite element approach within the flexible body description. The approach is further extended to handle material nonlinearities of flexible multibody systems also [17.12]. Using an updated Lagrangean formulation, the equations of motion obtained for the flexible bodies are general and allow modeling most of the geometric and material nonlinearities. In the sequel of the applications presented an alternative contact model is presented for cases involving the finite element nodal contact and impact.

17.2   Equations of motion for a nonlinear flexible body

The motion of a flexible body, depicted by Figure 17.1, is characterized by a continuous change of its shape, due to internal or external forces, and by large displacements and rotations, associated to the gross rigid body motion. The initial configuration of the flexible body, denoted as configuration 0, and the last configuration t are known equilibrium configurations. The actual equilibrium configuration $t+\Delta t$ is generally not known. Therefore, all measures of stress or strain in the actual configuration have to be referred to one of the known configurations.   In what follows, the last known equilibrium configuration t is chosen as the reference configuration, leading to an updated Lagrangean formulation.

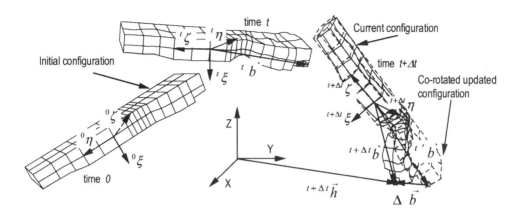

Figure 17.1 General motion of a flexible body

Let XYZ denote the inertial reference frame and $\xi\eta\zeta$ a body fixed coordinate frame. Let the principle of the virtual works be used to express the equilibrium of the flexible body in the current configuration $t+\Delta t$

$$\int_{t+\Delta t_V} (\delta \;_{t+\Delta t}\mathbf{e})^T \;^{t+\Delta t}\boldsymbol{\tau} \;^{t+\Delta t}dv = \;^{t+\Delta t}\mathbf{R} \qquad (17.1)$$

being the external virtual work given by

$$^{t+\Delta t}\mathbf{R} = \int_{t+\Delta t_V} {}^{t+\Delta t}\rho(\delta\mathbf{h})^T ({}^{t+\Delta t}\ddot{\mathbf{h}} + {}^{t+\Delta t}_{t+\Delta t}\mathbf{f}_b)^{t+\Delta t}dv + \int_{t+\Delta t_A} (\delta\mathbf{h})^T {}^{t+\Delta t}_{t+\Delta t}\mathbf{f}_s \;^{t+\Delta t}da \qquad (17.2)$$

In equations (17.1) and (17.2) the left superscript denotes the configuration in which a quantity is measured while the left subscript denotes the configuration to which the quantity is referred. The infinitesimal strains are denoted by **e** while $\boldsymbol{\tau}$ is a vector with the Cauchy stresses. The virtual displacements are denoted by $\delta\mathbf{h}$ and the body and surface forces by $\mathbf{f}_b$ and $\mathbf{f}_s$ respectively.

Being referred to an unknown configuration, equation (17.1) cannot be solved in its current form [17.13]. The choice of reference configuration, on the updated Lagrangean formulation, is the last known equilibrium configuration. As the Cauchy stresses are always referred to the configuration in which they exist another measure of stress, the 2nd Piola-Kirchoff, is adopted in this formulation. The Green-Lagrange strains are energy conjugate of the 2nd Piola-Kirchoff stress measure, and they are consequently adopted here.

Several researchers have shown that referring the equilibrium equations to a co-rotational coordinate system is advantageous with respect to the traditional updated Lagrangean formulation [17.14-15]. A natural choice is the body fixed coordinate system $\xi\eta\zeta$, already defined in the framework of rigid multibody systems as body fixed coordinate system. Such co-rotated updated configuration is denoted here by t'. The Green-Lagrange strain and the 2nd Piola-Kirchoff stress vectors measured with respect to the co-rotated updated configuration and expressed in the body fixed coordinate system are denoted by $^{t+\Delta t}_{t'}\boldsymbol{\varepsilon}'$ and $^{t+\Delta t}_{t'}\mathbf{S}'$ respectively.

Equation (17.1) is now referred to the co-rotated updated configuration t' and the strain and stress measures substituted, leading to [17.11]

$$\int_{t'V}\left(\delta_{t'}\boldsymbol{\varepsilon}'\right)^{T}\,{}^{t'}\mathbf{C}_{t'}\boldsymbol{\varepsilon}'\,{}^{t'}dv + \int_{t'V}\left(\delta_{t'}\boldsymbol{\eta}'\right)^{T}\,{}^{t'}\boldsymbol{\tau}\,{}^{t'}dv = {}^{t+\Delta t}\mathbf{R} - \int_{t'V}\left(\delta_{t'}\mathbf{e}'\right)^{T}\,{}^{t'}\boldsymbol{\tau}\,{}^{t'}dv \qquad (17.3)$$

where $_{t'}\mathbf{e}'$ and $_{t'}\boldsymbol{\eta}'$ are the linear and nonlinear terms of the Green-Lagrange strain increments

The nonlinear equations (17.3) are not linear in the displacement increment, implying that their direct solution is not possible. An approximate solution can be obtained by assuming $\delta_{t'}\boldsymbol{\varepsilon}' = \delta_{t'}\mathbf{e}'$ and an approximate constitutive equation $_{t'}\mathbf{S}' = _{t'}\mathbf{C}_{t'}\mathbf{e}'$. The substitution of these approximate relations in equation (17.3) leads to the linearization of the equilibrium equations given by [17.11,17.13].

$$\int_{t'V}\left(\delta_{t'}\mathbf{e}'\right)^{T}\,{}_{t'}\mathbf{C}_{t'}\mathbf{e}'\,{}^{t'}dv + \int_{t'V}\left(\delta_{t'}\boldsymbol{\eta}'\right)^{T}\,{}^{t'}\boldsymbol{\tau}^{t'}dv = -\int_{t'V}\left(\delta_{t'}\mathbf{e}'\right)^{T}\,{}^{t'}\boldsymbol{\tau}^{t'}dv$$

$$-\int_{t'V}{}^{t'}\rho\left(\delta\mathbf{h}\right)^{T}\,{}^{t+\Delta t}\ddot{\mathbf{h}}\,{}^{t'}dv + \int_{t'V}{}^{t'}\rho\left(\delta\mathbf{h}\right)^{T}\,{}^{t+\Delta t}_{t'}\mathbf{f}_{b}\,{}^{t'}dv + \int_{t'A}\left(\delta\mathbf{h}\right)^{T}\,{}^{t+\Delta t}_{t'}\mathbf{f}_{s}\,{}^{t'}da \qquad (17.4)$$

17.2.1 Finite element equations of motion

Let the finite element method be used to represent the equations of motion of the flexible body. Referring to figure 17.1, the assembly of all finite elements used in the discretization of the flexible body results in its equations of motion written as [17.11]

$$
\begin{bmatrix} \mathbf{M}_{rr} & \mathbf{M}_{rf} & \mathbf{M}_{rf} \\ \mathbf{M}_{\phi r} & \mathbf{M}_{\phi\phi} & \mathbf{M}_{\phi f} \\ \mathbf{M}_{fr} & \mathbf{M}_{f\phi} & \mathbf{M}_{ff} \end{bmatrix} \begin{bmatrix} \ddot{\mathbf{r}} \\ \dot{\omega}' \\ \ddot{\mathbf{u}}' \end{bmatrix} = \begin{bmatrix} \mathbf{g}_r \\ \mathbf{g}'_\phi \\ \mathbf{g}'_f \end{bmatrix} - \begin{bmatrix} \mathbf{s}_r \\ \mathbf{s}'_\phi \\ \mathbf{s}'_f \end{bmatrix} - \begin{bmatrix} \mathbf{0} \\ \mathbf{0} \\ \mathbf{f} \end{bmatrix} - \begin{bmatrix} \mathbf{0} & 0 & 0 \\ 0 & 0 & 0 \\ 0 & 0 & \mathbf{K}_L + \mathbf{K}_{NL} \end{bmatrix} \begin{bmatrix} \mathbf{0} \\ \mathbf{0} \\ \mathbf{u}' \end{bmatrix} \quad (17.5)
$$

where $\ddot{\mathbf{r}}$ and $\dot{\omega}'$ are respectively the translational and angular accelerations of the body fixed reference frame and $\ddot{\mathbf{u}}'$ denotes the nodal accelerations measured in body fixed coordinates. The local coordinate frame $\xi\eta\zeta$, attached to the flexible body, is used to represent the body's gross motion and its deformation. Vector $\mathbf{u}'$ denotes the displacements increments from a previous to the current configuration, measured in body fixed coordinates.

Equation (17.5) describes thoroughly the motion of the flexible body. Even if only small elastic deformations occur, this equation is highly nonlinear due to a variant mass matrix, changing external applied forces, gyroscopic and centrifugal forces and non-constant stiffness matrices.

The variant mass matrix for the flexible body results from the assembly of the individual contributions of each finite element. The sub-matrices describing the contribution of a single finite element for the mass matrix are given by

$$
\mathbf{M}_{rr_j} = \mathbf{I} \int_{V_j} \rho \ dv \qquad \mathbf{M}_{r\phi_j} = -\mathbf{A} \int_{V_j} \rho \ \tilde{\mathbf{b}}' dv
$$

$$
\mathbf{M}_{rf_j} = \mathbf{A} \int_{V_j} \rho \ \mathbf{N} \ dv \qquad \mathbf{M}_{\phi\phi_j} = -\int_{V_j} \rho \ \tilde{\mathbf{b}}'\tilde{\mathbf{b}}' dv \qquad (17.6)
$$

$$
\mathbf{M}_{\phi f_j} = \int_{V_j} \rho \ \tilde{\mathbf{b}}' \ \mathbf{N} \ dv \qquad \mathbf{M}_{ff_j} = \int_{V_j} \rho \ \mathbf{N}^T \mathbf{N} dv
$$

Here, $\mathbf{A}$ is the transformation matrix from the body fixed coordinate system to the inertial frame, $\mathbf{N}$ is the matrix of shape functions of element j and ρ is the mass density. Submatrices $\mathbf{M}_{rr}$ and $\mathbf{M}_{ff}$ are constant and represent respectively the mass of the entire element and the standard finite element mass matrix. $\mathbf{M}_{rf}$ and $\mathbf{M}_{\phi f}$ are the time variant matrices responsible for the inertia coupling between the gross motion of the body and its deformations. In the numerical implementation special attention must be paid to the evaluation of $\mathbf{M}_{\phi\phi}$ when large deformation develop. This sub-matrix, representing the inertia tensor of the flexible body is approximately constant if the body deformations are small, otherwise its time variance cannot be neglected. All other submatrices in equation (17.5) are either null or constant, provided that a proper choice is made for the location and orientation of the body fixed coordinate frame.

The right-hand side of equation (17.5) contains, the vector of gyroscopic and centrifugal forces **s** and the vector of generalized forces **g**, which are evaluated for each element j as

$$
\begin{bmatrix} \mathbf{s}_r \\ \mathbf{s}'_\phi \\ \mathbf{s}'_f \end{bmatrix}_j = \begin{bmatrix} \mathbf{A}\,\widetilde{\boldsymbol{\omega}}'\widetilde{\boldsymbol{\omega}}' \int_{V_j} \rho\, \mathbf{b}'\, dv \\ \int_{V_j} \rho\, \widetilde{\mathbf{b}}'\,\widetilde{\boldsymbol{\omega}}'\widetilde{\boldsymbol{\omega}}'\mathbf{b}'\, dv \\ \int_{V_j} \rho\, \mathbf{N}^T\, \widetilde{\boldsymbol{\omega}}'\widetilde{\boldsymbol{\omega}}'\mathbf{b}'\, dv \end{bmatrix} + 2 \begin{bmatrix} \mathbf{A}\,\widetilde{\boldsymbol{\omega}}' \int_{V_j} \rho\, \mathbf{N}\, dv \\ \int_{V_j} \rho\, \widetilde{\mathbf{b}}'\,\widetilde{\boldsymbol{\omega}}'\mathbf{N}\, dv \\ \int_{V_j} \rho\, \mathbf{N}^T\, \widetilde{\boldsymbol{\omega}}'\mathbf{N}\, dv \end{bmatrix} \dot{\mathbf{u}}'_j \qquad (17.7)
$$

The vector of the external generalized applied forces $\mathbf{g}_j$ for each element is:

$$
\begin{bmatrix} \mathbf{g}_r \\ \mathbf{g}'_\phi \\ \mathbf{g}'_f \end{bmatrix}_j = \begin{bmatrix} \int_{A_j} \mathbf{f}_s\, da \\ \int_{A_j} \widetilde{\mathbf{b}}'\,\mathbf{A}^T \mathbf{f}_s\, da \\ \int_{A_j} \mathbf{N}^T \mathbf{A}^T \mathbf{f}_s\, da \end{bmatrix} + \begin{bmatrix} \int_{V_j} \rho\, \mathbf{f}_b\, dv \\ \int_{V_j} \rho\, \widetilde{\mathbf{b}}'\,\mathbf{A}^T \mathbf{f}_b\, dv \\ \int_{V_j} \rho\, \mathbf{N}^T \mathbf{A}^T \mathbf{f}_b\, dv \end{bmatrix} \qquad (17.8)
$$

In equation (17.8), $\mathbf{f}_b$ and $\mathbf{f}_s$ are respectively the body and surface forces.

Matrices $\mathbf{K}_{Lj}$ and $\mathbf{K}_{NLj}$ are the linear and nonlinear stiffness matrices respectively, and $\mathbf{f}_j$ denotes the vector of equivalent nodal forces due to the state of stress

$$
\mathbf{K}_{L_j} = \int_{V_j} \mathbf{B}_L^T \mathbf{C}\, \mathbf{B}_L\, dv \qquad (17.9)
$$

$$
\mathbf{K}_{NL_j} = \int_{V_j} \mathbf{B}_{NL}^T\, \boldsymbol{\tau}'\, \mathbf{B}_{NL}\, dv \qquad (17.10)
$$

$$
\mathbf{f}_j = \int_{V_j} \mathbf{B}_L^T\, \boldsymbol{\tau}'\, dv \qquad (17.11)
$$

In these equations $\mathbf{B}_L$ and $\mathbf{B}_{NL}$ denote the linear and nonlinear strain matrices respectively and $\boldsymbol{\tau}'$ is the Cauchy stress tensor for the updated configuration. Note that the reference to the linearity of the stiffness matrices $\mathbf{K}_L$ and $\mathbf{K}_{NL}$ is concerned to their relation with the displacements. For a constitutive tensor $\mathbf{C}$ not constant both $\mathbf{K}_L$ and $\mathbf{K}_{NL}$ are not linear. This is the case when a multibody system experiences elasto-plastic deformations of one or more of its components. For these problems, an elasto-plastic constitutive tensor $\mathbf{C}$ must be used in the equation (17.9).

17.2.2 Generalized coordinates of the flexible bodies

Equation (17.5) is not efficient for numerical implementation due to the need to invert the variant mass matrix every time step, during the integration process. A simpler form of the equations of motion for a flexible body is obtained when a lumped mass formulation is used and the accelerations $\ddot{\mathbf{u}}'$ are substituted by the nodal accelerations relative to the inertial frame $\ddot{\mathbf{q}}'_f$.

Let the vectors of nodal accelerations be partitioned into translational and angular accelerations as:

$$\ddot{\mathbf{u}}' = \left[\ddot{\boldsymbol{\delta}}'^T , \ddot{\boldsymbol{\theta}}'^T \right]^T \quad \text{and} \quad \ddot{\mathbf{q}}'_f = \left[\ddot{\mathbf{d}}'^T , \ddot{\boldsymbol{\alpha}}'^T \right]^T \tag{17.12}$$

with relation to this partition of the flexible coordinates, the mass matrix of the flexible body is now evaluated using lumped masses and inertias at the nodal points [17.16], leading to

$$\mathbf{M}_{rr_j} = \sum m_k \mathbf{I} \qquad\qquad \mathbf{M}_{r\phi_j} = -\sum m_k \mathbf{A} \tilde{\mathbf{b}}'_k$$

$$\mathbf{M}_{rf_j} = \left[\sum m_k \mathbf{A} \underline{\mathbf{I}}_k^T \quad \mathbf{0} \right] \qquad \mathbf{M}_{\phi\phi_j} = -\sum m_k \tilde{\mathbf{b}}'^T_k \tilde{\mathbf{b}}'_k + \sum \mu_k \mathbf{I} \tag{17.13}$$

$$\mathbf{M}_{\phi f_j} = \left[\sum m_k \tilde{\mathbf{b}}'_k \underline{\mathbf{I}}_k^T \quad \sum \mu_k \underline{\mathbf{I}}_k^T \right] \quad \mathbf{M}_{ff_j} = \begin{bmatrix} \sum m_k \underline{\mathbf{I}}_k \underline{\mathbf{I}}_k^T & \mathbf{0} \\ \mathbf{0} & \sum \mu_k \underline{\mathbf{I}}_k \underline{\mathbf{I}}_k^T \end{bmatrix}$$

where m_k and μ_k are the lumped mass and inertia of node k respectively, and $\underline{\mathbf{I}}_k$ is a Boolean matrix, which associates node k to the finite element node numbering scheme. The lumped mass formulation of the centrifugal and gyroscopic forces leads to

$$\mathbf{s} = \begin{bmatrix} -\sum m_k \mathbf{A} \tilde{\boldsymbol{\omega}}' \tilde{\boldsymbol{\omega}}' \mathbf{b}'_k - 2\sum m_k \mathbf{A} \tilde{\boldsymbol{\omega}}' \dot{\boldsymbol{\delta}}'_k \\ -\sum m_k \tilde{\mathbf{b}}'_k \tilde{\boldsymbol{\omega}}' \tilde{\boldsymbol{\omega}}' \mathbf{b}'_k - 2\sum m_k \tilde{\mathbf{b}}'_k \tilde{\boldsymbol{\omega}}' \dot{\boldsymbol{\delta}}'_k \\ -\sum m_k \underline{\mathbf{I}}_k \tilde{\boldsymbol{\omega}}' \tilde{\boldsymbol{\omega}}' \mathbf{b}'_k - 2\sum m_k \underline{\mathbf{I}}_k \tilde{\boldsymbol{\omega}}' \mathbf{A}^T \dot{\boldsymbol{\delta}}'_k \\ \mathbf{0} \end{bmatrix} \tag{17.14}$$

The relation between the relative and absolute nodal accelerations for node k is given by:

$$\ddot{\mathbf{q}}'_{fk} \equiv \begin{bmatrix} \ddot{\mathbf{d}}' \\ \ddot{\boldsymbol{\alpha}}' \end{bmatrix}_k = \ddot{\mathbf{u}}'_k + \begin{bmatrix} \mathbf{A}^T & -\left(\tilde{\mathbf{x}}_k + \tilde{\boldsymbol{\delta}}_k\right)' \\ \mathbf{0} & \mathbf{I} \end{bmatrix} \begin{bmatrix} \ddot{\mathbf{r}} \\ \dot{\boldsymbol{\omega}}' \end{bmatrix} + \begin{bmatrix} \tilde{\boldsymbol{\omega}}'\tilde{\boldsymbol{\omega}}'\left(\mathbf{x}_k + \boldsymbol{\delta}_k\right)' + 2\tilde{\boldsymbol{\omega}}'\dot{\boldsymbol{\delta}}'_k \\ \tilde{\boldsymbol{\omega}}'\dot{\boldsymbol{\theta}}'_k \end{bmatrix} \quad (17.15)$$

where $\mathbf{x}_k$ is the vector containing the node position in the reference configuration. Equations (17.13) through (17.15) are evaluated for all finite elements and n nodes and substituted into equation (17.5). The result is simplified yielding [17.11]

$$\mathbf{A} \sum_{k=1}^{n} m_k \ddot{\mathbf{d}}'_k = \mathbf{g}_r \quad (17.16a)$$

$$\sum_{k=1}^{n} m_k \left(\tilde{\mathbf{x}}' + \tilde{\boldsymbol{\delta}}'\right)_k \ddot{\mathbf{d}}'_k = \mathbf{g}'_\theta \quad (17.16b)$$

$$\mathbf{M}_{ff} \ddot{\mathbf{q}}'_f = \mathbf{g}'_f - {}^t_t\mathbf{f} - \left({}^t_t\mathbf{K}_L + {}^t_t\mathbf{K}_{NL}\right)\mathbf{u}' \quad (17.16c)$$

Equations (17.16a) and (17.16b) describe the motion for the center of mass of a system of particles [17.17]. Assuming that the origin of the body fixed coordinate system is coincident with the center of mass of the flexible body, equations (17.16a) and (17.16b) describe the motion of the origin of the $\xi\eta\zeta$ referential. Equation (17.16c) is the equation of motion for the nodes of the flexible body, expressed in the body fixed coordinate system. Due to the use of the lumped mass formulation the mass matrix $\mathbf{M}_{ff}$ is diagonal.

17.2.3 Partially rigid-flexible body

Let a rigid body with mass and inertia similar to those of the flexible body, as described by equations (17.16b) and (17.16c) respectively, be defined. Moreover, assume that the body fixed coordinate system associated to such rigid body has the same location and orientation of the flexible body floating frame. This situation is shown in Figure 17.2, where the flexible body is represented as having rigid and a flexible parts.

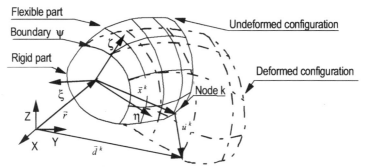

Figure 17.2   Flexible body with a rigid part

The equations of motion of a rigid body i are given by

$$m_i \ddot{\mathbf{r}}_i = \mathbf{f}_{r_i} \tag{17.17a}$$

$$\mathbf{J}'_i \dot{\omega}'_i = \mathbf{n}'_i - \tilde{\omega}'_i \mathbf{J}'_i \omega'_i \tag{17.17b}$$

where $\mathbf{f}_{r_i}$ and $\mathbf{n}'_i$ are the external forces applied over the center of mass of the rigid part and moments applied over the body respectively. In order for equations (17.17a) and (17.17b) to be equivalent to equations (17.16a) and (17.16b) it is necessary that the mass and inertia tensor of rigid body i are equal to those of the flexible body, as given in equations (17.16a) and (17.16b) respectively. Furthermore, the generalized external forces applied over the rigid body are $\mathbf{g}_r$ and $\mathbf{g}'_\theta$ given by the first two rows of equation (17.8).

Equations (17.17a), (17.17b) and (17.16c) represent the equations of motion of the partially flexible body provided that proper reference conditions are set. A suitable set of reference conditions is introduced by enforcing that the flexible and rigid parts are attached by the boundary nodes ψ. This is achieved by imposing kinematic constraints, which ensure that the nodes in the boundary ψ have null displacements, velocities and accelerations with respect to the body fixed coordinate frame, this is

$$\mathbf{u}'_k = \dot{\mathbf{u}}'_k = \ddot{\mathbf{u}}'_k = \mathbf{0} \tag{17.18}$$

Before the constraints implied by equations (17.18) can be applied they must be expressed in terms of the generalized flexible coordinates $\mathbf{q}'_f$. For a constrained node k equations (17.18) are used in the nodal acceleration equations (17.15) leading to

$$\begin{bmatrix} \ddot{\mathbf{d}}' \\ \ddot{\boldsymbol{\alpha}}' \end{bmatrix}_k = \begin{bmatrix} \mathbf{A}^T & -\tilde{\mathbf{x}}'_k \\ 0 & \mathbf{I} \end{bmatrix} \begin{bmatrix} \ddot{\mathbf{r}} \\ \dot{\boldsymbol{\omega}}' \end{bmatrix} + \begin{bmatrix} \tilde{\boldsymbol{\omega}}'\tilde{\boldsymbol{\omega}}'\mathbf{x}'_k \\ 0 \end{bmatrix} \qquad (17.19)$$

The acceleration equation of the constrained node is rearranged to obtain a form similar to that of equation (16.13), which is

$$\begin{bmatrix} -\mathbf{A}^T & \tilde{\mathbf{x}}'_k & \mathbf{I} & 0 \\ 0 & -\mathbf{I} & 0 & \mathbf{I} \end{bmatrix} \begin{bmatrix} \ddot{\mathbf{r}} \\ \dot{\boldsymbol{\omega}}' \\ \ddot{\mathbf{d}}'_k \\ \ddot{\boldsymbol{\alpha}}'_\kappa \end{bmatrix} = \begin{bmatrix} \tilde{\boldsymbol{\omega}}'\tilde{\boldsymbol{\omega}}'\mathbf{x}'_k \\ 0 \end{bmatrix} \qquad (17.20)$$

The nodal constraints are applied to the rigid part equations of motion, given by equations (17.17a) and (17.17b), and to the flexible part equations, represented by equation (17.16c), leading to

$$m\ddot{\mathbf{r}} + \mathbf{A}\boldsymbol{\lambda}_\delta = \mathbf{f}_r \qquad (17.21\text{a})$$

$$\mathbf{J}'_i\dot{\boldsymbol{\omega}}'_i - \tilde{\mathbf{x}}'_k\boldsymbol{\lambda}_\delta - \boldsymbol{\lambda}_\theta = \mathbf{n}'_i - \tilde{\boldsymbol{\omega}}'_i\mathbf{J}'_i\boldsymbol{\omega}'_i \qquad (17.21\text{b})$$

$$\mathbf{M}_{ff}\ddot{\mathbf{q}}'_f + \begin{bmatrix} \boldsymbol{\lambda}_\delta \\ \boldsymbol{\lambda}_\theta \end{bmatrix} = \mathbf{g}'_f - {}^t_t\mathbf{f} - \left({}^t_t\mathbf{K}_L + {}^t_t\mathbf{K}_{NL} \right)\mathbf{u}' \qquad (17.21\text{c})$$

Using equations (17.20) and (17.21), the Lagrange multipliers are eliminated from the equations of motion resulting in the dynamic equations for the rigid-flexible body [17.11]

$$\begin{bmatrix} m\mathbf{I} + \bar{\mathbf{A}}\underline{\mathbf{M}}^*\bar{\mathbf{A}}^T & -\bar{\mathbf{A}}\underline{\mathbf{M}}^*\mathbf{S} & 0 \\ -\left(\bar{\mathbf{A}}\underline{\mathbf{M}}^*\mathbf{S}\right)^T & \mathbf{J}' + \mathbf{S}^T\underline{\mathbf{M}}^*\mathbf{S} & 0 \\ 0 & 0 & \mathbf{M}_{ff} \end{bmatrix} \begin{bmatrix} \ddot{\mathbf{r}} \\ \dot{\boldsymbol{\omega}}' \\ \ddot{\mathbf{q}}'_f \end{bmatrix} = \begin{bmatrix} \mathbf{f}_r + \bar{\mathbf{A}}\mathbf{C}'_\delta \\ \mathbf{n}' - \tilde{\boldsymbol{\omega}}'\mathbf{J}'\boldsymbol{\omega}' - \mathbf{S}^T\mathbf{C}'_\delta - \bar{\mathbf{I}}^T\mathbf{C}'_\theta \\ \mathbf{g}'_f - \mathbf{f} - \left(\mathbf{K}_L + \mathbf{K}_{NL}\right)\mathbf{u}' \end{bmatrix} \quad (17.22)$$

where for notation convenience, auxiliary matrices are introduced to represent the existence of more than one constrained node. These are defined as

$$\bar{\mathbf{A}}^T = \begin{bmatrix} \mathbf{A} & \mathbf{A} & \cdots & \mathbf{A} \end{bmatrix}^T ; \quad \bar{\mathbf{I}} = \begin{bmatrix} \mathbf{I} & \mathbf{I} & \cdots & \mathbf{I} \end{bmatrix}^T ;$$

$$\mathbf{S} = \begin{bmatrix} \left(\tilde{\mathbf{x}}'_1 + \tilde{\boldsymbol{\delta}}'_1\right)^T & \left(\tilde{\mathbf{x}}'_2 + \tilde{\boldsymbol{\delta}}'_2\right)^T & \cdots & \left(\tilde{\mathbf{x}}'_n + \tilde{\boldsymbol{\delta}}'_n\right)^T \end{bmatrix}^T$$

Vectors $\mathbf{C}'_\delta$ and $\mathbf{C}'_\theta$, appearing in equation (17.22), represent the reaction force and moment of the flexible part of the body over the rigid part, given by

$$\mathbf{C}'_\delta = \mathbf{g}'_{\underline{\delta}} - \mathbf{F}_{\underline{\delta}} - \left(\mathbf{K}_L + \mathbf{K}_{NL}\right)_{\underline{\delta}\delta} \delta' - \left(\mathbf{K}_L + \mathbf{K}_{NL}\right)_{\underline{\delta}\theta} \theta' \qquad (17.23a)$$

$$\mathbf{C}'_\delta = \mathbf{g}'_{\underline{\theta}} - \mathbf{F}_{\underline{\theta}} - \left(\mathbf{K}_L + \mathbf{K}_{NL}\right)_{\underline{\theta}\delta} \delta' - \left(\mathbf{K}_L + \mathbf{K}_{NL}\right)_{\underline{\theta}\theta} \theta' \qquad (17.23b)$$

In these equations, the subscripts δ' and θ' refer to the partition of the vectors and matrices with respect to the translation and rotational nodal degrees of freedom. The underlined subscripts are referred to nodal displacements of the nodes fixed to the rigid part.

17.3   Contact model

Let the flexible body approach a surface during the motion of the multibody system, as represented in Figure 17.3. Without lack of generality, let the impacting surface be described by a mesh of triangle patches. In particular, let the triangular patch, where node k of the flexible body will impact, be defined by points i, j and l. The normal to the outside surface of the contact patch is defined as $\vec{n} = \vec{r}_{ij} \times \vec{r}_{jl}$.

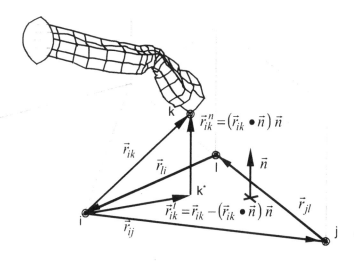

Figure 17.3 Contact detection between a finite element node and a surface

Let the position of the structural node k with respect to point i of the surface be

$$\mathbf{r}_{ik} = \mathbf{r}_k - \mathbf{r}_i \qquad (17.24)$$

This vector is decomposed in its tangential component, which locates point k^* in the patch surface, and a normal component, given respectively by

$$\mathbf{r}_{ik}^t = \mathbf{r}_{ik} - \left(\mathbf{r}_{ik}^T \, \mathbf{n}\right)\mathbf{n} \qquad (17.25)$$

$$\mathbf{r}_{ik}^n = \left(\mathbf{r}_{ik}^T \, \mathbf{n}\right)\mathbf{n} \qquad (17.26)$$

A necessary condition for contact is that node k penetrates the surface of the patch, i.e.

$$\mathbf{r}_{ik}^T \, \mathbf{n} \leq 0 \qquad (17.27)$$

In order to ensure that a node does not penetrate the surface through its 'interior' face a thickness e must be associated to the patch. The thickness penetration condition is

$$-\mathbf{r}_{ik}^T \, \mathbf{n} \leq e \qquad (17.28)$$

The condition described by equation (17.28) prevents that penetration is detected when the flexible body is far away, behind the contact surface.

The remaining necessary conditions for contact results from the need for the node to be inside of the triangular patch. These 3 extra conditions are

$$\left(\tilde{\mathbf{r}}_{ik}^t \, \mathbf{r}_{ij}\right)^T \mathbf{n} \geq 0 \; ; \; \left(\tilde{\mathbf{r}}_{ik}^t \, \mathbf{r}_{jl}\right)^T \mathbf{n} \geq 0 \; \text{ and } \; \left(\tilde{\mathbf{r}}_{ik}^t \, \mathbf{r}_{ki}\right)^T \mathbf{n} \geq 0 \qquad (17.29)$$

Equations (17.27) through (17.29) are necessary conditions for contact. However, depending on the contact force model actually used, they may not be sufficient to ensure effective contact.

The contact detection algorithm is also applicable to rigid body contact by using the position of a rigid body point P instead of the position of node k in equation (17.24). The global position of point P is given by $\mathbf{r}_i^P = \mathbf{r}_i + \mathbf{A}_i \mathbf{s}_i^{P'}$, where $\mathbf{s}_i^{P'}$ denotes the point location in body i frame.

If contact between a node and a surface is detected, a kinematic constraint is imposed. Assuming fully plastic nodal contact, the normal components of the node k velocity and acceleration, with respect to the surface, are null during contact. Therefore the global nodal velocity and acceleration of node k, in the event of contact, become

$$\ddot{\mathbf{q}}_k = \ddot{\mathbf{q}}_k^{(-)} - \left(\ddot{\mathbf{q}}_k^{(-)T} \mathbf{n} \right) \mathbf{n} \tag{17.30}$$

$$\dot{\mathbf{q}}_k = \dot{\mathbf{q}}_k^{(-)} - \left(\dot{\mathbf{q}}_k^{(-)T} \mathbf{n} \right) \mathbf{n} \tag{17.31}$$

where $\dot{\mathbf{q}}_k^{(-)}$ and $\ddot{\mathbf{q}}_k^{(-)}$ represent respectively the nodal velocity and acceleration immediately before impact.

The kinematic constraint implied by equations (17.30) and (17.31) is removed when the normal reaction force between the node and the surface becomes opposite to the surface normal. Representing by $\mathbf{f}_k$ the resultant of forces applied over node k, except for the surface reaction forces but including the internal structural forces due to the flexible body stiffness, the kinematic constraint over node k is removed when

$$\mathbf{f}_k^n = -\mathbf{f}_k^T \mathbf{n} > 0 \tag{17.32}$$

The contact forces between the node and the surface include friction forces modeled using Coulomb friction. The friction force, in the presence of sliding, is given by

$$\mathbf{f}^{friction} = -\mu_d \, f_d \left(\left| \mathbf{f}_k^n \right| / \left| \dot{\mathbf{q}}_k \right| \right) \dot{\mathbf{q}}_k \tag{17.33}$$

where μ_d is the dynamic friction coefficient and f_d is a dynamic ramp correction coefficient expressed as

$$f_d = \begin{cases} 0 & \text{if} \quad \left| \dot{\mathbf{q}}_k \right| \leq v_0 \\ \left(\left| \dot{\mathbf{q}}_k \right| - v_0 \right) / \left(v_1 - v_0 \right) & \text{if} \quad v_0 \leq \left| \dot{\mathbf{q}}_k \right| \leq v_1 \\ 1 & \text{if} \quad \left| \dot{\mathbf{q}}_k \right| \geq v_1 \end{cases} \tag{17.34}$$

The dynamic correction factor represented by equation (17.34) prevents that the friction force changes its direction for almost null values of the node tangential velocity. Otherwise, the rapidly changing friction forces are perceived by the integration algorithm, used to integrate the system equations of motion, as high frequency response contents forcing it to dramatically reduce the time step size.

17.4   Frictionless impact of an elastic beam

The continuous contact force model application is exemplified with the oblique impact of a hyperelastic beam against a rigid wall. The impact scenario, proposed by Orden and Goicolea [17.18], is described in Figure 17.4. The beam, with a mass of 20 kg, length of 1 m and a circular cross-section, is made of a material with an elastic modulus of $E=10^8$ Pa, Poisson's ration of $v=0.27$ and a mass density of $\rho=7850$ kg/m^3. For this geometry and material properties the equivalent stiffness coefficient used in equation (16.33) is $K=1.2$ 10^8 kg/m$^{2/3}$. Both contact models, represented by equations (16.33) and (17.30) and designated respectively by continuous force and kinematic constraint models.

The motion predicted for the beam model using 10 elements is similar to that presented by Orden and Goicolea [17.18] for a model made of 20 elements and using an energy-momentum formulation to describe contact. In Figure 17.5, the contact forces developed during nodal impact are presented for both contact models described.

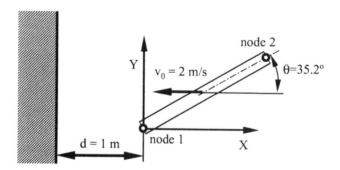

Figure 17.4 Impact scenario for an oblique elastic beam

The simulations results, summarized in Figure 17.5, show the similarity of the system response for both contact models, suggesting that they capture the most important features of the contact phenomena. The beam response shows high sensitivity to the friction forces developed during contact, demonstrating the need for an accurate modeling of these effects. The results suggest that by fine-tuning the continuous force model parameters it would be possible to obtain identical responses.

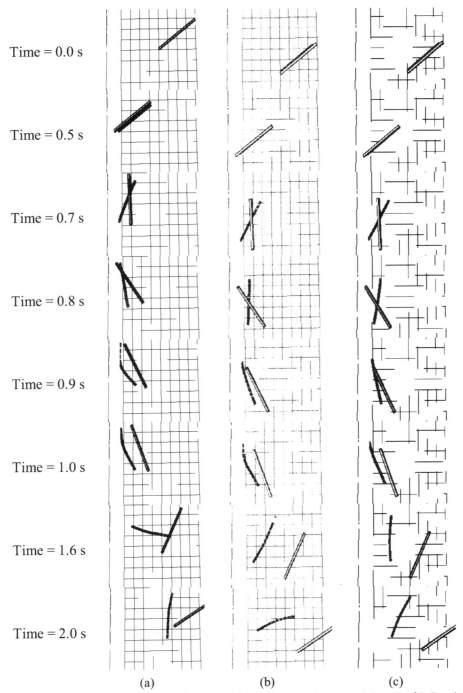

Time = 0.0 s

Time = 0.5 s

Time = 0.7 s

Time = 0.8 s

Time = 0.9 s

Time = 1.0 s

Time = 1.6 s

Time = 2.0 s

(a)   (b)   (c)

Figure 17.5 Impact of a hyperelastic beam: (a) Continuous force model, μ=0; (b) Continuous force model, μ=0.35; (c) Contact model with kinematic constraints, μ=0.35.

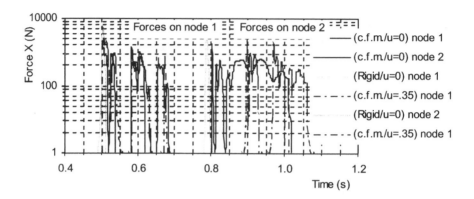

Figure 17.6 Forces developed between the (10 elements) beam end nodes and the rigid surface (c.f.m. – continuous force model; k.c. – kinematic constraint model)

Observing the contact forces, shown in Figure 17.6, it is clear that the impact phenomenon occurs with multiple contacts. Each of these contacts, for a beam model made of 10 finite elements, lasts for 0.02 s in average, which is similar to the contact duration of 0.018 s estimated by Orden and Goicolea [17.18] using the elastic wave travel time across the bar length. Moreover, these results confirm that both models predict similar contact forces.

17.5   Study of the crash-box of a sports car

The formulation presented in this chapter is applied to the redesign of the front crash-box of the sports car presented in Figure 17.7. The vehicle, a replica of the original Lancia Stratos, was assembled in-house at IDMEC/IST, providing an opportunity to obtain through direct measurements the structural and dynamic characteristics of all car components used.

Figure 17.7   Prototype of the sports car

The multibody model of the vehicle is composed of 16 rigid bodies and a nonlinear partially flexible body. The system components include the front double A-arm suspension system, the rear McPherson suspension system, wheels and chassis as depicted by Figure 17.8

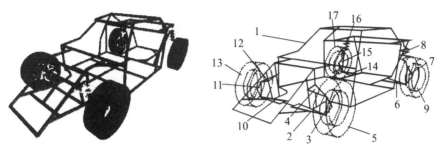

Figure 17.8 Multibody model of the sports vehicle

The mass and inertia characteristics and the initial positions of the vehicle components are presented in reference [17.19]. To model the tire interaction with the ground an analytical tire model with comprehensive slip is used [17.20]. A partially flexible body where the front part is considered flexible while the remaining structure and shell is considered rigid, as depicted by Figure 17.9, models the chassis of the vehicle. This modeling assumption is valid if plastic deformations are to occur in the parts of the vehicle modeled as flexible regions, such as the front, and no significant deformation takes place in the passenger compartment. This model is suitable to simulate frontal impacts. The flexible part of the chassis is composed of 36 nodal points and 38 nonlinear beam elements made of E24 steel and having hollow rectangular cross-sections. The complete system has 15 degrees of freedom associated with the rigid bodies and 186 nodal degrees of freedom.

The front crash-box of the vehicle is modeled and simulated for a frontal impact with an initial velocity of 50 Km/h. It is observed that deformations extend beyond the allowable limit compromising the survivable space. This is due to the mechanism of deformation that forces the crash-box to bend down without absorbing the energy that it is supposed to. A new design for the crash-box, compatible with the functional geometry of the vehicle front is proposed. The original and new designs for the crash-box are shown in Figure 17.10.

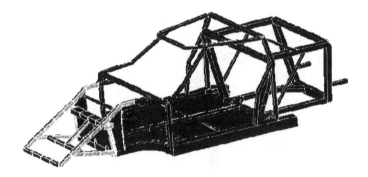

Figure 17.9 Model the car chassis with a nonlinear flexible front crash-box

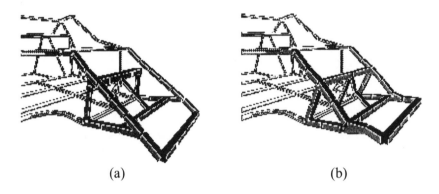

(a)   (b)

Figure 17.10 (a) Original and (b) modified crash-box of the sports vehicle.

For the same initial conditions, the contact forces generated between the crash-box and the rigid surface are depicted by figure 17.11. There, it is observed that for the new design the level of energy absorbed by the structure is higher and the force levels, of the structural fuses, are maintained for larger periods of time. The sharp peaks of the reaction forces are observed when nodal points of the crash-box come in contact with the rigid surface. When a node starts to penetrate the contact surface, a kinematic constraint, equivalent to setting the pseudo-velocity and pseudo-acceleration of penetration to zero, is applied. Because a lumped mass formulation is being used, the reaction forces corresponding to such kinematic constraint rise sharply when the constraints are activated, and almost immediately decrease to their sustained level.

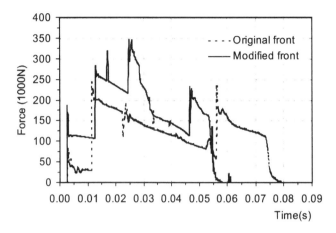

Figure 17.11 Forces developed between the crash-box and the rigid surface interface for a 50 Km/h crash.

The deformation mechanisms resulting from the simulations of the original and improved crash-boxes are presented in Figure 17.12. In the original crash-box the deformation progressed by bending the crash-box down and actually moving the structure out of the way of the remaining of the chassis without forcing the material to reach its limit for energy absorption. Therefore, the deformation progresses into the occupant safety envelope. In the improved design, the deformation of the crash-box is such that all structural components fold, as if they are kept inside the original volume, exploring the energy absorption characteristics of the structure.

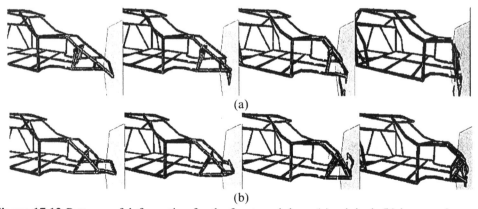

Figure 17.12 Patterns of deformation for the front crash-box: (a) original; (b) improved

The new design of the vehicle with the improved crash-box is simulated in different impact situations accounting for several relative orientations of the impact surfaces, various road profiles and several friction characteristics for the chassis-surface contact. For a constant initial velocity of the vehicle of 50 Km/h the impact scenarios are summarized in Figure 17.13.

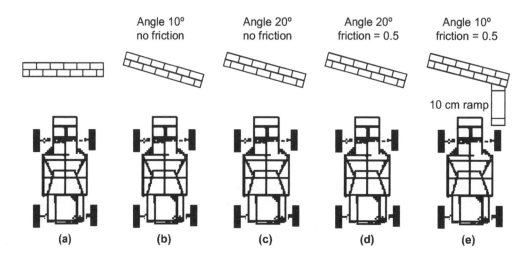

Figure 17.13 Impact scenarios: a) Frontal impact; b) Impact with a 10° oblique surface, no contact friction; c) Impact with a 20° oblique surface, no friction; d) Impact with a 20° oblique surface, contact friction; e) Same as (d) but the vehicle travels over a ramp with a 10 cm height.

The vehicle motion for the different impact scenarios is presented in Figure 17.14. For the frontal crash scenario it is observed that the deformation of the crash-box progresses with a pattern similar to that predicted for the modified structure shown in Figure 17.12b. At the simulated impact speed the influence of the car suspension elements over the deformation mechanism is minimal.

In Figure 17.15 the velocity and acceleration of the center of mass of the vehicle are plotted for the frontal impact scenario. It is observed that all kinetic energy of the vehicle is absorbed, mainly by plastic deformation of the crash-box, leading to the vehicle complete stop 0.09 s after the start of the analysis. Note that in the initial conditions the front of the vehicle crash-box is 0.570 m from contact surface, when the motion starts.

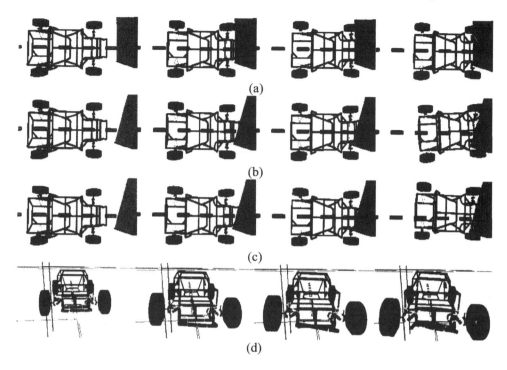

Figure 17.14 Motion of the vehicle in different crash scenarios: (a) Frontal impact; (b) 20°
Oblique impact without contact friction; (c) 20° Oblique impact with contact
friction; (d) Impact with an oblique surface for a vehicle traveling over a ramp

It is observed that for the frontal crash the front structure dissipates all
kinetic energy. The efficiency of the crash-box to dissipate the kinetic energy
of the vehicle decreases with the increase of the angle value between surface
normal and vehicle heading. The vehicle motion is deflected, and would
continue if no other component of the car entered in contact with the surface.
In figure 17.16 the forces developed between vehicle and surface are plotted.

For the crash scenarios where the vehicle collides with an oblique
surface, and no contact friction is modeled, the front of the vehicle slides on the
surface forcing the chassis to have a counterclockwise rotation on the
horizontal plane. The presence of friction forces between the impacting vehicle
and surface is very important for the efficiency of the crash-box, as an energy
management component in crash events. For the crash scenarios where friction
is modeled, the deflection of the vehicle motion does not occur, enabling the
structure to deform with a crushing mechanism similar to that of the frontal

impact. Only a slight sideways translation of the vehicle is observed. This result clearly emphasizes the importance of a correct model for the friction forces developed during contact.

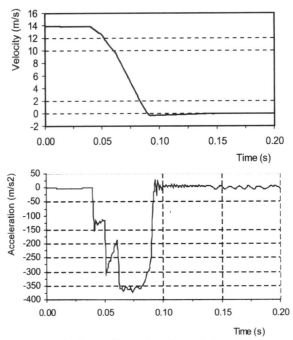

Figure 17.15 Displacement, velocity and acceleration of the chassis center of mass for the frontal impact scenario

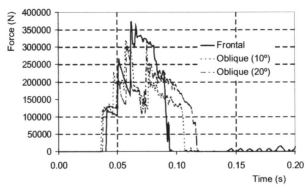

Figure 17.16 Contact forces between the vehicle front and impact surface

REFERENCES

[17.1] Erdman, A. G. and Sandor, G. N., Kineto-elastodynamics - a review of the state of the art and trends, *Mech. Mach. Theory,* **7,** 19-33, 1972.

[17.2] Lowen, G.G. and Chassapis, C., Elastic behavior of linkages: an update, *Mech. Mach. Theory,* **21,** 33-42, 1986.

[17.3] Thompson, B.S. and Sung, G.N., Survey of finite element techniques for mechanism design, *Mech. Mach. Theory,* **21,** 351-359, 1986.

[17.4] Song, J.O. and Haug, E.J., Dynamic analysis of planar flexible mechanisms, *Computer Methods in Applied Mechanics and Engineering,* **24,** 359-381, 1980.

[17.5] Shabana, A.A. and Wehage, R.A., A coordinate reduction technique for transient amalysis os spatial structures with large angular rotations, *Journal of Structural Mechanics,* **11** , 401-431, 1989.

[17.6] Shabana, A.A., *Dynamics of Multibody Systems,* John Wiley & Sons, New York, New York, 1989

[17.7] Belytschko, T. and Hsieh, B.J., Nonlinear transient finite element analysis with convected coordinates, *Int. J. Nume. Methods in Engng.,* **7** , 255-271, 1973.

[17.8] Simo, J.C. and Vu-Quoc, L. , On the dynamics in space of rods undergoing large motions – a geometrically exact approach, *Comp. Methods Appl. Mech. Eng.,* **66,** 125-161, 1988

[17.9] Bathe, K.-J. and Bolourchi, S., Large displacement analysis of three-dimensional beam structures, *Int. J. Nume. Methods in Engng.,* **14,** 961-986, 1979

[17.10] Cardona,A. and Geradin,M., A beam finite element non linear theory with finite rotations, *Int.J. Nume Methods in Engng.,* **26,** 2403-2438, 1988.

[17.11] Ambrósio, J. and Nikravesh, P. , Elastic-plastic deformations in multibody dynamics, *Nonlinear Dynamics,* **3** , 85-104, 1992.

[17.12] Ambrósio, J.A.C. and Pereira, M.S. Multibody dynamic tools for crashworthiness and impact. In *Crashworthiness Of Transportation Systems: Structural Impact And Occupant Protection* (J.A.C. Ambrósio, M.S. Pereira and F.P. Silva, Eds.), NATO ASI Series E. Vol. 332, 475-521. Dordrecht: Kluwer Academic Publishers, 1997

[17.13] Bathe, K.-J., *Finite Element Procedures In Engineering Analysis*, Prentice-Hall, Englewood Cliffs, New Jersey, 1982.

[17.14] Nygard, M.K. and Bergan, P.G., Advances on treating large rotations for nonlinear problems, In *State Of The Art Surveys On Computational Mechanics* (A.K. Noor and J.T. Oden, Eds.), ASME, New York, 305-333, 1989.

[17.15] Wallrapp, O. and Schwertassek, R. , Representation of geometric stiffening in multibody system simulation, *Int. J. Nume. Methods Engng.* **32,** 1833-1850, 1991.

[17.16] Hinton, E., Rock, T. and Zienckiewicz, O.C. , A note on the mass lumping and related processes in the finite element method, *Earthquake Engineering and Structural Mechanics*, **4,** 245-249, 1976.

[17.17] Greenwood, D.T. , *Principles of Dynamics*, Prentice-Hall, Englewood-Cliffs, New Jersey, New Jersey, 1965.

[17.18] Orden, J.C.G. and Goicolea, J.M., Conserving properties in constrained dynamics of flexible multibody systems, *Multibody System Dynamics*, **4** , 221-240, 2000.

[17.19] Ambrósio, J.A.C., Geometric and material nonlinear deformations in flexible multibody systems, In *Proceedings of the NATO-ARW on Computational Aspects Of Nonlinear Structural Systems With Large Rigid Body Motion,* (J.A.C. Ambrósio and M. Kleiber, Eds.), Pultusk, Poland, July 2-7, 91-115, 2000.

[17.20] Gim, G., *Vehicle Dynamic Simulation With A Comprehensive Model For Pneumatic Tires*, Ph.D. Dissertation, University of Arizona, Tucson, Arizona, 1988.

18. VEHICLE AND OCCUPANT INTEGRATED SIMULATION

by
J.A.C. AMBRÓSIO and M.P.T SILVA[1]

18.1   Introduction

The safety of occupants and their potential survival in crash events in transportation systems involve topics as different as interior trimming of the passenger compartment structural crashworthiness, or restraint systems efficiency. The analysis of such aspects is currently done during the initial design stages.   Current design methodologies entail the use of different computer simulations of increasing complexity ranging from simplified lumped mass models [18.1-2], multibody models [18.3] to complex geometric and material nonlinear finite element based representations of occupant [18.4] and vehicle structures [18.5].   Some well-known simulation programs are now available: PAM CRASH [18.5], WHAMS-3D [18.6] and DYNA 3D [18.7] for structural impact and CAL3D [18.8] and MADYMO [18.9] for occupant dynamics. These programs are able to simulate with relative detail frontal, rear and side impact scenarios. However, in most cases, the structural impact and the occupant dynamics are treated separately.   The structural crash analysis provides the relevant accelerations pulses, which are used in the occupant analysis phase.   The injury indices such as HIC and chest accelerations [18.10-11] can then be evaluated to access design performances.

[1] Instituto de Engenharia Mecânica, Instituto Superior Técnico, Av. Rovisco Pais, 1049-001 Lisboa, Portugal

The procedure described is valid for most impact scenarios but it has shortcomings for more complex crash situations. In vehicle simulations that involve complex gross motion during the crash events, such as for vehicle rollover, a combined model with the ability to analyze in a coupled manner complex crash events and occupant dynamics is necessary, specially for small cars where the occupant masses become comparable to the vehicle mass. The multibody methodologies have been successfully in modeling these types of problems [18.12-13]. These computer models include features such as tire-terrain characteristics, steering and suspension sub-systems representations and nonlinear flexibility effects to model structural deformation [18.14].

A fully coupled vehicle-occupant model and the interaction between the compartment and the occupant is described within the framework of a multibody dynamics formulation. Both vehicle and occupants are assumed to be composed of rigid bodies linked by different kinematic joints. The vehicle model includes a detailed suspension system with a tire-terrain description, steering capabilities and a spatial description of the chassis with appropriate force-deflection characteristics for the potential points of contact between the chassis and the ground. The occupant model includes a set of nonlinear actuators, which develop forces and moments between the different biomechanical segments controlling in an appropriate manner its relative motion. The different body segments are spatially described as ellipsoids in order to treat the contact problem between the occupant and the complex geometry of the surrounding passenger compartment. Shoulder and lap safety belt systems are also included.

18.2   Biomechanical model of the occupant

The methodology presented is applied to the representation of a three-dimensional, whole body response, biomechanical model of the human body suitable for impact simulations, based on the occupant model of SOMLA [18.15]. The model is general and accepts data for any individual. The information required to assemble the equations of motion of the model includes the mass and inertia of the biomechanical segments, their lengths, location of the body-fixed coordinate frames and the geometry of the potential contact surfaces, as pictured in Figure 18.1. The data hold within a database can be expanded for different individuals.

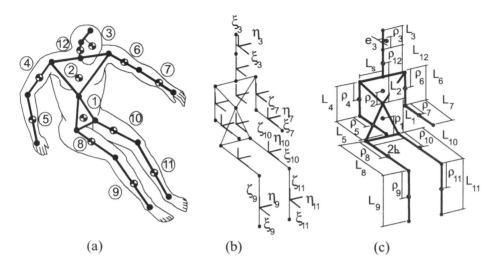

Figure 18.1 Three-dimensional biomechanical model for impact: (a) actual model; (b) local referential locations; (c) dimensions of the biomechanical segments.

In contact/impact simulations the relative kinematics of the head-neck and torso are important to the correct evaluation of the loads transmitted to the human body. Consequently, the head and neck are modeled as separate bodies and the torso is divided in two bodies. The hands and feet do not play a significant role in this type of problems and consequently are not modeled. The model is described using 12 rigid bodies defined by 16 basic points and 17 unit vectors located at the articulations and extremities. The resulting biomechanical model has 29 degrees of freedom. In Table 18.1, the description and location of the kinematic joints is presented.

Table 18.1   Kinematic joint description for biomechanical model.

Joint	Type	Description
1	spherical	Back, (12th thoracic and 1st lumbar vertebrae).
2	spherical	Torso-Neck (7th cervical and 1st thoracic vertebrae).
3-5	spherical	Shoulder.
4-6	revolute	Elbow.
7-9	spherical	Hip.
8-10	revolute	Knee.
11	revolute	Head-Neck, (at occipital condyles).

The principal dimensions of the model are represented in Figure 18.1(c). In most cases, the effective link-lengths between two kinematic joints are used instead of standard anthropometric dimensions based on external measurements. The set of data for the models referred are described in reference [18.16].

18.2.1 Joint resisting moments

In the biomechanical model, no active muscle force is considered but the muscle passive behavior is represented by joint resistance torques. Applying a set of penalty torques when adjacent segments of the biomechanical model reach the limit of their relative range of motion prevents physically unacceptable positions of the body segments. A viscous torsional damper and a non-linear torsional spring, located in each kinematic joint, describe the joint torques. Take the elbow of the model, represented in Figure 18.2 for instance. The axis for the relative rotation of the lower and upper arm is represented. The torsional damper has a small constant coefficient j_i, being the total damping torque at each joint given by

$$\mathbf{m}_{d_i} = -j_i \, \dot{\boldsymbol{\beta}}_i \qquad (18.1)$$

where $\dot{\boldsymbol{\beta}}_i$ is the relative angular velocity vector between the two bodies interconnected by joint i.

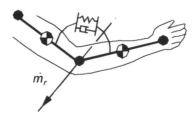

Figure 18.2 Joint resistance torque modeled with a non-linear spring and damper torsional element.

A constant torque m_{r_i} that acts resisting the motion of the joint is applied to the whole range of motion in the dummy model [18.16]. For the human joint this torque has an initial value, which drops to zero after a small angular displacement from the joint initial position. The torque has a direction opposite to that of the relative angular velocity vector between the two bodies interconnected in the joint

$$\mathbf{m}_{r_i} = -m_{r_i}\dot{\boldsymbol{\beta}}_i \left\| \dot{\boldsymbol{\beta}}_i \right\|^{-1} \tag{18.2}$$

A second moment, applied at the joint, is a penalty resisting torque m_{pi} which is null during the normal joint rotation but it increases rapidly, from zero to a maximum value, when the two bodies interconnected by that joint reach physically unacceptable positions. The curve for the penalty resisting moment is represented qualitatively in Figure 18.3.

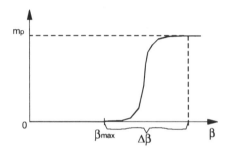

Figure 18.3 Penalty moment for the elbow.

The shoulder is a biomechanical joint modeled by a spherical joint as depicted by Figure 18.4. The calculation of the penalty torque requires the construction of the cone of feasible motion. This cone has its tip in the center of a sphere with a unit radius. While the upper arm moves inside the cone its motion does not imply displacements of the upper or lower torsos.

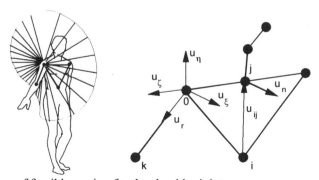

Figure 18.4 Cone of feasible motion for the shoulder joint.

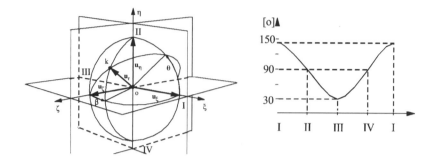

Figure 18.5 Angles and interpolation curve for the shoulder joint.

A local reference frame is constructed and rigidly attached to the shoulder joint (points 3 and 5). The local axes u_ξ and u_η are defined from the geometry of the upper torso. Referring to Figure 18.4, these are

$$u_\xi = u_n \quad ; \quad u_\eta = \frac{r_j - r_i}{\|r_j - r_i\|} \quad ; \quad u_\zeta = \tilde{u}_\xi u_\eta \qquad (18.3)$$

A fourth vector in the direction of the upper arm is calculated using the two anatomical points of this body

$$u_r = \frac{r_k - r_o}{\|r_k - r_o\|} \qquad (18.4)$$

The angles θ and β of the unit vector u_ρ are calculated in the local reference frame, as depicted in Figure 10(a). The angle of β_{max} is also calculated for a specified longitude θ, using a cubic spline interpolation curve, as shown in Figure 18.5(b). If the effective latitude β is larger than the maximum latitude β_{max}, a penetration on a zone of unfeasible motion occurs and a penalty resisting torque is applied in the direction of the cross-product between vector u_ζ and vector u_ρ. The penalty torque is given by

$$m_{p_i} = m_{p_i} \left[3 \left(\frac{\beta_i - \beta_{i_{max}}}{\Delta \beta_i} \right)^2 - 2 \left(\frac{\beta_i - \beta_{i_{max}}}{\Delta \beta_i} \right)^3 \right] \tilde{u}_\zeta u_r \qquad (18.5)$$

where the term between brackets is a third order polynomial with a behavior similar to that depicted by Figure 18.3. Table 18.2 describes the values for the limit angles for different joints of the human body.

18.2.2 Contact surfaces

A set of contact surfaces is defined for the calculation of the external forces exerted on the model when the bodies contact other objects or different body segments. These surfaces are ellipsoids and cylinders with the form depicted by Figure 18.6 and have the dimensions described in table 18.3.

Table 18.2   Joint limit angles and force data.

Joint	$\beta_{i\ I}[°]$	$\beta_{i\ II}[°]$	$\beta_{i\ III}[°]$	$\beta_{i\ IV}[°]$	$\Delta\beta_i[°]$	$m_i[g]$	$m_{pi}[Nm]$	$j_i[Nms]$
1	40	35.0	30.0	35.0	11.5	2.0	226.0	16.95
2	60	40.0	60.0	40.0	15.0	2.0	678.0	3.39
3-5	140	90.0	30.0	90.0	11.5	1.0	226.0	3.76
4-6	90	-	45.0	-	11.5	1.0	226.0	3.39
7-9	10	120.0	50.0	45.0	11.5	2.0	452.0	5.65
8-10	-	90.0	-	45.0	11.5	1.0	226.0	5.65
11	19	-	2.0	-	15.0	2.0	452.0	16.95

When contact between components of the biomechanical model are detected a contact forces, with the characteristics described by equation (16.33), are applied to such components in the points of contact. Friction forces are also applied to the contact surfaces using Coulomb friction. It must be noted that the characterization of the surfaces in contact is important for general applications of the biomechanical model.

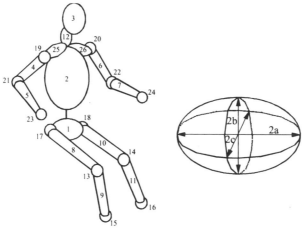

Figure 18.6   Representation of contact surfaces.

Table 18.3   Dimensions of contact surfaces.

Body	50%ile Dummy $R_i[m]$	95%ile Dummy $R_i[m]$
1	0.102	0.114
2	0.127	0.114
3	0.095	0.0874
4 - 6	0.053	0.050
5 - 7	0.042	0.047
8 - 10	0.083	0.079
9 - 11	0.057	0.058
12	0.051	0.051

18.3   All-Terrain vehicle multibody model

The vehicle model simulated here is an all-terrain truck M151A2 Jeep with ¼ ton and a rollbar cage for the occupant protection, as shown in Figure 18.7. The occupant model described before is positioned in the driver seat of the truck and it is restrained to the vehicle by a system of seat belts. The vehicle with different occupant sizes is then simulated for a rollover scenario. The coupling between the two systems is considered in the simulations.

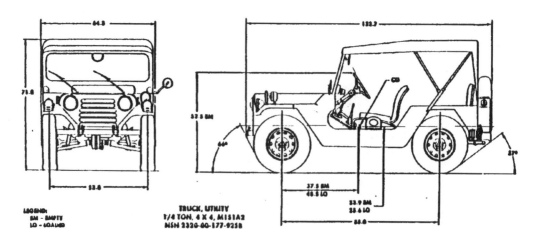

Figure 18.7   The M151A2 Jeep - General dimensions.

18.3.1 Vehicle model

The truck fully equipped has a total mass of 1470 Kg that corresponds to a weight of 799 Kg_f in the front axle and 670 Kg_f in the rear axle. The height of the vehicle is 1.803 m while its length and width are 1.633 m and 3.371 m respectively. The location of the center of mass is 0.650 m above the ground and 1.232 m from the rear centerline of the front axle. Moreover, the truck is equipped with a single A-arm suspension in the rear and a double A-arm for the front suspension [18.17].

The utility truck is modeled with thirteen rigid bodies. In Table 18.4, the mass and inertia for each system component are presented. It is assumed that bodies of the suspension system have negligible masses as compared to the mass of the wheels and chassis.

Table 18.4   Rigid bodies inertial properties

Body	Description	Mass [Kg]	$I_{xx}/I_{hh}/I_{zz}[Kg.M^2]$
1	Chassis	1368.00	1300/2500/2000
2	Left-Front-Upper A-Arm	1.00	1.0/1.0/1.0
3	Left-Front Hub	1.73	0.6/1.0/0.6
4	Left-Front-Lower A-Arm	1.00	1.0/1.0/1.0
5	Left-Rear A-Arm	1.73	0.6/1.0/0.6
10	Left-Front Wheel	22.70	1.0/1.9/1.0
11	Left-Rear Wheel	22.70	1.0/1.9/1.0

18.3.2 Springs, dampers and tires

The suspension system includes four spring-damper assemblies with nonlinear characteristics, as represented in Figure 18.8. The attachment points of the spring-damper assemblies in the vehicle model and their nonlinear characteristics are described in Tables 18.5 and 18.6 respectively.

In the vehicle model a comprehensive tire model is used to describe the interaction with the ground. This model, proposed by Gim [18.18] includes traction, braking and lateral forces due to steering depending on the normal force, slip angle and camber angle. The tire data for the truck is described in Table 18.7.

Table 18.5   Spring and damper attachment points.

	Body	$\xi^p/\eta^p/\zeta^p$ [m]
Left-Front	1	1.329/0.377/-0.094
	4	0.097/0.236/-0.038
Left-Rear	1	-0.840/0.476/-0.089
	5	0.093/-0.197/-0.090

Table 18.6   Spring and Damper non-linear characteristics

	L_0 [m]	d_1 [m]	d_2 [m]	f_1 [N]	f_2 [N]	s_1 [$^+$]	s_2 [$^+$]	s_3 [$^+$]
Fr.Spring	0.279	-0.099	-0.021	-9041	-1945	2521700	91600	1251
Re.Spring	0.328	-0.117	-0.052	-6893	-3052	898619	58766	2093
	v_1 [m/s]	v_2 [m/s]	f_1 [N]	f_2 [N]	$d_1[^*]$	$d_2[^*]$	d_3 [*]	$d_4[^*]$
Fr.Damp.	-0.339	0.339	-489	1757	1180	1443	5181	1082
Re.Damp.	-0.339	0.339	-556	2113	590	1640	6230	2164

*N.s/m; $^+$N/m

Table 18.7   Tire characteristics

Radius	0.406	(m)
Radial stiffness	5.84×10^5	(N/m)
Longitudinal stiffness	1.0×10^5	(N/slip)
Cornering stiffness	1.0×10^5	(N/rad)
Maximum friction coefficient	0.80	
Minimum friction coefficient	0.60	

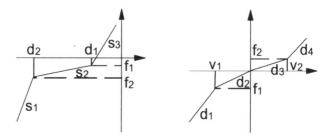

Figure 18.8   Non-linear a)Spring; b)Damper

18.3.3 Rollbar cage

A rollbar cage is installed in the truck in order to protect the vehicle occupants in case of rollover. This is a flexible frame, made of 1025-1030 steel, mounted over the chassis, as depicted by Figure 18.9. The cross-sectional area of each bar is annular with an outside radius of 2.54 cm. Though in other models of the vehicle rollover the deformation of this structure was modeled [18.19] it is not considered explicitly here. The effects of the deformation are lumped in the continuous force contact.

The interaction of the vehicle and/or the rollbars with the ground is described by controlling the coordinates of six points in the rollbar cage P_1 through P_6 and eight points (C_1 through C_8) for possible ground contact. The position of the control points is described in Table 18.8. When contact is detected, a force is applied on the contacting points of the vehicle in the direction normal to the ground surface. This force is calculated using the model described in equation (16.33). In addition, a friction force is also applied using the Coulomb friction model. The direction of this force is opposite to the direction of the projected velocity of the contact point on the ground. For the rollover simulation, the values of n=1.5, K=1.05×10^7 N/m$^{1.5}$, e=0.75 and a friction coefficient of μ=0.75 were used in the continuous force model referred.

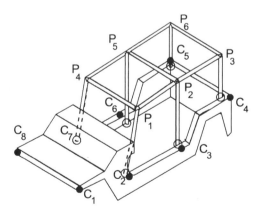

Figure 18.9 Rollbar cage computational model.

Table 18.8   Local coordinates of control points

Pt	$\xi^P/\eta^P/\zeta^P$ [m]	Pt	$\xi^P/\eta^P/\zeta^P$ [m]
P_1	0.303/0.708/1.001	C_1	1.457/0.692/0.319
P_2	-0.483/0.692/1.045	C_2	0.589/0.692/-0.231
P_3	-1.387/0.692/0.995	C_3	-0.432/0.692/-0.088
P_4	0.303/-0.708/1.001	C_4	-1.387/0.692/0.244
P_5	-0.483/-0.692/1.045	C_5	1.457/-0.692/0.319
P_6	-1.387/-0.692/0.995	C_6	0.589/-0.692/-0.231
		C_7	-0.432/-0.692/-0.088
		C_8	-1.387/-0.692/0.244

18.3.4 Rigid seat data

Before integrating the vehicle occupant in the model a seat needs to be introduced. The seat is formed by two planes, i.e., the seat pan and back cushions, as shown in Figure 18.10(a). Once penetration of a contact surface of the occupant is detected in one of the seat cushions, a contact force is calculated

$$f_s = A\left(e^{B\delta} - 1\right) + C\dot{\delta}$$

This force is applied at the contact point, with a direction normal to the seat pan/back cushion. A friction force is also calculated using Coulomb's friction and applied:

$$f_f = \mu f_s$$

The dimensions of the seat and the coefficients of the force model are shown in Table 18.9.

Table 18.9   Seat data

Seat	$\xi^P/\eta^P/\zeta^P$ [m]	θ_s/θ_b [°]	t [m]	A [N]	B [1/m]	C [N.s/m]	μ
1	0.414/0.419/0.062	10/20	0.07	3380.0	19.7	420.0	0.3

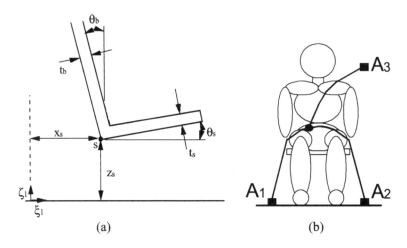

Figure 18.10 (a)Seat dimensions, and; (b) seat belts attachment points.

18.3.5 Seat belts

In order to include a seat belt in the integrated simulation of the rollover of vehicle and occupant, it is necessary to introduce the coordinates of the three attachment points of the seat belt to the car as pictured in Figure 18.10(b). The coordinates are shown in table 18.10.

Table 18.10   Seat belt attachment points.

Pt	$\xi^P/\eta^P/\zeta^P$ [m]
A_1	-0.477/0.113/-0.124
A_2	-0.477/0.725/-0.124
A_3	-0.663/0.092/0.580

18.4   Vehicle rollover with occupant simulation

The vehicle is simulated here in a rollover situation with the initial conditions described in Figure 18.11.   The vehicle rollover has been extensively analyzed with the purpose of studying the rollbar cage influence in the vehicle stability and its structural integrity [18.12]. There, the rollbar cage is modeled as a nonlinear flexible body experiencing large plastic deformations. The rollbar cage deformation is not included in the model.

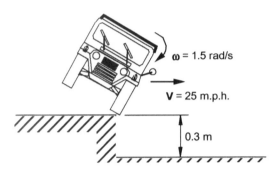

Figure 18.11 Rollover simulation scenario.

The initial conditions of the simulations correspond to experimental setup where the vehicle moves on a cart with a lateral velocity of 13.41 m/s until the impact with a water-filled decelerator system occurs. The vehicle is ejected with an roll angle of 23 degrees. The initial velocity of the vehicle, when ejected, is approximately 11.75 m/s in the Y direction while the angular roll velocity is 1.5 rad/s. The first 1.5 seconds of the simulations for the vehicles without and with occupants are represented in Figure 18.12. It is observed that the first contact of the wheels with the ground occurs at 0.3 s, for all simulations, causing the vehicle to bounce from the ground with an increasing roll velocity. At about 0.8 s the truck impacts the ground with the rollbar cage and continues its rolling motion with the contact of different points of the rollbar. It is noticeable that the vehicle with occupant rolls more than the empty vehicle.

The occupant motion is constrained by the seat belts and the contact of the biomechanical segments with the vehicle floor, firewall and dashboard. Figure 18.13 presents the head acceleration, showing that high acceleration peaks, resulting from the direct contact of the head with the vehicle interior, appear shortly after the vehicle contact with the ground is detected. HIC values over 5400 are obtained for the occupant in both simulations as a result of the extremely severe head contact with the vehicle side and top panel.

Observing the forces on the seat belts, presented in Figure 18.14, it is noticeable that it is the lap belt that plays the most important role in keeping the vehicle occupant in place during the rollover. Peaks of 8000 N and 9500 N are observed in the lap belt when the vehicle impact with the ground starts. While the upside down position is maintained forces on the lap belts increase to 35000 N with both simulations. During the impact, the shoulder belt plays a minor role comparatively to the role played by the lap belt.

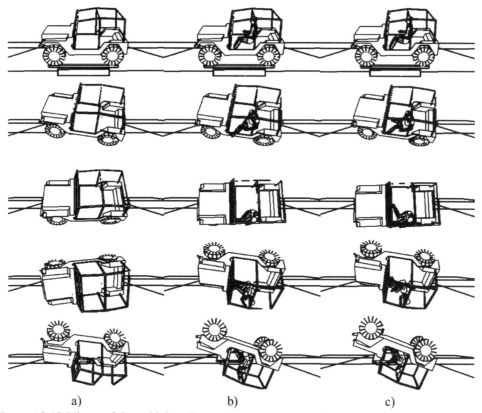

a)   b)   c)

Figure 18.12 Views of the vehicle rollover from 0 through 1.5s (steps of 0.3s). a) no occupant;
b) 50%ile ocupant; c) 95%ile occupant

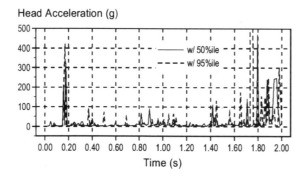

Figure 18.13   Occupant head and chest acceleration.

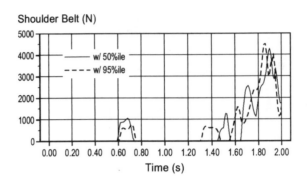

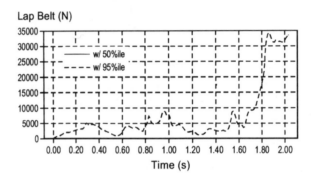

Figure 18.14   Forces on the seat belts.

REFERENCES

[18.1]   Kamal, M. M., Analysis and simulation of vehicle-to-barrier impact, SAE Paper N. 700414, In *International Automobile Safety Conference Compendium*, SAE, Inc., 1970.

[18.2]   Herridge, J. T. and Mitchell, R. K., *Development of a Computer Simulation Program for Collisions Car/Car and Car/Barrier Collision*, DOT-HS-800-645, 1972.

[18.3]   Ambrósio, J. A. C. and Pereira, M. S., Multibody dynamic tools for crashworthiness and impact, In *Proceedings of the NATO Advanced Study Institute on Crashworthiness on Transportation Systems* (J.A.C. Ambrósio and M.S. Pereira, Eds.), Tróia, Portugal, 1996.

[18.4]   King, H. Y., Pan, H., Lasry, D. and Hoffman, R., Finite element modelling of the hybrid III dummy chest, In *Crashworthiness and Occupant Protection in Transportation Systems*, AMD-Vol. 126/BED-Vol.19, ASME, 1991.

[18.5]   Haug, E. and Ulrich, D. The PAM-CRASH code as an efficient tool for crashworthiness simulation and design, In *Proceedings of the Second European Car/trucks Simulation Symposium*, Munich, 1988.

[18.6]   Belytschko, T. and Kenedy, J. M., *WHAMS-3D, An Explicit 3D Finite Element Program*, KBS2 Inc. P.O. Box 453, Willow Springs, IL 60480, 1988.

[18.7]   Halquist, J. O., *Theoretical Manual for DYNA-3D*, Lawrence Livermore Laboratory, 1982.

[18.8]   Fleck, J. T. and Butler, F. E., *Validation of the Crash Victim Simulator, Vol.1: Engineering Manual, Part I: Analytical Formulations*, NTIS, no.Pb86-21243-8, 1981.

[18.9]   *MADYMO User's Manual*, Software version 4.0, TNO, The Hague, Netherlands, 1986.

[18.10] *Motor Vehicle Safety Standard No. 208*, (MVSS 208), U. S. Federal Safety Standards, 1988.

[18.11] Viano, D. C. and Lau, I. V., A viscous tolerance criterion for soft tissue injury assessment, *J. Biomechanics*, **21**, 387-399, 1988.

[18.12] Ambrósio, J., Nikravesh, P. and Pereira, M. S. 'Crashworthiness analysis of a truck', J. Mathematical Computer Modelling, **14**, 959-964, 1990.

[18.13] Nikravesh, P. , Ambrósio, J. and Pereira, M. S., Rollover simulation and crashworthiness analysis of trucks, *J. of Forensic Engineering*, **2**(3), 387-401, 1990.

[18.14] Pereira, M.S., Ambrósio, J..A.. and Dias, J.M., Crashworthiness analysis and design using rigid-flexible multibody dynamics with application to train vehicles, *Int. J. Numerical Methods in Engng*, **40**, 655-687, 1997.

[18.15] Laananen, D., *Computer Simulation Of An Aircraft Seat And Occupant In A Crash Environment - Vol. 1: Technical report*, Technical Report DOT/FAA/CT-82/33-I, US Dep. of Transportation, F.A.A., 1983.

[18.16] Silva, M.S.T, Ambrósio, J.A.C. and Pereira, M.S., Biomechanical model with joint resistance for impact simulation, *Multibody System Dynamics*, **1**(1), 65-84, 1997.

[18.17] Gim G., Pereira, M., Lankarani, H.M. and *Nikravesh, P.E. Rollover Analysis of Vehicles With Safety Rollbars*, Technical Report CAEL-87-5, University of Arizona, 1987.

[18.18] Gim, G. *Vehicle Dynamic Simulation With a Comprehensive Model for Pneumetic Tyres*, Ph.D. Dissertation, University of Arizona, 1988.

19. ADVANCED DESIGN OF STRUCTURAL COMPONENTS FOR CRASHWORTHINESS

by
J.A.C. AMBRÓSIO, J.P. DIAS[1] and M. SEABRA PEREIRA[1]

19.1 Introduction

The passive safety of vehicle occupants and the crashworthy characteristics of the vehicle structural components are issues of growing importance for manufacturers, users and legislators. The demands put in the systems responsible for maximizing safety must be included in the vehicle design, during its early phases, without increasing the time necessary for the designer to come with a reasonably good result. In one hand, the design tools used cannot be so complex and time consuming that different design iteration steps take longer than what is reasonable and, in the other hand, the methodologies used must still be accurate and have low requirements of information. These objectives are achieved by using design methodologies allowing establishing accurate and versatile conceptual vehicle models and by integrating optimization procedures in the reanalysis process in order to support the decisions.

With the increase of computer power and availability of cheap machines the numerical optimization procedures have been introduced in the industrial design processes as a very efficient tool to support decisions at the design and analysis stages. These procedures, though very popular in different aspects of

---

[1] Instituto de Engenharia Mecânica, Instituto Superior Técnico, Av. Rovisco Pais, 1049-001 Lisboa, Portugal

the design process, only now start to be applied to complex nonlinear dynamic problems and in particular to crashworthiness [19.1-4]. The major problems that are addressed, to have efficient numerical tools, are the evaluation of the design variables sensitivities and the optimization algorithms used. The research reported addresses mostly the evaluation of the sensitivities and rely on public domain or commercial algorithms for the optimization process. Among these, the multi-purpose program ADS [19.5-6] is thoroughly used here.

In the methodologies discussed, the vehicle model is described using either the flexible multibody formulation, described in Chapter 17 or the plastic hinge technique, presented in Chapter 16. These methodologies, besides supporting the optimization procedure, serve also as analysis tools. The dynamic analysis program and the optimization methodology are kept as separate programs that run concurrently. The design procedure developed in this form is then applied to the crashworthy design of the end underframe of a railway car. In the problem, several objective functions and constraints are specified for different designs. The section geometry characteristics of components and structural parameters associated to the structure topology are used as design variables. The results of the optimization procedures are used in the final design of the vehicle structure, which in turn, is analyzed with a commercial nonlinear finite element program to demonstrate the validity of the numerical procedure developed.

19.2   Optimization strategy

The optimization strategy can be separated in three stages. In the first level the initial constrained problem is solved as a sequence of simpler problems, using for instance sequential convex programming. Then, an objective function and the constraints must be evaluated at every iteration step. Finally, the way in which the optimal point is to be reached, or approached, is defined. The optimization procedure must be supported by an efficient way of calculating the problem sensitivities to the design variables that will ultimately contribute to the decisions taken by the optimizer. The sensitivities are evaluated by using analytical expressions or by applying the finite difference method.

19.2.1 Statement of the optimization problem

The general statement of the optimization problem is given in the form:

$$\min \quad F(\mathbf{b}) \tag{19.1}$$

$$\text{subject to} \quad \mathbf{G}(\mathbf{b}) \leq \mathbf{0}$$
$$\mathbf{H}(\mathbf{b}) = \mathbf{0} \tag{19.2a}$$

$$\mathbf{b}_{\min} \leq \mathbf{b} \leq \mathbf{b}_{\max} \tag{19.2b}$$

where $F(\mathbf{b})$ is a given objective function, $\mathbf{H}(\mathbf{b})$ and $\mathbf{G}(\mathbf{b})$ are *ncon* functions representing the constraints and $\mathbf{b}$ is a vector of design variables. The limits $\mathbf{b}_{\max}$ and $\mathbf{b}_{\min}$ represent the acceptable limits of variation of the variables, being generally associated with technological or biological constraints or with ranges of uncertainty depending on the application.

The objective function is a measure of the system characteristic that is being optimized. In crashworthiness typical objective functions are the maximum acceleration reached by a system component during impact, the deformation energy of the structural or mechanical components or the injury measures. For the first case the objective function is of particular type and it is generally written as

$$F(\mathbf{b}) = \max \ f(\mathbf{q}, \dot{\mathbf{q}}, \ddot{\mathbf{q}}, \lambda, t, \mathbf{b}) \tag{19.3}$$

The energy and injury measures are represented by objective functions of the integral type, which are written in their general form as

$$F(\mathbf{b}) = \int_{t_0}^{t_1} f(\mathbf{q}, \dot{\mathbf{q}}, \ddot{\mathbf{q}}, \lambda, t, \mathbf{b}) \, dt \tag{19.4}$$

The optimization process requires that the sensitivities of the objective functions and constraints to the design variables are evaluated. These sensitivities are used by the optimizer algorithm to decide on the direction to take in the search for an optimal point. The sensitivities may be evaluated in an analytical form [19.4] or using the finite difference method [19.7]. The advantage of the analytical sensitivities is that once the expressions for them are established their numerical evaluation is very efficient. The effort is put in deriving the sensitivities analytical expressions and implementing them in a computer code. The finite difference method for the evaluation of the sensitivities, though is not as efficient as the evaluation of the analytical sensitivities, is much easier to implement and it does not require any change in the analysis numerical tool. The numeric sensitivity of a given function $u(\mathbf{b})$ is

$$\frac{du}{db_i} \cong \frac{\Delta u}{\Delta b_i} = \frac{u(b_i + \Delta b_i) - u(\mathbf{b})}{\Delta b_i} \qquad (19.5)$$

Both methods are described and used here to evaluate the sensitivities for the optimization problems.

19.2.2 Integration of the analysis and optimization tools

The first optimization strategy devised uses a general purpose multibody code based on the formulations presented in Chapter 17, which is designated by MBOSS_FLEX, as the system analysis tool and the ADS program [19.5] as the optimization tool. None of the programs is modified for the optimization problems that are solved. In turn a set of interface programs are developed to allow these two numerical tools to exchange information. The methodology implemented is described in Figure 19.1.

The procedure starts by invoking the optimizer that initializes the design variables in the first call and decides if an optimum for the objective function is reached in subsequent calls. If the optimum has not been reached the design variables are updated and the control of the problem transferred to the first interface program. This interface program reads the values of the design variables and constructs the input files for the evaluation of the geometric properties of the structural components and for the dynamic analysis program. The program GEOM [19.8] is used to evaluate the geometric properties of the multibody structural components that are required by the nonlinear finite element method implemented in the MBOSS program. In its present implementation the geometric properties are the cross-section characteristics of thin-walled beams used in mechanical construction and the corresponding integration points for the finite element formulation. With all information necessary generated the dynamic analysis of the system undergoing impact is performed by MBOSS for a specified period of time and an output file is generated. The second interface program is then invoked to analyze this output file and construct an input file for the ADS optimization code. The process continues until an optimum for the problem is reached.

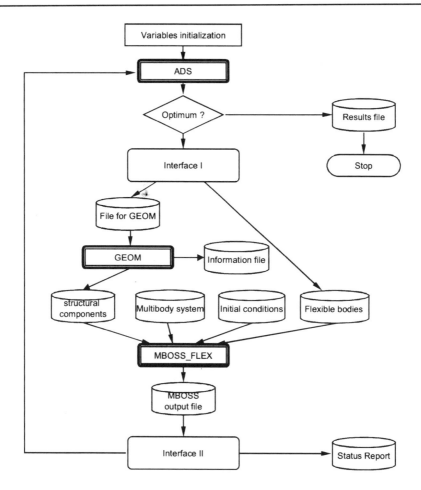

Figure 19.1   Optimization solution strategy

19.2.3 Application example

The methodology presented is tested in the optimization of the crashworthy characteristics of a crash-box typically used in automobiles. The original structure, proposed by Ni [19.9], and presented in Figure 19.2, is used as the starting point for the optimization. In this example two different objective functions are tested in order to demonstrate the methodology described. This structure is modeled using 10 nonlinear beam elements, for each side. It is made of A36 steel that has the following material properties: Young modulus E=210 Gpa; Poisson ratio v=0.3; mass density ρ=7912 Kg/m^3; Yield stress σ_y=248 Mpa; Plastic modulus E_T=E/10.

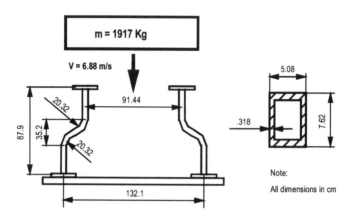

Figure 19.2   Crash-box structure

Two different objective functions are considered for two optimization problems. In the first problem the objective is to minimize the maximum acceleration of the points of the structure where the mass is lumped. For the second problem the objective is to minimize the Severity Index, i.e., an integral measure of the acceleration. In both problems the design variables are the thickness e, the height a and the width b of the cross-section. In both optimization problem the constraints used include a limit of 0.25 m for the displacement of the frontal node of the crash-box relative to the node fixed to the ground and technological constraints imposed in the beam cross-sectional dimensions, i.e., $0<e<40$ mm, $0<a<100$ mm and $0<b<100$ mm. The results of the optimization problems are synthesized in table 19.1.

Table 19.1   Results of the optimization of the crash-box

	Initial	Optimum	
Optimization problem	-	I	II
a (mm)	47.625	41.910	41.371
b (mm)	73.025	65.649	68.208
e (mm)	3.175	2.845	2.808
Objective function (m/s^2)	90.170	63.280	
Objective function (SI)	243.70		173.48
Constraint violation (m)	$-0.270 \ 10^{-2}$	0.00	0.00
Number of iterations	-	8	12

In both cases a reduction of about 30% of the objective function is achieved. The constraints set for the problem are completely used also. The evolutions of the objective function and displacement constraint are shown in Figure 19.3, for the optimization case where the maximum acceleration is minimized. For the second optimization case the evolution of the objective function and constraints is similar to that of the first case. Comparing the results of these two optimization problems it is observed that both objective function provide similar optimal designs.

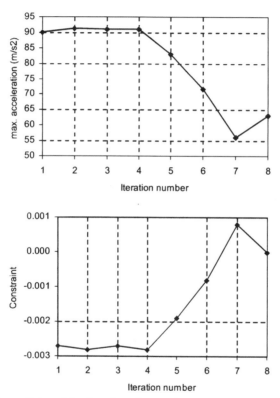

Figure 19.3 Evolution of the objective function and constraint for the minimization of the maximum acceleration.

It is visible for the evolutions of the objective function and constraint that in the 7th iteration the maximum acceleration is a minimum but the constraint is violated. The optimization proceeds by preventing the constraint violation and re-evaluating the objective function.

19.3   Design of the end underframe of a railway vehicle

The methodology is applied to the design of a railway vehicle end underframe. The objective is to get a design that ensures a better energy management of the train structural components in the event of a train crash. Resulting from statistic evaluation of train accidents [19.10] it is observed that most of the occupant injuries occur when the train speeds are below 30 Km h^{-1}. Consequently, in the analysis that supports the end underframe design optimization, the complete model of the railway vehicle has an initial velocity of 30 Km h^{-1}.

19.3.1 Model of the railway vehicle

The model of the train carbody for the computer simulations is composed of 13 moving bodies and 20 sets of spring-dampers for the suspension systems. The interaction between wheels and rails is modeled by nonlinear spring-damper elements and high friction coefficients. It must be noted that this is not suitable for high speed operating conditions, because the wheel/track interaction model is not exact. However, for operating speeds in the order of 30 Km/h, this model is acceptable. The train data, supplied by the manufacturer, corresponds to the main line carbody series B^{SYF} 1401 with 39 ton., depicted in Figure 19.4. For the computer model, the system components are described in table 19.2.

Table 19.2   Description of train carbody moving components

Body Number	Description
1	Complete carbody without boggies
2	Front plate between carbody and secondary suspension
3	Front boggie chassis
4	Front right wheel (front boggie)
5	Rear right wheel (front boggie)
6	Rear plate between carbody and secondary suspension
7	Rear boggie chassis
8	Front right wheel (rear boggie)
9	Rear right wheel (rear boggie)
10	Front left wheel (front boggie)
11	Rear left wheel (front boggie)
12	Front left wheel (rear boggie)
13	Rear left wheel (rear boggie)

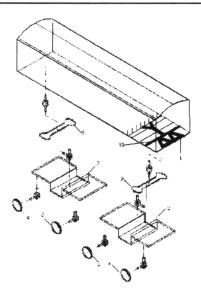

Figure 19.4 Representation of the mechanical elements that are included in the railway vehicle model

19.3.2 Initial design of the end underframe

The schematic model of the vehicle presented in Figure 19.4 shows the basic modeling assumptions made. The main body is considered to be rigid, except for the extremities that are flexible. This assumption restricts the modeling of crashworthy components to the region of end underframe. All other bodies of the system are considered to be rigid. The structure of the end underframe is made of stainless steel OX602 with a yield stress of σ_y=510 MPa. The finite element model, represented in Figure 19.5(a), is made of 45 nonlinear and uses thin-walled beams with the cross section shown in Figure 19.5(b).

The finite element structure represented is the flexible part of the carbody and it is fixed to the rigid part of the vehicle. The mesh of the end underframe reflects its expected behavior in a crash environment, i.e., the deformation energy is obtained essentially by crushing of the region of the component that is located around the V substructure. For optimization procedures where the design variables are associated to the geometric location of geometric features of the end underframe the re-meshing of the finite element model is a function of the design variables.

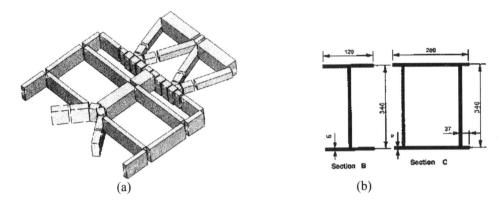

(a)   (b)

Figure 19.5   Representation of the end underframe: (a) finite element mesh; (b) Typical beam
cross sections used in the construction

19.3.3 Design optimization

For the optimization of the end underframe two different re-analysis are
carried with the same objective function and constraints but with different sets
of design variables. For the first optimization problem the topology of the
structure is maintained constant and the geometries of the cross sections,
represented in Figure 19.5(b), are used as design variables. For the second
optimization problem the location of the V shaped substructure is used as
design variable, as depicted by Figure 19.6. The location of this feature affects
directly the deformation mechanism of the end underframe. The cross section
geometry of the beams are kept constant in this case, being their dimensions
similar to those obtained in the first optimization problem.

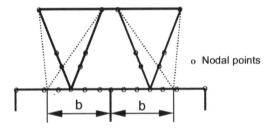

Figure 19.6   The location of the 'V' feature of the end-underframe is used as design variable
in the optimization process

In both optimization problems the objective function is the minimization of the maximum acceleration of the passenger compartment during the crash. It has been seen that the use of this discrete objective function provides results similar to those obtained by using an integral objective function such as an injury measure. In both optimization steps the maximum deformation of the structure of 0.560 m is used as constraint. Table 19.3 synthesizes the results of the optimization problems. The optimization procedure used for all cases has been the modified method of feasible directions [19.6].

Table 19.3   Optimization of the end underframe

Optimization problem	Case 1		Case 2	
	Initial	Optimum	Initial	Optimum
Height Section A (m)	0.350	0.345	0.340	-
Height Section B (m)	0.350	0.346	0.340	-
b (m)	0.500	-	0.300	0.454
Objective function (m/s^2)	106.14	86.46	97.82	80.449
Constraint violation (m)	$-3.60\ 10^{-2}$	$-2.80\ 10^{-2}$	$3.90\ 10^{-2}$	$-3.00\ 10^{-2}$
Number of iterations		42		17

In both cases, a reduction of about 20% of the maximum acceleration is observed. Due to the initial conditions specified in the second optimization problem the constraint starts by being violated. During the optimization the system is redesigned and the constraint is verified, making the design feasible. The evolution of the objective function and design variable of the second optimization problem are shown in Figure 19.7.

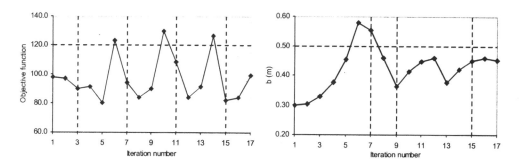

Figure 19.7   Evolution of the (a) Objective function, (b) Design variable

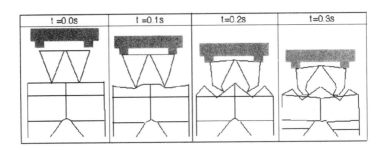

Figure 19.8 Predicted mechanism of deformation of the end underframe

The evolution of the pattern of deformation of the end underframe for the final design is depicted in Figure 19.8. It is observed that the deformation proceeds in a similar fashion for all the different designs attempted during the optimization process with relatively small variations of the maximum load that the structure can carry.

19.4 Optimization procedures using analytical sensitivities

19.4.1 Design Sensitivity Analysis

The general design problem has been stated in equations (19.1) and (19.2). There, the objective function and the constraint functions are functions of the nb design variables $\mathbf{b} = [b_1, b_2, ..., b_{nb}]^T$. The design variables vector $\mathbf{b}$ is also function of the state variables which include positions, velocities, accelerations and Lagrange multipliers, thus

$$\mathbf{\Psi} = [\mathbf{F}, \mathbf{G}, \mathbf{H}]^T = \mathbf{\Psi}(\mathbf{q}, \mathbf{v}, \dot{\mathbf{v}}, \lambda, t, \mathbf{b}) \qquad (19.6)$$

In the optimization process the cost and constraint functionals are evaluated as well as the corresponding gradients with respect to the design variables

$$F_{\mathbf{b}} = \frac{\partial F}{\partial \mathbf{b}} \quad \text{and} \quad \mathbf{G}_{\mathbf{b}} = \frac{\partial \mathbf{G}}{\partial \mathbf{b}} \; ; \; \mathbf{H}_{\mathbf{b}} = \frac{\partial \mathbf{H}}{\partial \mathbf{b}} \qquad (19.7)$$

A sensitivity matrix $\mathbf{L}$, can be defined such that

$$\partial \mathbf{\Psi} = \left[\frac{\partial F^T}{\partial \mathbf{b}} \quad \frac{\partial \mathbf{G}^T}{\partial \mathbf{b}} \quad \frac{\partial \mathbf{H}^T}{\partial \mathbf{b}} \right]^T \delta \mathbf{b} = \mathbf{L}^T \delta \mathbf{b} \qquad (19.8)$$

Using the direct differentiation method [19.11], a first step for the design sensitivity analysis requires the linearization of the functionals (19.3) and (19.4). Without loss of generality consider the particular case constraints

$$\delta \Psi = (f_q \delta q + f_v \delta v + f_{\dot{v}} \delta \dot{v} + f_\lambda \delta \lambda + f_b \delta b) \Big|_{t=t^j} \qquad (19.9)$$

The elimination of the dependency on the state variables is possible using the first variations

$$\delta q = q_b \delta b \qquad (19.10)$$
$$\delta v = v_b \delta b \qquad (19.11)$$
$$\delta \dot{v} = \dot{v}_b \delta b \qquad (19.12)$$
$$\delta \lambda = \lambda_b \delta b \qquad (19.13)$$

In design sensitivity analysis the kinematic constraint equations (16.11) are also functions of the design variables, thus

$$\Phi(q, t, b) = 0 \qquad (19.14)$$

Calculating the first variation and taking into account (19.9)

$$D q_b = -\Phi_b \qquad (19.15)$$

where D is the Jacobian matrix. Differentiating equation (19.15) twice with respect to time leads to

$$D v_b = -(Dv)_q q_b - \dot{\Phi}_b \qquad (19.16)$$
$$D \dot{v}_b = -\ddot{D} q_b - 2\dot{D} v_b - \ddot{\Phi}_b \qquad (19.17)$$

The generalized mass matrix and the generalized force vector are also functions of the design variables, thus

$$M = M(q, b) \qquad (19.18)$$

and

$$g = g(q, v, t, b) \qquad (19.19)$$

The first variation of the system dynamic equations of motion (16.25) is

$$M v_b + D^T \lambda_b = g_q q_b + g_v v_b + g_b - (Mv)_q q_b$$
$$- (Mv)_b - (D^T \lambda)_q q_b + (D^T \lambda)_b \qquad (19.20)$$

Equations (19.17) and (19.20) can be written in matrix form as

$$
\begin{bmatrix} \mathbf{M} & \mathbf{D}^{\mathrm{T}} \\ \mathbf{D}^{\mathrm{T}} & 0 \end{bmatrix} \begin{Bmatrix} \mathbf{v}_b \\ \lambda_b \end{Bmatrix} =
$$

$$
\begin{Bmatrix} \mathbf{g}_q \mathbf{q}_b + \mathbf{g}_v \mathbf{v}_b + \mathbf{g}_b - (\mathbf{Mv})_q \mathbf{q}_b - (\mathbf{Mv})_b - (\mathbf{D}^{\mathrm{T}}\lambda)_q \mathbf{q}_b + (\mathbf{D}^{\mathrm{T}}\lambda)_b \\ -\mathbf{Dq}_b - 2\mathbf{Dv}_b - \Phi_b \end{Bmatrix} \tag{19.21}
$$

which can be integrated in time with the initial conditions

$$
\begin{cases} \mathbf{q}_b(t^0) = \mathbf{q}_b^0 \\ \mathbf{v}_b(t^0) = \mathbf{v}_b^0 \end{cases} \tag{19.22}
$$

Since the equations of motion (16.26) have the same left hand side as equations (19.21), they can be solved simultaneously to obtain the state variables $\mathbf{q}$, $\mathbf{v}$, $\dot{\mathbf{v}}$, λ and the sensitivities $\mathbf{q}_b$, $\mathbf{v}_b$, $\dot{\mathbf{v}}_b$, λ_b. This allows a greater efficiency in terms of computational efforts.

19.4.2 *Application example - optimization of an end underframe*

The end underframe model of Figure 19.9 indicates the members that are to be optimized in their cross section dimensions which in turn are used as design variables. Due to symmetry, only the left side of this model was considered in the simulation of the frontal impact against a rigid wall. The structure is modeled by rigid bodies connected by plastic hinges located either in the zones of geometric discontinuities, or at points of concentrated loading or at points where the stresses are large. The multibody dynamics model is composed of 16 bodies and 16 plastic hinges. The buffers are represented by special nonlinear translational springs acting only in the compression phase. Interior contact between bodies and the contact of the structure with the rigid wall, is represented with a contact model based on the Hertz contact law, proposed by Lankarani [19.12].

Two colliding velocities have been considered in the simulation: V=30 Km/h (8.33 m/s) and V=50 Km/h (13.833 m/s). Plastic hinges are modeled by nonlinear torsional springs mounted over revolute joints. The nonlinear response of the torsional springs are described in terms of a quasi-static

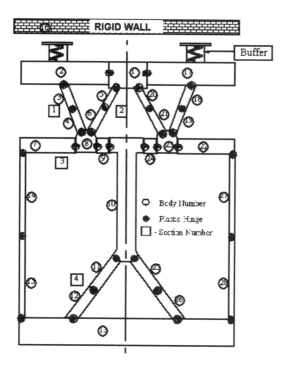

Figure 19.9   End-underframe model using plastic hinges

moment-angle relationship based on the curves for elastic-plastic behavior, where the bending stiffness of the elastic part is obtained from the elementary beam theory[19.13] as

$$k = \frac{EI}{l} \tag{19.23}$$

where E is the modulus of elasticity, I is the cross sectional moment of inertia and l is assumed to be the flange width. The limit plastic moment is obtained for each section. For sections 1 to 4, shown in figure 19.10, IPE profiles (European Profile) have been considered in the optimization process. In this case the plastic moment is calculated as

$$M_p = \sigma_y \left[\frac{1}{2} e \ell^2 + \frac{1}{4} a^2 (H - 2e) \right] \tag{19.24}$$

where σ_y is the yielding stress, and the other constants are the cross sections dimensions. The material is a cold-rolled steel with $\sigma_y = 500$ MPa.

In the present case, the design criterion for the crashworthiness has been based only on the structural response predicted by the crashworthiness analysis for each design problem. Consequently, an objective function is defined as the acceleration peak experienced at the end underframe in the compartment zone (body 13). Because the acceleration levels are related with human injury tolerance levels, its reduction will reduce the potential for human injuries.

A set of design variables has been considered corresponding to the different beam heights, H, of the cross sections 1 to 4. The other geometric parameters e, a, and ℓ, can be obtained as functions of H, considering the range of cross sections commercially available.

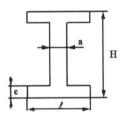

Figure 19.10 Beam cross sections

Longitudinal constraints are also defined to impose limits on the maximum structural crash deformation to avoid failure of the cross sections and overlapping of the structural components.

The objective function and constraints are summarized as:

Minimize $\quad |\ddot{y}_{13}(\mathbf{b},t)_{13}|$

Subject to: $\quad$ max. $\left|(y_{13} - y_1) - (y_{13}^0 - y_1^0)\right| - X_1 \leq 0$

$\qquad\qquad$ max. $\left|(y_{13} - y_{10}) - (y_{13}^0 - y_{10}^0)\right| - X_2 \leq 0$

The subscripts indicate the body number, y is the displacement of the body in the O_y direction and the superscript indicates the initial position. The maximum deformations allowed are respectively, $X_1 = 0.937$ m and $X_2 = 0.254$ m. The purpose of the second constraint is to avoid the overlapping of the structure members in the zone of sections 1, 2 and 3.

Four design variables are considered corresponding to the sections represented in Figure 19.10, as $\mathbf{b} = [H_1, H_2, H_3, H_4]$. The initial design vector is $\mathbf{b} = [300, 300, 400, 500]$ (mm) and the lower and upper bounds are, $\mathbf{b}_l = [60, 60, 60, 60]$ (mm) and $\mathbf{b}_u = [600, 600, 600, 600]$ (mm). The lower and upper vectors correspond respectively, to the minimum and maximum depth of the sections available commercially.

Table 19.4   Crashworthiness Design of The End Underframe of a Train

	Crash Situation			
Velocity	30 Km/h		50 Km/h	
	Initial	Optimal	Initial	Optimal
Objective Function (G)	27.98	17.79	130.8	31.18
Constraint 1 (m)	-3.77×10^{-1}	1.71×10^{-4}	1.64×10^{-1}	-5.7×10^{-3}
Constraint 2 (m)	-3.11×10^{-2}	-8.04×10^{-4}	-2.02×10^{-2}	-2.2×10^{-2}
$b_1 = H_1$ (mm)	300.0	99.3	300.0	352.6
$b_2 = H_2$ (mm)	300.0	82.0	300.0	319.0
$b_3 = H_3$ (mm)	400.0	120.9	400.0	469.0
$b_4 = H_4$ (mm)	500.0	459.0	500.0	600.0
CPU Time (s)[*]	6 h 46 m		10 h 4 m	

[*]Workstation HP 735

Table 19.4 shows the results obtained for the two colliding velocities. For the impact at 30 Km/h, the results for deformation and acceleration are presented in Figures 19.11 and 19.12, respectively. These results show that a decrease in the maximum acceleration is achieved with a larger deformation. This is obtained globally with a reduction of the sections size.

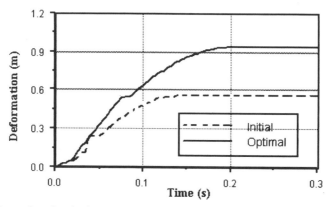

Figure 19.11 Deformation for the impact at V=30 Km/h

The results of the displacement and acceleration for the impact at V=50 Km/h, are shown in Figures 19.13 and 19.14 respectively. In this case the global deformation constraint (constraint 1) is violated in the initial design.

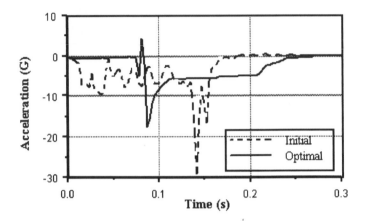

Figure 19.12 Acceleration for the impact at V=30 Km/h

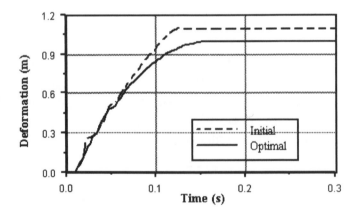

Figure 19.13 Deformation for the impact at V=50 Km/h

For this simulation the initial design yields a very large deformation of the structure and a very strong collision of the lower extremity of central beam (body 10) with the lower beam (body 13), resulting in a large acceleration peak. The optimal design is associated with an increase of the dimensions of all sections, corresponding to higher levels of energy absorption in sections 1, 2 and 3.

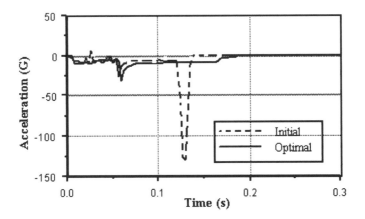

Figure 19.14 Acceleration for the impact at V=50 Km/h

19.5 Detailed analysis of the final design

The model of the structure of the end underframe obtained in the optimization procedure is a conceptual design. During the subsequent phase of the design process not only some of the results obtained in the optimized conceptual design have to be adjusted but also some further operating conditions, not considered in the optimization process, have to be verified. To ensure the reasonability of the final design a detailed analysis of the end underframe is performed using a complete finite element model in a nonlinear commercial finite element code. The methodology used in this verification step is completely independent from the procedure used in the optimization phase. Therefore, the results can be understood as a validation phase.

Due to the symmetry of the end underframe only half of the structure is modeled. The boundary conditions are set in order to account for this approach. Three-dimensional triangular and quadrangular plate/shell elements are used in the model shown in Figure 19.15(a).

The operating conditions for the end underframe impose that no plastic deformation can occur for two loading cases. In the first loading case, corresponding to the normal operating conditions of the train, a traction load of 1000 KN is applied to the center of the front of the underframe. The second loading case, representing the loads developed when a train is assembled or the braking of the train during its operation, corresponds to the application of a compressive load of 1000 KN in the buffer locations. The results, summarized respectively in Figures 19.15(b) and 19.15(c) for the two loading cases described, show that the maximum stresses are below the 510 MPa yield stress of the end underframe material.

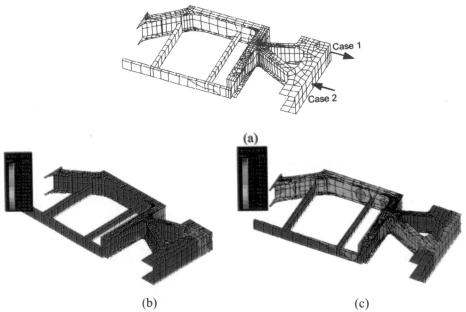

Figure 19.15 Finite element model: (a) loading conditions; (b) Von-Mises equivalent stresses for traction loading; (c) Von-Mises equivalent stresses for compressive loading.

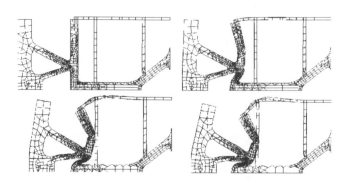

Figure 19.16 Pattern of deformation for the end underframe predicted by the finite element model

The elastic-plastic structural response of the end underframe carried on by imposing the displacement of the front nodes of the structure by a moving plane which corresponds to the compressive loading that develops during the crash events. In Figure 19.16 the kinematics of the end underframe deformation is presented.   In Figure 19.17 the comparison between the predicted deformation and the result of an experimental dynamic crash test [19.10] performed in a railway vehicle is presented.   It is observed here that the experimental deformation measured matches very closely the predicted results.

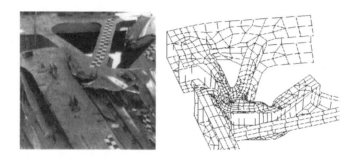

Figure 19.17 Comparison between the final deformation observed in the experimental test and predicted

In the actual design of the end underframe the structure has to resist wide variety of loads that are present during the normal operation of the vehicle without any plastic deformations.   The loads, compressive or traction, are established by the standardization agencies that regulate the railway operations. In a more sophisticated optimization procedure for this structure the ability to sustain such operating loads without plastic deformation can be modeled as design constraints.

REFERENCES

[19.1]   Bennett, J.A., Lin, K.H. and Nelson, M.F., The application of optimization techniques to problems of automobile crashworthiness, *SAE Transactions*, **86**, 2255-2262, 1976.

[19.2]   Song, J.O., An optimization method for crashworthiness design, In *Proceedings of the 6th Int. Conference on Vehicle Structural Mechanics*, Detroit, Michigan, 39-46, 1986.

[19.3]   Lust, R., Structural optimization with crashworthiness constraints, Structural Optimization, **4**, 85-89, 1992.

[19.4]   Dias, J. and Pereira, M.S., Design for vehicle crashworthiness using multibody dynamics, Int. *J. of Vehicle Design*, **15**(3/4), 563-577, 1994.

[19.5]   Vanderplaats, G., *ADS-A Fortran Program For Automated Design Synthesis Vs. 2.01*, Eng. Design Optimization Inc., California, 1987.

[19.6]   Vanderplaats, G. and Moses, F., Structural optimization by methods of feasible directions, *J. Comp. and Structures*, **3**, 739-755, 1973.

[19.7]   Arora, J.S., *Introduction To Optimum Design*, McGraw-Hill, New York, New York, 1989.

[19.8]   M S Pereira, J B Cardoso and J Ambrósio, *GEOM A Program For The Evaluation Of The Cross Section Geometric Characteristics Of Thin Walled Beams Vs. 3.*', IDMEC, Instituto Superior Técnico, Lisboa, Portugal, 1994.

[19.9]   C.-M. Ni, Impact response of curved box beam columns with large global and local deformations, *AIAA/ASME/SAE 14th Structures, Structural Dynamics and Material Conference*, March 20-22, Williamsburg, Virginia, 1973.

[19.10] SOREFAME, *Dynamic Crash Tests: Dynamic Test of a Vehicle Against an Obstacle*, Technical Report n. TRAINCOL/T7.2-F, SOREFAME, Portugal, 1993.

[19.11] Chang, C. O. and Nikravesh, P. E., Optimal Design of Mechanical Systems with Constraint Violation Stabilization Method, *ASME J. Mech., Trans., and Auto. in Design*, **107**, 493-498, 1985.

[19.12] Lankarani, H.M. and Nikravesh, P.E., Continuous contact force models for impact analysis in multibody systems, *Nonlinear Dynamics*, **5**, 193-207, 1994

[19.13] Shigley, J.E., and Mischke, C.R., *Mechanical Engineering Design*, 5th Edition, McGraw Hill, New York, 1989.

Part V

CRASHWORTHINESS BIOMECHANICS *

J. Wismans
TNO Automotive, Delft, The Netherlands

20.   INTRODUCTION IN INJURY BIOMECHANICS

20.1   Introduction

Injury of the human body can be caused by different loading situations including mechanical, chemical, thermic and electric load. The field of injury biomechanics deals with the effect on the human body of mechanical loads, in particularly impact loads. Therefore also often the term "impact biomechanics" is used instead of "injury biomechanics". Viano [20.1] defines the objectives and research methods of injury biomechanics as follows:

"The broad goal of injury biomechanics research is to understand the injury process and to develop ways to reduce or eliminate the structural and functional damage that can occur in an impact environment". "To achieve this goal, researchers must identify and define the mechanisms of impact injury, quantify the responses of body tissues and systems to a range of impact conditions, determine the level of response at which the tissues or systems will fail to recover, develop protective materials and structures that reduce the level of impact energy and force delivered to the body, and develop test devices and computer models that respond to impact in a human like manner, so that protective systems can be accurately evaluated"

Note that above description relates to impact type of loading conditions. But usually the response of biological tissue under *static* injury producing loading conditions is considered to be subject of injury biomechanics research as well.

In this Chapter, first a short discussion on the use of human body models in injury biomechanics research will be presented (Section 20.2).

Human body models are used to study the dynamical response of the human body under extreme loading conditions and more specifically, to determine injury mechanisms and injury tolerances. Furthermore human body models are applied for the assessment of the protection offered by safety provisions. Five types of models are distinguished: human volunteers, human cadavers, living and dead animals, mechanical (crash dummies) and mathematical models.

In Section 20.3. the load/injury model will be introduced, which can be of help in understanding and interpreting the methodology and terminology used in the field of injury biomechanics research. Some basic relevant anatomy and injury terminology used in this course is summarized in Appendix A.

An important aspect of injury biomechanics and trauma research is the classification of the injury severity. Section 20.4 deals with this subject, where various methods to assess the severity of injury will be introduced, including the Abbreviated Injury Scale (AIS) and scales which prescribe the severity of an injury in terms of societal costs.

Assessment of the (statistical) relationship between injury severity and injury criteria will be discussed in Section 20.5. The so-called injury risk function will be introduced which describes the probability of a certain injury or a fatality as function of the loading in an accident or a laboratory experiment.

The last Section of this Chapter discusses the tolerance of the whole human body in an impact environment. Injury mechanisms and tolerances of individual body segments will be presented in separate Chapters. Attention will be given to injury biomechanics of the head (Chapter 23), the spine (Chapter 24) and the thorax (Chapter 25).

20.2   Models in injury biomechanics

20.2.1 Introduction

In order to protect the human body against injuries in case of extreme loading conditions, a clear insight in the ways injuries arises and into loads at which they occur is needed. Accident analysis studies and biomechanical research are carried out for this purpose. In injury biomechanics research several types of models for the human being are used to study biomechanical response of the human body and to study injury mechanisms and tolerances. Also for the assessment of the protection offered by safety provisions, like seat belts and crushable vehicle structures, models for the human being are needed. In fact five types of human models can be distinguished: human volunteers, human cadavers, animals, mechanical models and mathematical models. In the next Sections the capabilities and limitations of the various models will be discussed.

20.2.2 Human volunteers

For obvious reasons it is impossible to experiment with human beings fitted with instrumentation under injury producing conditions. Only in low severity tests, i.e. below the pain thresholds, sometimes human volunteers are used. Such tests, which are limited by rigid regulations and guidelines, can contribute to general knowledge on the human body non-injurious response. The results are very important for the development of mechanical or mathematical models. The human subjects are mostly young, well-trained, military volunteers (Figure 20.1). Their pain tolerance is usually much higher than that of the general population. Therefore the test results are not representative of females, children and elderly people.

An advantage of the use of human volunteers is that the effect of muscle tone and pre-bracing on the dynamical response can be studied. But this influence, which might be relatively large at low impact levels, cannot be simply extrapolated to higher impact levels.

The first well-documented human volunteer test was conducted in 1954 in the dessert of New Mexico at Holloman Air Force Base. Test subject was Colonel John Paul Stapp who sustained, without serious complaints, a velocity change of 1000 km/h during 1.4 sec on a rocket propelled sled device. The maximum deceleration was 40g, in other words 40 times the deceleration of gravity. For comparison: the average deceleration during a 50 km/h vehicle crash test against a brick wall is about 25g. And this deceleration usually doesn't take longer than 0.1 sec.

Well-known for its research using human volunteers is the Naval Biodynamics Laboratory in New Orleans. One of the objectives of their tests is the analysis of head-neck motions. In Chapter 24 some of the Naval Biodynamics work will be presented.

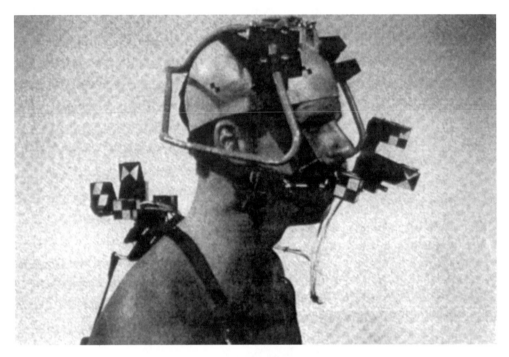

Figure 20.1   Human volunteer test at the Naval Biodynamics Laboratory in New Orleans

20.2.3 Human cadavers

The primary research tool to evaluate injurious biomechanical response is a human cadaver (sometimes referred to as PMHS: post-mortem human subject). The anthropometry of a human cadaver is almost identical to the living human and material properties are often close to living tissue. This latter aspect however is strongly dependent on the preparation techniques applied and the time duration since death. Very often the lungs of the cadavers are inflated during the test and the blood pressure is partly restored by infusion. Disadvantages of the use of human cadavers are the absence of muscle tone and that physiologic responses cannot be determined. Furthermore the age of cadavers is often high and since the mechanical strength of most tissues in the human body tends to decrease with age, the data obtained are not necessarily representative of the general population.

20.2.4 Animals

Research using anaesthetized animals as human surrogates, is vital to obtain information on physiologic responses, in injury producing loading conditions, for specific body areas like the brains and the spinal cord. Furthermore tests with animals can provide insight in the differences between dead and living surrogates and as such provide the information for correct interpretation of human cadaver testing. Due to differences in size, shape and also structural differences between humans and animals, quantitative scaling and extrapolation of results of animal testing to the human is very difficult.

20.2.5 Mechanical models

Mechanical models or crash dummies (sometimes also referred to as anthropomorphic test devices) normally consist of a metal or plastic skeleton, including joints, covered by a flesh-simulating plastic or foam. They are constructed such that dimensions, masses and mass-distribution, and therefore the kinematics in a crash are human like. The dummy is fitted with instrumentation to measure accelerations, forces and deflections during the tests that correlate with injury criteria for human beings. Dummies are often used in approval tests on vehicles or safety devices, in which the measured values should remain below certain (human tolerance) levels. Very important in this respect is a repeatable response of the dummy in identical tests. Furthermore crash dummies sometimes are used for the reconstruction of real accidents in order to study the correlation between real world injuries and actual loading conditions.

20.2.6 Mathematical models

Mathematical models can also simulate the behaviour of human beings in crashes. Together with a mathematical description of the environment (e.g. steering-wheel, dashboard, seat and belts) and the impact conditions (e.g. vehicle deceleration), the model provides a numerical description of the crash event. Like mechanical surrogates, mathematical models can be used to study the protection offered by safety devices and for accident reconstructions. Furthermore they can be used to analyze biomechanical experiments and to quantify mechanical parameters in the experiment which cannot be assessed experimentally. Another potential of computer models is the use in the design of vehicles and restraint systems as a computer aided design tool. They offer a very economical means of system optimisation. Finally mathematical models can be used to scale results of animal tests.

An important limitation of a computer model is that the accuracy and reliability of the model strongly depends on the (biomechanical) information available on the system(s) to be modeled as well as the assumptions made in the model formulation. Therefore detailed experimental verification is often a necessity. Chapter 21 deals with mathematical models.

20.3   The load-injury model

20.3.1 Introduction

Figure 20.2 schematically presents the accident/injury process in case of mechanical (over)loading and the various parameters affecting the outcome of this process. This schematic representation will be called the load/injury model. The model will be used here to define and explain some of the terminology used in injury biomechanics research.

The load/injury model shows that, as a consequence of an accident, the human body is exposed to a mechanical load condition. Through *injury control measures*, as discussed in this chapter, the magnitude of the load acting on the body can be reduced.

In the following Sections the reaction of the human body to the mechanical load will be discussed. First the so-called biomechanical response will be described, then the injury mechanisms and finally definitions like injury severity, injury criteria and injury tolerances will be introduced.

20.3.2 Biomechanical response

Due to the mechanical load a body region will experience mechanical and physiological changes: the so-called *biomechanical response*. Biomechanical response will be defined here as any change in time of the position and shape (due to deformation) of the human body, a body region or tissue and any physiological changes related to these mechanical changes. Examples of biomechanical response are the motion of the head-neck system in an accident, the brain motions and deformations in case of a head impact or thorax deformation in case of an impact with the steering wheel.

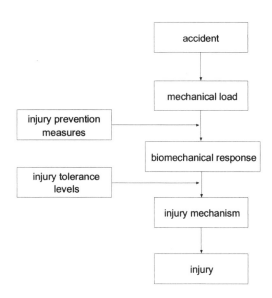

Figure 20.2   The load/injury model.

Examples of physiological changes are changes in reflexes, dizziness and headache. It is important to realize here that biomechanical response not necessarily has to result in injury. Human volunteers e.g. are tested in impact biomechanics research below pain levels as was discussed in Section 20.2.2. The information of such tests can be very important to develop mechanical or mathematical models of the human body.

20.3.3 Injury mechanisms

Injury will take place if the biomechanical response is of such a nature that the biological system deforms beyond a recoverable limit, resulting in damage to anatomical structures and alteration in normal function. The mechanism involved is called *injury mechanism*. Identification and understanding of injury mechanisms is one of the main objectives of research in the field of injury biomechanics.

Usually a distinction is made between penetrating and non-penetrating injuries. Penetrating injuries are caused by high-speed projectiles, such as bullets, or by sharp objects moving with a lower velocity such as knives. In this type of injury the (impact) energy is concentrated on a small area and injury will occur if local stresses exceed failure levels.

Non-penetrating injuries are more complicated and the corresponding injury mechanisms are not always easy to understand. Loads are usually acting on a larger contact area caused by contacts with blunt objects. Inertia, elastic and visco-elastic aspects will effect the load distribution within the body region involved. Bones like the ribs and the skull protect vital organs by absorbing some of the impact energy. Damage of an anatomical structure will occur if tensile, compressive or shear stresses exceed a tolerable level.

Three principal injury mechanisms are usually distinguished, as is illustrated in Figure 20.3 [20.2]:

- Compression of the body causing injury if elastic tolerances are exceeded. Injury can occur in case of slow deformation of the body (crushing) as well as due to high velocity impacts.

- Impulsive (shock) type of loading causing shock waves in the body, which results in internal injuries, if viscous tolerances are exceeded. Even without significant outside deformation of the body this type of injury is possible, for example in the chest cavity in case of an impact on the sternum.

- Acceleration type of loading causing tearing of internal structures due to inertia effects. In case of brain injuries, for example, this mechanism plays an important role.

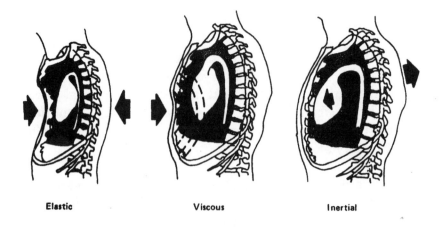

Elastic   Viscous   Inertial

Figure 20.3   Three principle injury mechanisms [20.2].

Above mechanisms also can occur in combination with each other depending on the loading situation. Increase of the load levels often will lead to more severe injuries while the same injury mechanism takes place. However, higher loads may also result in different mechanisms and as a consequence will show a different injury outcome.

20.3.4 Injury severity, criteria and tolerances

The severity of the resulting injury is indicated by the expression *injury severity*. It is defined as the magnitude of changes, in terms of physiological alterations and/or structural failure, which occur in a living body as a consequence of mechanical violence [20.3]. Section 20.4 will deal in detail with injury severity and various methods to assess the injury severity level including the widely used Abbreviated Injury Scale (AIS).

The next term that will be introduced here is the *injury criterion*, which is defined as a physical parameter or a function of several physical parameters which correlates well with the injury severity of the body region under consideration. Frequently used parameters are those quantities that relatively easy can be determined in tests with human substitutes like the linear acceleration experienced by a body part, global forces or moments acting on the body or deflection of structures. Later in this course several examples of injury criteria will be given.

In conjunction with the injury criterion the term *tolerance level* (or injury criterion level) will be defined here as the magnitude of loading indicated by the threshold of the injury criterion which produces a specific type of injury severity. It should be noted that there are large variations in tolerance levels between individuals. Tolerance levels for populations can therefore only be determined statistically. Section 20.5 deals with this subject.

Above terminology refers here to the living human body as well as to biological substitutes like human cadavers or animals. In some literature (e.g. Ref. [20.3]) above terminology is solely reserved for the living human body since the injury mechanisms in substitutes might differ from those in the living human body. For the injury process in human substitutes a corresponding terminology is proposed like damage severity instead of injury severity and damage criteria instead of injury criteria. In this course we will not adopt this terminology since it is seldom applied in practice up to now.

20.3.5 Protection criteria and biofidelity

Also for non-biological substitutes, i.e. the mechanical and mathematical models the terms injury severity, injury criterion and tolerance level are applied although these substitutes are not actually injured. For mechanical models (dummies) rather than the term injury criterion however also sometimes the term *performance or protection criterion* is used. In these models, levels of biomechanical response are determined and judged as if the model were a living human body.

The correct interpretation of injury criteria determined at models and the "translation" into injury in the living human body is one of the challenges in injury biomechanics research. Important here is whether the model responds to impact in a human like manner, the so-called *biofidelity* of the model. The biofidelity of a model can be judged by comparing the response (i.e. motions, displacements etc..) with average data generated in test with human volunteers or cadavers. Response data from tests with biological substitutes used for the purpose of biofidelity judgement are often defined as *biofidelity performance requirements*.

20.4   Injury scaling

20.4.1 Introduction

Injury scaling is defined here as the numerical classification of the type and severity of an injury. Many schemes have been proposed for ranking and quantifying injuries. They can be grouped into three main types:

- *Anatomic scales* which describe the injury in terms of its anatomical location, the type of injury and its relative severity. These scales rate the injuries itself rather than the consequences of injuries. The most well-known scale, which is accepted worldwide, is the Abbreviated Injury Scale (AIS). Section 20.4.2 deals with anatomic scales.

- *Physiologic scales* which describe the physiological status of the patient based on the functional change due to injury. This status and consequently its numerical assignment, may change over the duration of the injury's treatment period, in contrast to anatomical scales where a single numerical value is assigned to a certain injury. A well-known example is the Glasgow Coma Scale (GCS) which was specifically developed for head injuries. It is a way of quickly assessing the nature and severity of brain injuries on the basis of three indicators: eye opening, verbal response and motor response. Physiological scales are particularly important in a clinical environment; they will not be further discussed here.

- *Impairment, disability and societal loss scales.* Here not the injury itself or the functional changes due to the injury are rated, but the long term consequences and in relation to this the "quality of life". Examples are the Injury Cost Scale ICS, the Injury Priority Rating, IPR and the HARM concept, which all are attempts to assign an economic value to the various injuries. These scales are discussed in 20.4.3.

Note that above types of injury scales basically relate to injuries in the living human body. Anatomical scales like the AIS however also can be applied to rate the injury severity in human cadavers after dissection of the body.

In the Netherlands none of the above injury scales are applied, except for the AIS which was used in the SWOV accident investigation mentioned earlier (see 21.3.5). In the LMR/SIG hospital data base (see 21.3.3) the ICD (International Classification of Diseases) is used, however this classification only codes the location and nature of the injury and not its severity.

20.4.2 Anatomical scales

20.4.2.1   The Abbreviated Injury Scale (AIS)

The need for a standardized system for injury severity rating arose in the mid nineteen-sixties in the USA with the first generation of multi-disciplinary motor vehicle crash investigation teams. In 1971 the first Abbreviated Injury Scale (AIS) was published and has since then been revised four times (1976, 1980, 1985 and recently in 1990). This last update [20.4] will be referred to as "AIS 90". Although originally intended for impact injuries in motor vehicle accidents, the several updates of the AIS allow its application now also for other injuries like burns and penetrating injuries (gun shots). The AIS distinguishes between the severities of injury described in Table 20.1.

Table 20.1   The Abbreviated Injury Score (AIS) [20.4].

AIS	Severity code
0	no injury
1	minor
2	moderate
3	serious
4	severe
5	critical
6	maximum injury (virtually unsurvivable)
9	unknown

The information for AIS scaling is contained in the AIS manual, which is organized into nine Sections dealing with several body regions (head, face, neck, thorax, abdomen and pelvic contents, spine, upper extremity, lower extremity, external and other). Within each Section, injury descriptions are provided by specific anatomical part. In case of the lower extremities for instance, these anatomical parts are: the whole area, vessels, nerves, skeletal-bones, skeletal-joints and muscles-tendons-ligaments. For each specific injury the manual provides a 7-digit coding, where the digit right of the decimal point is the AIS score. The other digits are used to specify body region, anatomic structure, type of injury etc. Table 20.2 illustrates examples of injuries in the AIS manual for several body regions.

The AIS is a so-called "threat to life" ranking. Higher AIS levels indicate an increased threat to life. The numerical values have no significance other than to designate order. They do not indicate relative magnitudes, in other words an AIS 2 level is not twice as severe as an AIS 1 level. Several attempts have been made to establish a quantitative relationship between the various AIS levels. One attempt is the calculation of a fatality rate for each AIS value. Table 20.3 summarizes the range of results of several studies [20.5]. Another attempt is the HARM concept [20.6], which assigns an average economic value to each of the AIS injuries. See Table 20.3 and Section 20.4.3 for more details.

Table 20.2 AIS examples by body region [20.5].

AIS	Head	Thorax	Abdomen and pelvic contents	Spine	Extremities and bony pelvis
1	headache or dizziness	single rib fracture	abdominal wall: superficial laceration	acute strain (no fracture or disl.)	toe fracture
2	unconscious less than 1 hour; linear fracture	2-3 rib fracture; sternum fracture	spleen kidney or liver: laceration or contusion	minor fracture without any cord involve-ment	tibia, pelvis or patella: simple fracture
3	unconscious 1-6 hours; depressed fracture	≥4 rib fracture; 2-3 rib fracture with hemoth. or pneumoth.	spleen or kid-ney: major laceration	ruptured disc with nerve root damage	knee disloca-tion; femur fracture
4	unconscious 6-24 hours: open fracture	≥4 rib fracture with hemoth. or pneumoth.; flail chest	liver: major laceration	incomplete cord syndrome	amputation or crush above knee; pelvis crush (closed)
5	unconscious more than 24 hours; large hematoma (100 cc)	aorta laceration (partial transection)	kidney, liver or colon rupture	quadriplegia	pelvis crush (open)

Table 20.3   AIS vs. fatality rate and vs. Economic Costs (HARM) [20.5].

Injury Severity AIS	Fatality rate (range %)	Costs (HARM) ($1000)
1	0.0	0.4
2	0.1-0.4	2.7
3	0.8-2.1	7.1
4	7.9-10.6	38.8
5	53.1-58.4	186.6
6	...	165.0

20.4.2.2   Multiple injury scales

The AIS does not assess the effect of multiple injuries in patients. One possibility is to take the highest AIS score for a certain body region as a measure for the overall injury severity: the M(aximum)AIS. The value of the MAIS in trauma research is considered limited due to its nonlinear relationship with the probability of death.

As mentioned before the AIS is not applied in the Netherlands. The only exception is the SWOV accident investigation conducted in the mid-seventies (see Chapter 21). As a measure for the injury severity of the complete body the O(verall)AIS was used which is comparable with the MAIS. Results of the AIS findings in this SWOV study for various accident configurations are summarized in Table 20.4.

A more general accepted approach for rating multiple injuries is the ISS or Injury Severity Score [20.4]. The ISS distinguishes 6 body regions: Head or Neck, Face, Chest, Abdominal or Pelvic contents, Extremities or Pelvic girdle and External, where external injuries include lacerations, abrasions, contusions and burns, independent of their location on the body surface. For each of these regions the most severe injury on the basis of the AIS code is determined. The ISS is the sum of the squares of the three largest AIS values. The maximum value for the ISS is 75 (three AIS 5 injuries). An AIS 6 automatically becomes an ISS 75 score according to the ISS specifications. The ISS has been shown to correlate quite well with the probability of death as is illustrated in Figure 20.4.

Table 20.4 Severity of injuries in SWOV accident investigation for some collision types, drivers only [20.7].

OAIS	Lateral (%)	Front (%)	Rear (%)	Rollovers (%)	Total (%)
1	18.0	22.4	21.4	18.0	18.6
2	10.9	16.1	3.0	14.1	11.2
3	4.1	4.4	0.7	5.3	3.1
4	1.0	0.7	0.4	0.3	0.5
5	0.3	0.2	-	-	0.2
6	2.2	2.7	-	1.9	1.5
Total	100% (2044 cases)	100% (2138 cases)	100% (743 cases)	100% (640 cases)	100% (8173 cases)

20.4.3 Injury cost scales

In case of injury severity scales which rate the long term consequences of injuries, numerical values are much more difficult to determine and the status of the accident victim has to be monitored during a long period. The interest in, and importance of these scales is growing since they provide a much better means of establishing priorities for injury prevention measures than the anatomical based scales since much more factors are taken into account and since injury severities usually are expressed in term of economic costs. Examples of these scales are the Injury Priority Rating, IPR, the HARM concept and the Injury Cost Scale ICS, which all are attempts to assign an economic value to the various injuries.

The HARM concept is an attempt to assign an average economic cost to an AIS value [20.6]. Table 20.3 illustrates this relationship. Since the AIS rates the injury severity itself and not the long term consequences a large variation in actual economic costs per injury within an AIS level will exist and therefore useful application of the average costs for each AIS level as rated by HARM is rather limited. The US government applies the HARM score for instance to calculate the effect of various governmental interventions in the field of vehicle safety.

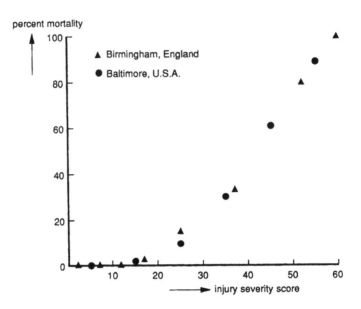

Figure 20.4 ISS as function of mortality rate [20.8].

A more reliable and detailed effort to weigh the various injuries in terms of economic costs, is the IPR [20.9]. The analysis, resulting in the IPR, was based on the NASS files discussed in Chapter 21. The information in the NASS files was first augmented with estimates for the impairment of the accident victims. Then the loss of expected lifetime earnings were estimated, had the victim not been injured. In this way for the various injuries within NASS "costs" values became available, the so-called Injury Priority Rating (IPR). The analysis presented in Ref. [20.9] dealt with the NASS data for 1980 and 1981. IPR's per injury were summed per body region. Results for each body region were expressed as a percentage of the total "costs" for all body regions (Table 20.5). Clearly the importance of head, face and neck injuries is illustrated: together more than 60%.

Since the NASS injury surveillance system only takes into account tow away motor vehicle accidents the IPR information only applies to occupants of motor vehicles. So lower leg injuries e.g. are underrepresented compared to the all over traffic accident situation. The last scale to be discussed here, i.e. the Injury Cost Scale (ICS) [20.10], is more general in this respect since it also considers other road user categories. Moreover the ICS is based on actual cost figures rather than estimates like the IPR.

Table 20.5   IPR Distribution by body region [20.9].

Body Region	Distribution (%)
Head	44.6
Face	10.5
Neck	5.1
Shoulder	0.3
Chest	18.9
Back	1.6
Abdomen	7.5
Pelvis	1.1
Thigh	2.1
Knee	1.6
Lower leg	1.0
Ankle/Foot	0.6
Lower Limb	0.0
Upper Arm	1.3
Elbow	0.5
Forearm	1.3
Wrist/Hand	0.4
Upper Limb	0.3
Whole Body	0.9
Unknown	0.2
TOTAL	100.0

The ICS is based on data analysis on road accidents of the working population in Germany, available in a data base of the German Workman's Compensation (an insurance type of data base). The analysis concerns accidents in 1985 with 1026 traffic fatalities and 15.407 injured persons. For each type of injury the average costs were determined taking into account costs of medical treatment and rehabilitation, loss of income, disability benefits etc. Also the societal costs due to a fatality were estimated. This analysis resulted in two injury cost scales: the ICS and the ICSL. The difference between both scales is that the ICS only considers injured persons while the ICSL (Injury Cost Scale Lethal) also takes fatalities into account. For example: the costs of a Contusio Cerebri are 129.000 DM according to the ICS scale and 288.000 DM according to the ICSL scale. For a cervical spine fracture these numbers are 217.000 DM (ICS) and 464.000 DM (ICSL), respectively.

Table 20.6 Most cost-intensive injuries arranged according to the resultant social costs of the sample (ICSL) [20.10].

Injuries	Resultant social costs in million $
1. contusion cerebri	225
2. closed fracture of hip joint, pelvis or femur	120
3. spine fracture	98
4. closed tibia/fibula fracture	79
5. closed foot fracture	48
6. closed glenohumeral fracture	45
7. open tibia/fibula fracture	45
8. commotio cerebri	39
9. closed ulna/radius fracture	34
10. knee laceration	19

Apart from the costs per injury type in this study also the frequency of injuries was taken into account. An "expensive" injury type with a very low frequency is, from the societal point of view, less relevant than an injury with lower costs and a high frequency. By multiplication of frequency and costs a ranking of the various injuries in the data base was obtained. Table 20.6 shows the 10 most cost-intensive injuries according to the ICSL scale. The most expensive injury is the contusio cerebri followed by a closed fracture of the upper leg/hip joint/pelvis.

20.5   Injury risk function

This Section deals with probably the most important problem in injury biomechanics research: assessment of the correct relationship between injury severity and a mechanical load which causes this injury. It is a statistical problem which usually cannot be solved directly by analyzing accident data since the load parameters which are assumed to cause, or to correlate well with, the injury can not be measured during the accident. A few attempts varying from simple methods to assess the belt elongation in a crash to complicated crash recorders have not been introduced yet on a large scale. Moreover the information obtained from these methods is limited to parameters like the vehicle acceleration or the belt load.

Therefore laboratory experiments have to be conducted using human substitutes which allow extensive measurements to be made. For a number of reasons the transformation of the findings from these laboratory experiments to the real world however is not trivial, like:

- differences in biomechanical response between surrogates, (e.g. human cadavers, animals) and the living human body.
- the small number of tests and differences between the population of test subjects and the real world population of risk.
- large spread in test results due to, for instance, variations in test conditions and variations in anthropometry, age and sex of the subjects.
- the problem of finding the appropriate parameter(s), i.e. injury criteria which cause or correlate well with the injury sustained.
- the large number of injury mechanisms and resulting injuries that can occur.

Moreover injury biomechanical tests often are, what statisticians call, "censored". This means that the actual parameter (load) measured in the test is usually not the magnitude that exactly causes the injury. It is either too small if no injury occurs or too large in case of injury.

Let us consider a hypothetical experiment dealing with bone fractures. Figure 20.5 illustrates the results of a set of n experiments. The outcome of the experiments is either a fracture or a non-fracture. For small loads all tests appear to result in a non-fracture. For medium level loads there is a transition zone where both fractures and non-fractures can be observed. In this zone some randomness can be observed. Finally for high loads only fractures occur.

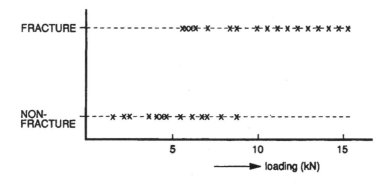

Figure 20.5   A hypothetical example of a bone fracture experiment.

Statistical methods are available which can deal with this type of observations. One of the methods which has been recommended for this purpose is the so-called "Maximum Likelihood Method", developed already in the twenties. Another method is "Probit analysis". Detailed treatment of such methods here would be out of the scope of this course, however some background information is considered useful for an engineer working in this field. For more details it is suggested to study Ref [20.11] and statistical handbooks dealing with probability and reliability analysis. We will concentrate here on the Maximum Likelihood Method.

Let us first introduce for the above example the so-called "cumulative frequency distribution" $F(z)$, illustrated in Figure 20.6. It can be interpreted as the probability, or risk, that a loading of the magnitude z would cause a fracture. It rises from zero to 100 % over the range of loadings. The objective of the statistical method is to find the best possible representation for $F(z)$, i.e. to find a curve which best fits the experimental test series.

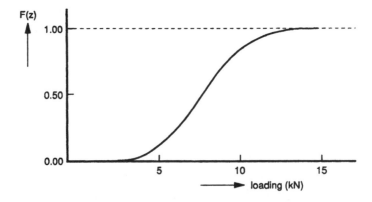

Figure. 20.6 Cumulative frequency distribution: Risk for fracture as function of loading.

The probability that, complementary, a non-fracture would occur at a given loading z is identical to $(1 - F(z))$. We will index now the non-fractures by $i = 1, 2, ...p$, where p is the total number of non-fractures, and the fractures by $j = 1, 2,.....q$, where q is the total number of fractures. Each test in our test series has had an occurrence probability which is $F(z_j)$ for a fracture and $1 - F(z_i)$ for a non-fracture. The likelihood, denoted by L, for all observations in this test series to occur is:

$$L = \prod_{j=1}^{q} F(z_j) \times \prod_{i=1}^{p} (1 - F(z_i)) \qquad (20.1)$$

where Π is the product sign. The Maximum Likelihood Method tries to find a distribution F(z) such that L gets a maximum value. The resulting F(z) is assumed to be the best model for the recorded observations. The candidate distributions F(z) should be selected from the many S-shaped (sigmoid) functions that are available. Such distributions have a value close to 0 % for low values of z and close to 100 % for high values. A function that successfully has been applied in injury biomechanics research for this purpose is the Weibull cumulative distribution function, which has the form:

$$W(z, \alpha, \beta, \gamma) = 1 - e^{-((z-\gamma/\alpha)^\beta} \qquad (20.2)$$

where α is the scale parameter ($\alpha > 1$), β is the shape parameter ($\beta > 1$) and γ is the location parameter, which are yet to be determined. Standard optimization routines can be applied, like available for instance in the software library NAG, to solve this optimization problem. Such routines allow additional constraints to be added in the search for a maximum of L, e.g. based on physical reasoning. Application of the Weibull function has shown to result often in higher likelihood values than other candidate distributions.

Assuming that the test sample is representative for the population of risk (which is however often not the case) and that the loading parameter is considered suitable to be used as an injury criterion, one might propose a tolerance level (or injury criteria level) from the resulting distribution for instance at the 25 or 50% risk level.

20.6   Whole body tolerance to impact

The term "whole body tolerance" originates from the earliest laboratory studies with human substitutes, in which the complete body was exposed to acceleration or impact type of loading. Results of such tests, together with information obtained from analysis of real accidents, like falls have contributed to a great extent to our present knowledge on the tolerance of the human body. The term "whole body response" was used, since in these tests the physiological response, which concerns the complete body, is evaluated. If injury occurred like in a rare accidental case with human volunteer tests or in tests with animals, actually one or more specific body parts like the head will be injured, rather than the "whole body".

In order to specify the direction of the load acting on the body first an orthogonal body fixed axis system (x,y,z) is introduced. In most injury biomechanics literature the axis system shown in Figure 20.7 is employed, although there are a few exceptions like the axis system recommended by a Committee of the Society of Automotive Engineers (SAE) for use in mechanical human substitutes [20.12]. Throughout this part we will adopt the axis system in Figure 20.7. The positive x-axis is forward (anterior direction), the z-axis upwards (superior) and the y-axis to the left. The axis system proposed by SAE is rotated 180° about the y-axis, in other words the positive z-axis is directed downward.

The direction of a load vector acting on the human body is usually denoted as plus (+) or minus (-) G_i, where the subscript i refers to the coordinate axis. For example in a $-G_X$ impact, the load vector acts from front to back on the body, causing the body to accelerate in backward direction. In air force crash-safety literature often the motion of the eyeballs relative to the human body is used to define the impact direction. A $-G_X$ impact is defined in this literature as "eyeballs out", a $+G_X$ as "eyeball in", a $+G_Z$ as "eyeball down" etc. According to these definitions, a normal seated occupant (i.e. forward facing), which is exposed to a pure frontal impact in a motor vehicle will experience the equivalent of a $-G_X$ impact.

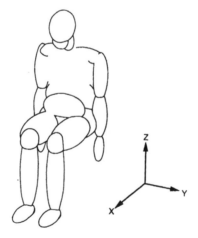

Fig. 20.7   Human body with coordinate system.

The earliest studies of the effects of impact loads on the human body are primarily from the field of aviation and aerospace medical impact research. They were initiated from the need to study the levels of acceleration pilots can sustain during high speed manoeuvres or in case of ejection from the aircraft in case of an emergency. Particularly the increase in aircraft accidents during the second world war stimulated the research in crash injuries. To study tolerance against long duration accelerations, tests with human volunteers in centrifuges were widely conducted. The "injury" in such tests usually took the form of unconsciousness due to decreased blood flow to the brain [20.13].

For short duration impacts swing type of impact devices or deceleration sleds were used. Well known are the rocket sled test with human volunteers and animals conducted by Stapp at the Aeromedical Research Laboratory at Holloman Airforce Base in New Mexico (USA), where thousands of tests have been conducted [20.14].

The earlier experiments with volunteers have shown that a $+G_X$ impact, where the body is accelerated in the forward direction, is the most survivable one of the principal directions, provided that the body and head are fully supported. This is mainly due to the fact that the load on the body can be distributed under this condition over a large surface. The most severe impact reported for a human volunteer (with reversible injuries), occurred in such a configuration. In this test an acceleration of 82.6 G was recorded at the sternum and 40.6 g at the sled for about 40ms. The subject complained of severe lower back pain and lost consciousness 10 seconds after the test. He was hospitalized for 3 days to recover from headache and backpain [20.13]. It should be realized that the measurement technology in these early days was still rather primitive and consequently less accurate than what is possible today.

Numerous of above types of tests have been documented extensively in literature. Several studies have attempted to bring together these data, where the most important ones are Eiband in 1959 [20.15] and Snyder in the early seventies [20.14]. Two figures which summarize some of the findings will be presented here. Figure 20.8 shows a graph originally developed by Roth in 1967 and adapted by Snyder [20.14] which summarizes human impact experience as function of deceleration distance and impact velocity. Most of the data concern free fall situations. The free fall distance is included in the graph. Assuming constant deceleration over the stopping distance, the average acceleration as well as stopping time was calculated and included in the graph.

From this data an approximated survival limit of 175-200 G was estimated. This limit should be used with much caution due to the various assumptions made. Moreover it is a survival limit which means that below this limit serious injuries can occur.

Figure 20.9 is from Eiband [20.15] who summarized human and animal impact experience for various impact directions. Injury occurrence is presented as function of mean acceleration level and duration of this acceleration in case of $+G_X$ impacts. Three categories are shown: voluntary human exposure, moderate injury and severe injury.

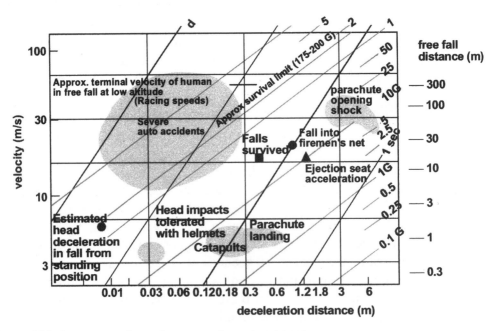

Figure. 20.8 Impact experience from several sources [20.14].

The most interesting finding from this figure is that the tolerance decreases with increasing pulse duration. The limit of voluntary exposure is about 40 G. Also such values should be handled with much caution since many factors will affect the actual tolerance like the restraint system used, the shape of the acceleration pulse applied and the individual condition of the subject.

For $-G_X$, $+G_X$ and $-G_Z$ impacts similar graphs were prepared by Eiband. The tolerance in $-G_X$ impacts was found to be from the same order of magnitude. For z-direction impacts the tolerable levels were found to be lower, in particular the level of voluntary exposure (about 15 G in case of $+G_Z$ and 10 G for $-G_Z$). Tolerance in $-G_Z$ impacts has had very limited study since the human body is hardly exposed to accidents in this loading direction. Also in lateral direction (G_y) much less data appear to be available. Impact tolerances in this direction are reported to be much lower than in G_X or $+G_Z$ direction [20.13].

The whole body response data presented here represents the basis of the current knowledge on tolerance values for thorax and head. In the next Chapters we will address in detail head injury tolerances, neck injury tolerances and thoracic injury tolerances.

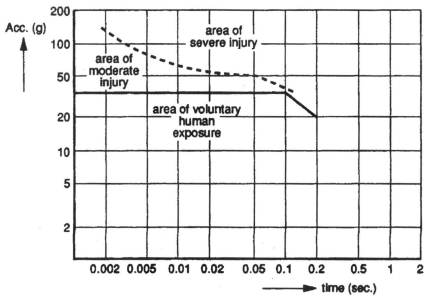

Figure. 20.9 Summary of impact experience in $+G_X$ impacts as function of mean acceleration and pulse duration (after Eiband in [20.13]):

REFERENCES

[20.1]   Viano, C., King, A.I., Melvin, J.W. and Weber, K., Injury biomechanics research: an essential element in the prevention of trauma, *J. Biomechanics*, **22**(5), 403-417, 1989.

[20.2]   Committee on Trauma Research, Commission on Life Sciences, National Research Council and the Institute of Medicine: *Injury in america. a continuing public health problem,* National Academy Press, Washington, 1985.

[20.3]   Aldman, B., Mellander, H. and Mackay, M., The structure of the European research into the biomechanics of impacts, In *Proceedings of the 27th Stapp Car Crash Conference*, San Diego, SAE paper 831610, Soc. of Automotive Engineers, Inc., 1983.

[20.4]   *The abbreviated injury scale - 1990 revision*, Association for the Advancement of Automotive Medicine, 2340 Des Plaines River Road, Suite 106, Des Plaines, Il 60018, USA, 1990.

[20.5]   Pike, J.A., *Automotive safety*. Society of Automotive Engineers, Inc., 400 Commonwealth Drive, Warrendale, PA 15096-0001, USA, 1990.

[20.6]   Malliaris, A.C. et. al., *Harm causation and ranking in car crashes*, SAE paper 85090, Society of Automotive Engineers, Inc., 400 Commonwealth Drive, Warrendale, PA 15096-0001, USA, 1985.

[20.7]   *Lateral car collisions*, Report R-79-48, Institute for Road Safety Research, SWOV, The Netherlands, 1979.

[20.8]   Baker, S.P. and O'Neill, B.O., The injury severity score: an update, *J. of Trauma*, **11**, 882-885, 1976.

[20.9]   Carsten, O. and Day, J., Injury priority analysis. Task A, In *Technical report DOT HS 807 224*, U.S. Department of Transportation National Highway Traffic Safety Administration (NHTSA) Washington, D.C. 20590, February 1988.

[20.10] Zeidler, F., Pletschen, B., Mattern, R., Alt, B., Miksch, T., Eichendorf, W., and Reiss, S., Development of a new injury cost scale, In *33rd*

Annual Proceedings, Association for the Advancement of Automotive Medicine, 1989.

[20.11] Ran, A., Koch, M. and Mellander, H., Fitting injury versus exposure data into a risk function, International IRCOBI conference, Delft 1984.

[20.12] *Instrumentation for impact test*, SAE Recommended Practise, SAE J211 OCT88, 1988.

[20.13] McElhaney, J.H., Roberts, V.L. and Hilyard, J.F., *Handbook of human tolerance*, Japan Automobile Research Institute, Inc. JARI, 1976.

[20.14] Snyder, R.G., Human impact tolerance, Paper 700398, In *1970 International Automobile Safety Conference Compendium*. Society of Automotive Engineers, Warrendale, PA, 1970.

[20.15] Eiband, A.M., Human tolerance to rapidly applied accelerations - A summary of the literature, *NASA Memo-5-19-59E*, 1959.

[20.16.] Human tolerance to impact conditions as related to motor vehicle design, *SAE information report J885 JUL86*, Society of Automotive Engineers, Warrendale, PA, 1991.

21.   DESIGN TOOLS: HUMAN BODY MODELLING

21.1 Introduction

In crash safety research and development, like in many other engineering disciplines, a strong increase could be observed during the last thirty years in the use of computer simulations. This is due in part to the fast developments in computer hardware and software, but in case of automotive safety, probably even more due to emphasis which has been placed on the development of reliable models describing the human body in an impact situation, as well as numerous validation studies which have been conducted using these models. Mathematical models of the human body in conjunction with a mathematical description of the vehicle structure and the various safety provisions and restraint systems, appear to offer a very economical and versatile method for the analyses of the crash response of complex dynamical systems.

In the crash safety field mathematical models can be applied in practically all area's of research and development including:

- reconstruction of actual accidents
- design (CAD) of the crash response of vehicles, safety devices and roadside facilities
- human impact biomechanics studies

Furthermore computer simulations may be part of safety regulations, much like standard crash tests.

Dependent on the nature of the problem, several types of crash analyses programs have been developed, each with their own, but often overlapping, area of applicability. Most of the models are of the deterministic type, that is, based upon measured or estimated parameter values, representing characteristics of the human body, safety devices, the vehicle and its surroundings and using well established physical laws, the outcome of the crash event is predicted. Another category of models used in the crash safety field is of a statistical nature. They are used for instance in injury biomechanics research to assess the correct relationship between loading conditions and resulting injuries by means of regression type of equations (the so-called injury risk function). This Chapter will concentrate on deterministic models for the human body.

Although the various deterministic models may differ in many aspects, they are all dynamic models. They account for inertial effects by deriving equations of motions for all movable parts and solving these equations. The mathematical formulations used for these models can be subdivided into lumped mass models, multi-body models and finite element models. Lumped mass models are usually one- or two-dimensional, multi-body models two- or three-dimensional and finite element models usually three-dimensional.

Lumped mass models

In a lumped mass model a system is represented by one or more rigid elements often connected by mass-less elements like springs and dampers. An example of a lumped-mass model is shown in Figure 21.1. It is a one-dimensional model of the human thorax developed by Lobdell in 1973 [21.1]. This model simulates the thorax response in case of a loading by an impactor. The model consist of 3 rigid bodies with masses m_1, m_2 and m_3 connected by springs and dampers. Mass m1 represents the impactor mass and masses m_2 and m_3 the sternal and vertebral effective mass. Spring k_{12} represents the skin and flesh between impactor and sternum. The internal spring and dampers represent the connection between sternum and thoracic spine. The response of this model was shown to correlate well with human cadaver tests.

Multi-body models

The most important difference between a lumped mass model and a multi-body model is that elements in a multi-body formulation can be connected by various joint types due which the number of degrees of freedom between the elements can be constrained. A lumped mass model in fact can be considered as a special case of the more general multi-body model formulation. The motion of the joint-connected elements in a multi-body model is caused by external forces generated by so-called force-interaction models. Examples of force-interaction models in a multi-body model for crash analyses are the model to account for an acceleration field, spring-damper elements, restraint system models and contact models. Another difference with lumped mass models is that in a multi-body formulation instead of rigid bodies also flexible bodies can be specified.

A first historical example of a multi-body model is presented in Figure 21.2. The model was developed in 1963 by Mc Henry [21.2]. The model which represents the human body together with restraint system and vehicle is 2-dimensional and has 7 degrees of freedom. The human body part of the model is represented by 4 rigid bodies representing thorax/head, upper arms, upper legs and lower legs. The rigid elements were connected by simple pin joints. The values for the parameters in the model were estimated. Mc. Henry compared his model calculations with experimental data to demonstrate the potential of this type of calculations. He was able to show quite good agreement for quantities like hip displacements, chest acceleration and belt loads.

A second more recent example of a multi-body model is presented in Figure 21.3. It is 3-dimensional MADYMO model of a Chrysler Neon vehicle suitable for frontal collisions. The model was developed as part of a co-operative effort between the European Community and the NHTSA (DOT-USA) to study vehicle compatibility issues [21.3]. The vehicle part of the model has more than 200 elements and includes a description of interior, restraint system, suspension, steering wheel, bumper, engine, hood etc. The human body part of the model is a 32-segment model of the Hybrid III dummy. Model results were compared with rigid wall test and offset deformable barrier tests and quite realistic results were obtained.

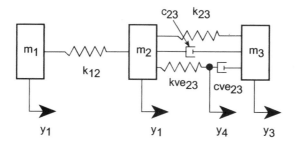

Figure 21.1 Example of lumped mass model: the Lobdell thorax model [21.1]

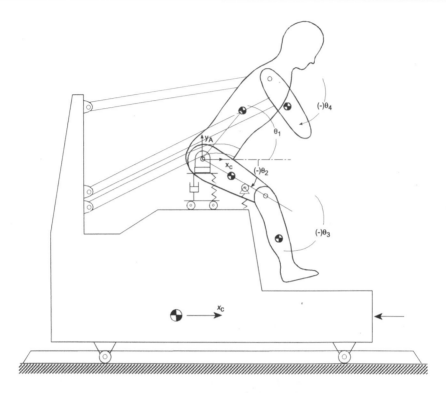

Figure. 21.2 Example of multi-body model: 7 degrees of freedom model for frontal collisions by Mc. Henry [21.2]

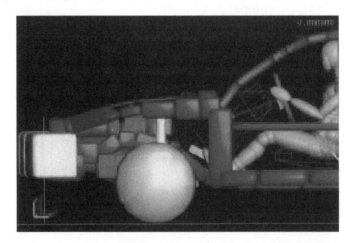

Figure 21.3 Example of multi-body model: 3-dimensional model for frontal collisions of Chrysler-Neon with Hybrid III dummy [21.3]

Finite element models

In a finite element model the system to be modelled is divided in a number of finite volumes, surfaces or lines representing an assembly of finite elements. These elements are assumed to be interconnected at a discrete number of points: the nodes. In the displacement-based finite element formulation, which is applied in practically all major finite element software packages, the motion of the points within each finite element is defined as a function of the motion of the nodes. The state of stress follows from the deformations and the constitutive properties of the material modelled. Figure 21.4 shows one of the earlier examples of using the finite element method for human body impact modelling: a model of the human head. The model was developed in the seventies by Shugar [21.4]. The model was 3-dimensional and included a representation of the skull and brain. Linear elastic and linear visco-elastic material behaviour was assumed. Skull bone response and brain response from the model was compared with experimental results of head impact tests with primates.

Multi-body models versus finite element models

Both the multi-body method and the finite element method offer their specific advantages and disadvantages in case of crash analyses. The multi-body approach is particularly attractive due to its capability of simulating in a very efficient way complex kinematical connections like they are present in the human body and in parts of the vehicle structure like the steering assembly and the vehicle suspension system. The finite element method offers the capability of describing (local) structural deformations and stress distribution, due to which it becomes possible to study for instance injury mechanism in the human body parts. Usually much larger computer times are required to perform a finite element crash simulation than a multi-body crash simulation, making the finite element method less attractive for optimisation studies involving many design parameters.

Figure 21.5 shows an example in which both the multi-body approach and the finite element approach are used. This integrated approach is sometimes referred to as "hybrid" approach. It is model of a car occupant interacting with a passenger airbag developed in the eighties by Bruijs [21.5]. The airbag (and airbag straps) was modelled in the PISCES 3D-ELK program (now MSC-DYTRAN) using almost 2000 triangular membrane elements. For the gas inside the airbag was described using the perfect gas law. The model takes into account leakage through airbag material and exhaust orifices. The most important assumption in the airbag model is that the pressure and temperature in the airbag are constant and that inertia effects of the gas are neglected. The finite element airbag model interacts with a multi-body model of the Hybrid III crash dummy modelled in MADYMO 3D.

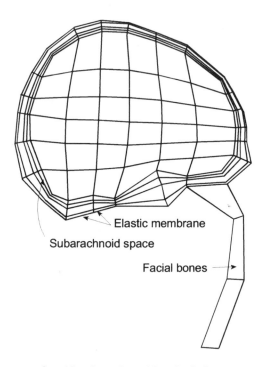

Combined mesh - midsagittal plane

Figure 21.4   Example of a finite element model simulating the human body: a head model by Shugar [21.4]

Set-up of this Chapter
      Multi-body models in which the complete human body is simulated for the purpose of crash analyses are often referred to as Crash Victim Simulation models (CVS), human body gross-motion simulators or whole body response models. The theoretical basis for this type of models will be presented in section 21.2. The finite element method as such is assumed to be known here, so this method will not be treated here further except for the aspect of integrated multibody-finite element simulations (section 21.2.5).

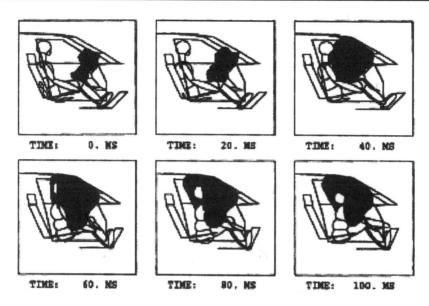

TIME: 0. MS   TIME: 20. MS   TIME: 40. MS

TIME: 60. MS   TIME: 80. MS   TIME: 100. MS

Figure 21.5 Example of integrated multi-body finite element model: occupant-airbag interaction by Bruys [21.5]

Human body models for crash analyses can be subdivided into models of crash dummies and of the real human body. Section 21.3. discusses aspects of crash dummy models. Real human body models will be presented in section 21.4. In the final section 21.5 some concluding remarks will be made and future trends will be discussed. Among others attention shall be given to the aspect of model validation in view of the increasing importance of virtual testing.

21.2   The multi-body method for crash analyses

21.2.1 Introduction

One of the first human body gross-motion simulation models, shown in Figure 21.1, was developed in the early 1960s by McHenry [21.2]. The results obtained were so encouraging that since then many, more sophisticated, models have been developed. The most well known are the two-dimensional 8-segment MVMA-2D model and the three-dimensional 6-segment HSRI occupant model both developed by Robbins et. al. [21.6,7], the three-dimensional 12-segment UCIN model [21.8] and the three-dimensional Calspan 3D CVS (CAL3D) which allowed up to 20 elements [21.9]. For a review of the status of human body gross-motion simulators up to 1975 see King and Chou [21.10].

Two of the reviewed models, i.e. MVMA 2D and the CAL3D, have gone through an extensive further validation and development effort since then and are still frequently used at present. An example of these developments is the ATB (Articulated Total Body) program, which is a special version of CAL3D for aircraft safety studies, developed by the Air Force Aerospace Medical Research Laboratory in Dayton [21.11].

This section will be mainly related to a more recent program, i.e. MADYMO [21.12], which is a general multibody/finite element program with a number of special features for crash analyses. The program was developed in Europe by TNO Automotive in Delft, The Netherlands. For a comparison of the basic features between MADYMO, MVMA2D and CAL3D the reader is referred to reviews of gross-motion simulation programs by Prasad [21.13,21.14] and Prasad and Chou [21.15]. MADYMO has a two- and a three-dimensional version referred to as MADYMO 2D and MADYMO 3D, respectively. This section will concentrate on the multibody part and the force interactions in MADYMO 3D.

21.2.2 MADYMO set-up

MADYMO consists of a number of modules (Figure 21.6). The multi-body module of the program calculates the contribution of the inertia of bodies to the equations of motion; the other modules calculate the contribution of specific force elements such as springs, dampers, muscles, interior contacts and restraint systems or the effect of systems which are represented as a finite element model. Special models are available for vehicle dynamic applications including tyre models. A control module is available which offers the possibility to apply loads to bodies dependent on the body's motions. For this purpose motion quantities can be extracted from the bodies by sensors. The sensor signals can be manipulated with summers, transformers and controllers and used as input for actuators which apply forces and torques to the bodies. In the next section 21.2.3 the multi-body algorithm will be presented and in section 21.2.4 the most important force elements as used for crash analyses. The aspect of integrated multi-body/finite element analyses will be introduced in section 21.2.5.

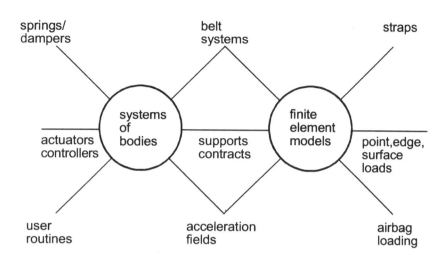

springs/
dampers

belt
systems

straps

systems
of
bodies

finite
element
models

actuators
controllers

supports
contracts

point,edge,
surface
loads

user
routines

acceleration
fields

airbag
loading

Figure 21.6 MADYMO modules

21.2.3 MADYMO multi-body algorithm

The multi-body module analyses the motion of systems of bodies. Bodies can be interconnected by kinematic joints such as spherical joints, revolute joints, universal joints, and translational joints. More over, new kinematic joints can be added by user subroutines.

Topology of a system of bodies

Consider a system of N bodies with a tree structure (Figure 21.7). A system of bodies has a tree structure if one can proceed from one arbitrary body to another arbitrary body along a unique sequence of bodies and joints. Systems with closed chains of bodies in it (like for instance a four bar linkage system) are permitted in the later MADYMO versions. For topology specification they have to be reduced to a tree structure however. The bodies are numbered in order to be able to specify which bodies are interconnected. One of the bodies is chosen as reference body and is given number 1. The other bodies are numbered from 2 till N in such a way that the numbers of the bodies on the path from the reference body to any other body are lower than the number of that specific body. The configuration of a system is completely defined by specifying for all bodies at the end of the tree (the peripheral bodies), the numbers of the bodies on the path to the reference body. The bodies on such a path form a branch. In the MADYMO input file the configuration of a system is defined by entering for each branch the numbers of the bodies in decreasing order. This table with branch numbers is called configuration table. For the example in Figure 21.7 the configuration specification is:

Branch 1: 3 2 1
Branch 2: 7 2 1
Branch 3: 6 5 1
Branch 4: 8 4 1

Kinematics of a rigid body

Consider the rigid body i shown in Figure 21.8. In order to describe its motion relative to an inertial space, a right-handed body-fixed base $\{e\}_i$ is introduced. Its origin is chosen coincident with the centre of mass since then the equations of motion of a single body have their simplest form: the Newton-Euler equations. The motion of the body is defined by the position of the origin and the orientation of the body-fixed base relative to an inertial base $\{E\}$.

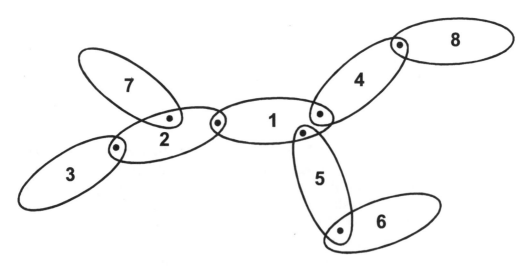

Figure 21.7 Example of body numbering

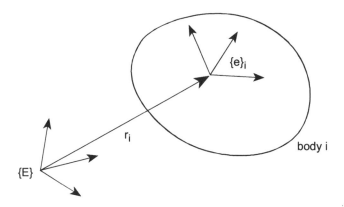

Figure 21.8 Specification of the motion of a rigid body

The position of the origin of the body-fixed base relative to the origin of the inertial base is given by the vector $\mathbf{r}_i$. The rotation matrix $\mathbf{A}_i$ defines the orientation of the body-fixed base relative to the inertial base. The elements of this matrix consist of scalar products of the inertial base vectors and the body-fixed base vectors.

Kinematics of a flexible body

In later versions of MADYMO 3D, bodies which experience small deformations can be modelled as flexible bodies. For a flexible body the motion of a point on a body is considered to be composed of a rigid body motion and a superimposed motion due to the deformation. The motions of the body due to the deformations are approximated by a linear combination of predefined displacement and rotation fields (deformation modes). Only at certain pre-defined points in the body (the nodes) the deformations are defined.

Kinematics of pair of bodies interconnected by a joint

Consider the pair of bodies interconnected by an arbitrary kinematic joint shown in Figure 21.9. (A kinematic joint as defined here can connect only two bodies.) The bodies are numbered i and j. The number of the lower numbered body i can be obtained from the configuration table. In MADYMO the motion of a body j is described relative to the corresponding lower numbered body i. This is done in terms of quantities that define the motion within the joint, the so called joint co-ordinates. Their number, n_{ij} equals the number of degrees of freedom of the joint. They are put in vector $\mathbf{q}_{ij}$.

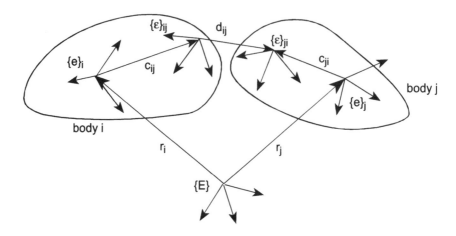

Figure 21.9 Pair of interconnected bodies

On each body a body-fixed joint base $\{\varepsilon\}$ is introduced in order to describe the relative motion of body j relative to body i. The origins of these joint bases are chosen such that the mathematical expression for the relative translation of the joint base origins is as simple as possible. As an example, for a spherical joint the origins will be chosen coincident with the articulation points on the bodies. Then, the relative translation vector is identically $\mathbf{0}$.

The selection of the orientations of the joint bases is made on the same ground. As an example, for a revolute joint one of the base vectors of each joint base will be chosen parallel to the rotation axis. The best choice for the origin and the orientation of joint bases depends on the specific joint which is considered.

Let the orientation of the joint base on body i (j) relative to the body base on body i (j) be specified by the time-independent rotation matrix $\mathbf{C}_{ij}$ ($\mathbf{C}_{ji}$). Let the orientation of the joint base on body j relative to the joint base on body i be specified by the rotation matrix $\mathbf{D}_{ij}$. This matrix is a function of the joint co-ordinates. Using these rotation matrices, the rotation matrix of the base of body j can be written as:

$$\mathbf{A}_j = \mathbf{A}_i \mathbf{C}_{ij} \mathbf{D}_{ij} \mathbf{C}_{ji}^{T} \tag{21.1}$$

Let $\mathbf{c}_{ij}$ and $\mathbf{c}_{ji}$ be the position vectors of the origins of the joint bases on body i and j, respectively, relative to the origin of the base of the corresponding body.

For a rigid body, the components of these vectors relative to the corresponding body base are constant. The vector from the origin of the joint base on body i to the joint base on body j is given by the vector $\mathbf{d}_{ij}$. The components of this vector relative to the joint base on body i, $\mathbf{d}_{ij}$, is a function of the joint co-ordinates. The position vector of the origin of the body base of body j can be written as:

$$\mathbf{r}_j = \mathbf{r}_i + \mathbf{c}_{ij} + \mathbf{d}_{ij} - \mathbf{c}_{ji} \tag{21.2}$$

Applying equation (21.1) and (21.2) successively for body 1 till body N yields the positions and orientations of all body bases relative to the inertial base.

Taking the first time derivative of (21.1) and (21.2) yields the following expressions for the angular and linear velocity [21.16]:

$$\omega_j = \omega_i + \omega_{ij} \tag{21.3}$$

$$\dot{\mathbf{r}}_j = \dot{\mathbf{r}}_i + \omega_i \times \mathbf{c}_{ij} + \dot{\mathbf{d}}_{ij} - \omega_j \times \mathbf{c}_{ji} \tag{21.4}$$

where ω_i and ω_j are the angular velocity of body i and j, respectively and ω_{ij} the angular velocity of the joint.

Taking the second time derivative of (21.1) and (21.2) yields similar expressions for the angular and linear acceleration [21.16].

Example of a kinematic joint
Consider the joint shown in Figure 21.10. This joint allows a relative translation s and a relative rotation φ. These joint co-ordinates are assembled in the column matrix $\mathbf{q}_{ij} = [s\ \varphi]^T$. The first base vectors of both joint bases are chosen parallel to the rotation/translation axis. The origin of the joint bases are chosen on the rotation/translation axis and are initially coincident. Then the components of the relative translation vector with respect to the joint base $\{\boldsymbol{\varepsilon}\}_{ij}$ and the rotation matrix are given by:

$$\mathbf{d}_{ij} = \begin{pmatrix} s \\ 0 \\ 0 \end{pmatrix} \tag{21.5}$$

$$\mathbf{D}_{ij} = \begin{pmatrix} 1 & 0 & 0 \\ 0 & \cos\varphi & -\sin\varphi \\ 0 & \sin\varphi & \cos\varphi \end{pmatrix} \tag{21.6}$$

The relative velocity of the origins of the joint bases is obtained by taking the first time derivative of (21.5). This yields:

$$\dot{\mathbf{d}}_{ij} = \begin{pmatrix} \dot{s} \\ 0 \\ 0 \end{pmatrix} \tag{21.7}$$

The relative angular velocity of the joint bases is the axial vector of the skew-symmetric matrix $\dot{\mathbf{D}}_{ij}\mathbf{D}_{ij}^{T}$. This yields:

$$\boldsymbol{\omega}_{ij} = \begin{pmatrix} \dot{\varphi} \\ 0 \\ 0 \end{pmatrix} \tag{21.8}$$

In MADYMO 3D a library of kinematic joints is available. Figure 21.11 summarises a number of the most frequently used joint types within MADYMO.

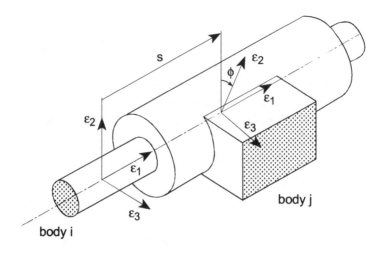

Figure 21.10 Translational-rotational joint

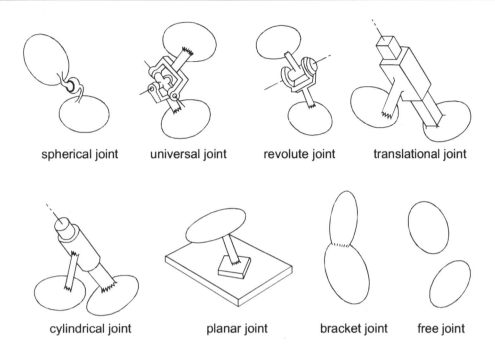

| spherical joint | universal joint | revolute joint | translational joint |

| cylindrical joint | planar joint | bracket joint | free joint |

Figure 21.11 Examples of kinematic joints

Equations of motion

The equations of motion (Newton-Euler) of a rigid body *i* referred to its centre of mass are [21.16]:

$$m_i \ddot{\mathbf{r}}_i = \mathbf{F}_i \tag{21.9}$$

$$\mathbf{J}_i \circ \dot{\boldsymbol{\omega}}_i + \boldsymbol{\omega}_i \times \mathbf{J}_i \circ \boldsymbol{\omega}_i = \mathbf{T}_i \tag{21.10}$$

where m_i is the mass, $\mathbf{J}_i$ is the inertia tensor with respect to the centre of mass, $\mathbf{F}_i$ is the resultant force vector, and $\mathbf{T}_i$ is the resultant torque vector relative to the centre of mass. For a body in a multibody system, $\mathbf{F}_i$ and $\mathbf{T}_i$ include the effect of the applied forces and torques acting on the rigid bodies due to the force-interaction models as well as the constraint forces and torques resulting from kinematic constraints in the joints. These constraint forces and torques cannot be determined until the motion of the system is known. This in contrast with the applied forces and torques that depend only on position and velocity.

The constraint forces and torques can be eliminated using the principle of virtual work. First equations (21.9) and (21.10) are multiplied by a variation of the position vector, δr_i, and a variation of the orientation, $\delta \pi_i$, and the resulting equations are summed for all bodies of the system:

$$\sum \left[\delta r_i \circ \{ m_i \ddot{r}_i - F_i \} + \delta \pi_i \circ \{ J_i \circ \dot{\omega}_i + \omega_i \times J_i \circ \omega_i - T_i \} \right] = 0 \qquad (21.11)$$

In case the variations δr_i and $\delta \pi_i$ of connected bodies are such that the constraints caused by the joint are not violated, the unknown joint forces and torques will cancel (principle of virtual work). Such variations can be obtained from equations (21.1) and (21.2) by varying the joint co-ordinates. These expressions can be substituted in (21.11). Starting with bodies at the end of the tree, expressions for the second time derivative of the joint co-ordinates are obtained:

$$\ddot{q}_{ij} = M_{ij} \, \dot{Y}_i + Q_{ij} \qquad (21.12)$$

$\dot{Y}_i$ is a 6×1 column matrix which contains the components of the linear and angular acceleration of the base of the lower numbered body i. The $n_{ij} \times 6$ matrix M_{ij} and the $n_{ij} \times 1$ column matrix Q_{ij} depend on the inertia of the bodies and the instantaneous geometry of the system. Q_{ij} depends additionally on the instantaneous velocity of the system and the applied loads. The matrices M_{ij} and Q_{ij} are calculated successively starting with body N to body 1. Then starting with the joint between the inertial space and body 1 the second time derivatives of the joint co-ordinates can be calculated from (21.12). Note that for this joint, i equals 0 and j equals 1 and the acceleration of the inertial space, $Y_0 = 0$.

This algorithm yields the second time derivatives of the joint co-ordinates in explicit form. The number of computer operations is linear in the number of bodies in case all joints have the same number of degrees of freedom. This leads to an efficient algorithm for large systems of bodies.

Time integration of the second time derivatives of the joint co-ordinates gives the joint co-ordinates and there first time derivatives at a new point in time. These are used to calculate the motion and velocity of the body bases relative to the inertial base using equations (21.1) and (21.2) and their first time derivatives (21.3) and (21.4). At the start of the integration the joint co-ordinates and their first time derivatives have to be specified (initial conditions).

For the time integration two explicit numerical integration methods are available, namely a fourth order Runge-Kutta method which uses a constant time step, and a fifth order Runge-Kutta-Merson method which uses a variable time step that is controlled by the local truncation error.

21.2.4 Force interaction models

Applied forces cause the motion of a system of joint-connected bodies. MADYMO offers a set of standard force-interaction models. The various types of force-interaction models are summarised below, and in Figure 21.6:

- Acceleration field model
- Spring-damper elements
- Muscle models
- Contact models
- Belt model
- Dynamic joint models

In addition the user can make and link his own routines to the MADYMO multibody module.

Forces (and torques) are specified as a (non-linear) function of parameters like deflections, elongations, penetrations, joint rotations etc. Such functions are defined by means of a set of function pairs, which internally in the program are approximated either by a spline function or by a piece-wise linear interpolation. Generally, quasi-static tests have to be carried out to obtain these characteristics. Various hysteresis models can approximate differences between loading and unloading responses.

Often function characteristics will be rate dependent in a highly dynamical environment like a crash. For this purpose in a number of the force-interaction models, velocity dependent damping can be introduced. Moreover there is a possibility to prescribe a so-called dynamic amplification factor which multiplies a statically determined force with a rate dependent factor in order to approximate the dynamic response. For details on the physical background of dynamic amplification in a crash environment see Prasad and Padgaonkar [21.17]. In MADYMO several dynamic amplification factors have been implemented including polynomial and logarithmic function of the rate of deformation (or penetration, elongation etc.).

Acceleration field model

The acceleration field model calculates the forces at the centres of gravity of bodies in a homogeneous time-dependent acceleration field **a** (Figure 21.12). This model can be applied for the simulation of the acceleration forces on a vehicle occupant during an impact. Consider as an example the impact of a vehicle against a rigid barrier. If the vehicle does not rotate during the crash, the actual recorded accelerations at the vehicle can be prescribed as an acceleration field acting on the occupant, while the vehicle is connected to the inertial space. The motion of the occupant relative to the vehicle is the same as when the actual recorded motion of the vehicle is prescribed. The components of vector **a** are defined as a function of time relative to the inertial co-ordinate system.

Spring-damper elements

Three different types of massless "spring-damper" models can be specified: a Kelvin-, a Maxwell- and a Point-restraint element (Figure 21.13). The Kelvin element is a uniaxial element which simulates a spring parallel with a damper. The Maxwell element is an uniaxial element which simulates a spring and damper in series. The point-restraint model can be considered as a combination of three Kelvin elements with infinite spring length, each parallel to one of the axes of an orthogonal co-ordinate system. All spring-damper models can be attached to arbitrary points of any two bodies or between a body and the inertial space.

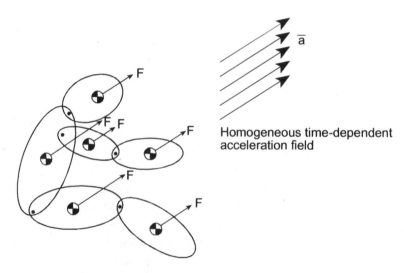

Homogeneous time-dependent acceleration field

Figure 21.12 A homogeneous acceleration field

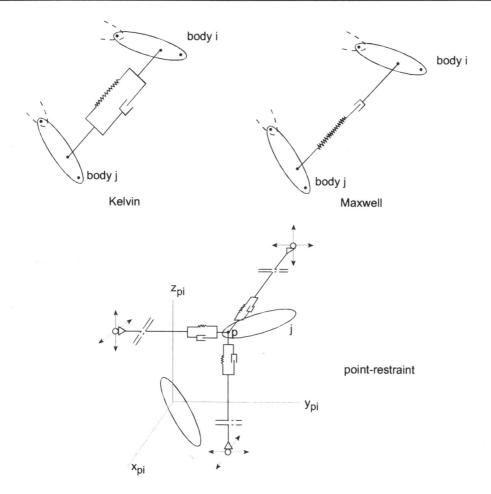

Figure 21.13 A Kelvin-, Maxwell- and Point-restraint element

Muscle models

The most common muscle model in biomechanical research is the Hill model (Figure 21.14). The model consists of a contractile element (CE) which describes the active force generated by the muscle, a parallel elastic element (PE) which describes the elastic properties of muscle fibers and surrounding tissue, 2 elastic elements (SE1 and SE2) which describe the elastic properties of tendons and aponeurosis and 2 masses (M1 and M2) to account for the muscle mass. The basic muscle model implemented in MADYMO consists of the combined contractile element CE and parallel elastic element PE. Muscles with varying complexity can be formulated using this basic model in combination with the standard MADYMO elements.

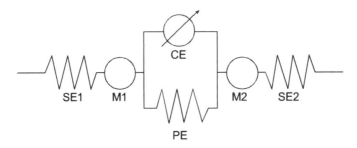

Figure 2.14 The Hill type muscle model

Contact models

Planes, cylinders, ellipsoids and in later versions of MADYMO also arbitrary shaped surfaces (called facet surfaces) are used to model contact with other bodies or the surroundings. The contact surfaces are of major importance in the description of the interaction of the human body and an impacting surface like the vehicle interior. Apart from standard ellipsoids, which are of the degree 2, also higher order ellipsoids can be specified:

$$\left(\frac{|x|}{a}\right)^n + \left(\frac{|y|}{b}\right)^n + \left(\frac{|z|}{c}\right)^n = 1 \qquad (21.13)$$

where a, b and c are the semi-axes of the (hyper)ellipsoid and n is the degree. If the degree n increases the (hyper)ellipsoid will approximate more and more a rectangular shape (Figure 21.15).

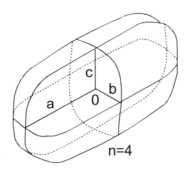

Figure 21.15 A hyper-ellipsoid of degree 4

The surfaces cannot deform themselves; instead they are allowed to penetrate into each other. The basic principle of the contact models in MADYMO (and other human body gross-motion simulators) is that a contact force is generated between two colliding surfaces, which is a function of the penetration of the two surfaces as well as of the relative velocity in the contact area (Figure 21.16). In this way elastic (including hysteresis and dynamic amplification), damping and friction forces can be specified in the contact. If a facet surface is involved the contact force instead of "force-penetration" based also can be based on a "stress-penetration" function.

The belt model

The belt model consists of a chain of connected, massless, spring type of segments (Figure 21.17). The end points of these segments are connected to rigid bodies or the inertial space at so-called attachment points. These attachment points cannot change during a simulation. An important feature is that belt material can slip through an attachment point from one segment to another. The belt model accounts for initial belt slack or pre-tension and rupture of belt segments. Elastic characteristics can be specified separately for each belt segment. Furthermore a retractor can be specified which is either of the vehicle sensitive type or of the webbing sensitive type. A vehicle sensitive reel can lock at a specified time or if a specific component of the calculated linear acceleration at the retractor location exceeds a prescribed level during a certain time interval. A webbing sensitive reel locks if the belt feed rate exceeds a specified limit. Finally also a pretensioner model is available.

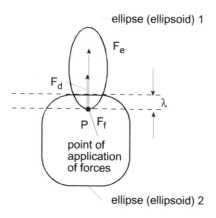

Figure 21.16 Contact loads in an ellipsoid-ellipsoid contact (only forces acting on the upper ellipsoid are shown).

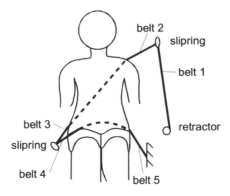

Figure 21.17 A 3-point belt with retractor.

Dynamic joint models

The last type of force-interaction models to be discussed here is the category of dynamic joint models. As described earlier, in a joint two types of loads are acting, the internal forces and torques caused by the kinematic joint constraints, and the applied loads representing for instance passive type of loads due to friction or elastic resistance or active loads caused for instance by muscle activity. The applied joint loads are taken into account by the dynamic joint force models. The most simple type generates a force (or torque) as function of a single joint co-ordinate. An example is the torque as function of the rotation in a revolute joint (Figure 21.18). In this model, elastic (including hysteresis and dynamic amplification), as well as damping forces (torques) can be prescribed, much like e.g. in a spring-damper model. Moreover in this model a friction torque can be introduced.

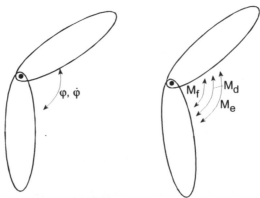

Figure 21.18 A revolute joint with joint torques

A more complicated dynamic joint is the so-called "flexion-torsion restraint" model which can be applied for instance in case spherical joints are used to represent flexible rubber type structures like the neck and spine in crash dummies. In this model the relative joint position of the joint is considered to be the result of two successive rotations, i.e. a "bending" and a "torsion" motion of segment j relative to segment i. Both for the "bending" and "torsion" motion an elastic torque has to be defined where the bending torque can be defined directional dependent, this means that for instance the "bending" stiffness in forward direction can differ from the stiffness in backward or lateral "bending".

21.2.5 Integrated multi-body finite element simulations

In the MADYMO finite element module, truss, beam, shell, brick and membrane elements are implemented and material models like elastic, visco-elastic, elasto-plastic, hysteresis and Moonley-Rivlin can be applied. Also models are available for sandwich material, solid foam and for honeycomb material. A MADYMO model can be made of only multi-body systems, only finite element structures or combinations.

Figure 21.19 illustrates the interaction between the multi-body and the finite element module. Two kind of interactions generate forces between the multi-body and the finite element model: support and contacts. A support is a finite element node rigidly connected to a body of a multi-body system (or to a belt segment of the belt model in order to model belt parts by membrane or truss elements). The necessity of these two types of interactions can be illustrated using the example of a finite element driver airbag simulation. The airbag unit is connected to the steering column which often is modelled as a multibody system. The airbag model can be attached to the multi-body steering column using supports while the interaction of the airbag with a multi-body occupant can be handled through contacts.

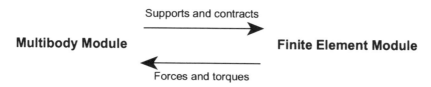

Figure 21.19 Interaction between multi-body and finite element module

This approach allows the use of different integration methods for the equations of motion of the finite element module and the multi-body module. For an integrated MADYMO analyses the 4th order Runge-Kutta or Euler method is used for the time integration of the equations of motion of the multi-body system. The central difference method is used for the equations of motion of the finite element model. Actual body positions and velocities at each time step of the central difference method determine the support and contact forces. The forces acting on the multi-body system are accounted for in each mean time point of the 4t^h order Runge-Kutta and each time step of the Euler method. Due to the fine spatial discritization often required in a finite element model a much smaller time step is required in finite element model compared to a multi-body model. To improve the efficiency of the integrated analyses the finite element analysis is subcycled with respect to the multi-body analyses.

Apart from the finite element module within MADYMO using above principles also external interfaces between MADYMO and finite element programs like PAMCRASH, LSDYNA, RADIOSS and MSC DYTRAN have been developed.

21.3 Crash Dummy modelling

21.3.1 Introduction

Well-validated human body crash models presented in literature so far mainly have been developed for mechanical models of the human body, i.e. crash dummies, rather than for real human beings themselves. The rationale for this is that most model input data in case of crash dummy models can be measured relatively easy. Moreover, results of experiments with crash dummies often are available for model validation and if not, such experiments, unlike tests with biological models, can be carried out in many well-equipped crash laboratories. Another reason for the emphasis on modelling crash dummies so far is the need, particular from the design departments in the automotive industry, for well-validated design tools which can reduce the number of regulatory tests with crash dummies in order to shorten and optimise the development process of a new car model. In the next section (section 21.3.2) the modelling process of a (multi-body) crash dummy model will be described. In section 21.3.3 typical examples of crash dummy models will be presented.

21.3.2 Modelling methodology

The first step in the modelling process of a crash dummy is the division of the dummy in segments and the specification of the parts that belong to each segment. The segments are selected by dividing the dummy into functional components. Each part of the dummy having significant mass and a flexible connection with other parts is considered as a segment. Dummy parts which do not show any relative motion are usually considered to be part of another segment except if load information is required at the interface between the two segments. In present dummy designs usually four types of kinematic connections between segments can be distinguished: revolute (or pin joints), translational joints, universal joints and universal joints.

Furthermore, often flexible structures are present in a dummy. Some are partly or completely made of rubber like the lumbar spine and neck. They are usually modelled by two universal joints located in the centres of the end planes of these structures or by a 6-degree of freedom joint. Flexible parts like the ribs in the dummies are usually represented by a number of rigid bodies, flexible bodies or by finite elements.

If the general model set-up has been specified the geometrical parameters have to be determined from technical drawings, CAD files and/or they have to be measured directly at the dummy. Information needed includes the joint locations within the individual segments, the joint axes orientations and the outside surface geometry. Three-dimensional measurements are conducted usually at a disassembled dummy. Some of the geometrical joint data have to be determined in an indirect way since the requested joint data may not be directly accessible a measuring device. Additional measurements of landmarks specifying segment local co-ordinate systems have to be conducted in order to express the data in a common body-fixed base.

The outside surfaces of the dummy segments are usually represented by means of ellipsoids or arbitrary shaped surfaces. The ellipsoids or arbitrary surfaces are used for visual presentation of the occupant kinematics as well as for the contact interactions between dummy segments and environment (e.g. the vehicle interior).

The next step is the determination of the inertial properties. The mass, location of the centre of gravity, the principal moments of inertia and the orientation of the principal axes must be determined for each dummy segment. In addition the position of segment landmarks have to be determined in order to express the inertia data in a body-fixed co-ordinate system. Experimentally the moments of inertia can be determined with a torsional vibration table [21.18]. The object which is fixed in a box is measured in several positions in order to get the complete inertia tensor.

The stiffness of the connections (joints) between the different segments is one of the parameters having a major effect on the motions of the dummy segments in a crash environment. These joint resistive properties are determined using various static and dynamic test methods. In these tests the range of motion corresponding to a joint co-ordinate is determined as function of the externally applied load. If a joint has more than one degree of freedom, like in a universal joint, for each degree of freedom separate measurements are conducted, keeping the others fixed. Since the actual joint resistance often will depend on the value of multiple joint co-ordinates, large test series may be required. In practice up to now this dependency on more than one degree of freedom is neglected and joints are tested with the other degrees of freedom fixed.

The last step is the specification of the surface compliance properties. Static as well as dynamic measurements with several penetrating surfaces must be performed at different locations on the dummy segments. The surface compliance is dependent on the skin covering thickness and density as well as the compliance of the underlying structure. If the dummy part to be modelled will be represented by a finite element model, material parameters describing the involved dummy materials have to be determined. The surface compliance tests can be used for model validation purposes in this case.

On the basis of these measurements a model of the dummy can be compiled now. After formulating a model, verification simulations are carried out to insure that the model adequately represents the complete dummy. For this purpose well controlled impactor tests and sled tests with the assembled dummy at different acceleration levels are used. If results are not completely satisfying further model refinements with corresponding input measurements may be required. A well validated computer model allows the user to apply the model for predictive simulations of events outside the range of validated simulations.

21.3.3 Examples of crash dummy models

The need for well-validated databases of crash dummies has been recognised by many organisations in the past and has resulted in a number of (co-operative) research efforts to develop such databases. A detailed presentation of these efforts is out of the scope of this Chapter. However worthwhile to mention here are the activities in the mid-eighties concerning the Hybrid III crash dummy. A series of frontal sled tests using a Hybrid III dummy at a rigid seat at 3 different severity levels was conducted in 1985 by Prasad [21.19]. The results are available for a SAE subcommittee for the purpose of validation of dummy databases of the ATB and MADYMO programs. These validation efforts were presented at the 1998 SAE congress by several authors [21.20-22] which resulted in a number of recommendations for further improvement of the quality of the Hybrid III dummy database.

Figs. 21.20 illustrates some of the validated "standard" multi-body dummy databases currently available with the MADYMO program. In addition to the dummy models shown in Figure 21.20 also models for the Hybrid II dummy, the various side impact dummies (EUROSID, BIOSID and DOT-SID) and various headforms and other impactors are available. Figure 21.21 shows 2 different MADYMO models for the EUROSID dummy: a model with arbitrary (facet) surfaces (left) and a finite element model.

2.4 Modelling the real human body

2.4.1 Introduction

A model of the real human body is much more difficult to develop than a model of a physical crash dummy. Mathematical modelling of the real human body potentially offers improved biofidelity compared to crash dummy models and allows the study of aspects like body size, body posture, muscular activity and post fracture response. Detailed human body models potentially allow analysis of injury mechanisms on a material level.

A large number of models describing specific parts of the human body have been published but only a few of these models describe the response of the entire human body in impact conditions. Models simulating the response of car occupants have been published for lateral loading by Huang [21.23,21.24] and

Irwin [21.25], frontal loading by Ma et. al. [21.26] and rearward loading by
Jakobsson et. al. [21.27] and van den Kroonenberg et. al. [21.28]. A model for
vertical loading has been published by Prasad and King [21.29], pedestrian
models have been published by Ishikawa et al. [21.30] and Yang et al. [21.31]
and a child model in a child restraint system by Wismans et. al. [21.32].

Figure 21.20 Examples of multi-body crash dummy models: Hybrid III dummy family(top);
5th, 50th and 95th%, USA child dummies(middle); Hybrid III 3yr and 6yr, and
Crabi 12 month and TNO child dummies (bottom); P3/4, P3, P6 and P10

21.4.2. Anthropometry

In occupant crash simulations the program GEBOD is often used to generate models representing arbitrary human body sizes. GEBOD produces geometric and inertia properties of human beings [21.33]. Joint resistance models for an adult male GEBOD model are described by Ma et al. [21.26]. GEBOD generates a model consisting of 15 segments: head, neck, upper arms, lower arms, thorax, abdomen, pelvis, upper legs, lower legs and feet. Computations for the geometrical parameters and mass distribution are based on a set of 32 body measurements. From these 32 parameters body segment sizes and joint locations are derived. Segments are described by ellipsoids except for the thorax and feet where more complex approximations (so-called elliptical solids) are used. Inertial properties are estimated by calculating the inertial properties of each segment ellipsoid or elliptical solid, assuming homogeneous body density. The 32 body parameters can be measured at a subject or can be generated by GEBOD using regression equations on the basis of body height and/or weight for both adult males and females. For children, regression equations are available on the basis of height, weight, age and combinations of these parameters. A major limitation of GEBOD is the approximation of body segments by simple geometrical volumes.

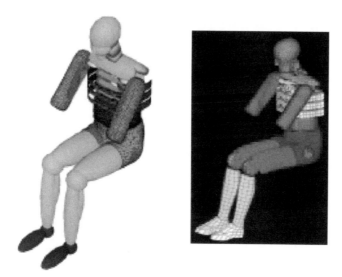

Figure 21.21 MADYMO multi-body facet model (left) and finite element model (right) of the EUROSID dummy

More recently the RAMSIS model [21.34] has been used anthropometry source for human body models [21.35]. RAMSIS has primarily been developed for ergonomic analyses and allows the generation of models with a wide range of anthropometry parameters. The RAMSIS model describes the human body as a set of rigid bodies connected by kinematic joints and the skin is described as a triangulated surface. RAMSIS provides a detailed geometric description of the body segments based on extensive anthropometric measurements on various civilian populations including automotive seated postures. The skin of the entire body is described as one "continuous" surface. Segment mass and centres of gravity are derived in RAMSIS using this realistic geometric description. RAMSIS provides a mathematical prediction for the increase of the average body height of the entire population during a given time period ("secular growth").

Anthropometric studies have shown that the body dimensions of each individual can be classified according to three dominant and independent features. These features are body height, the amount of body fat, and body proportion, i.e. the ratio of the length of the limbs to the length of the trunk. Using this classification scheme RAMSIS describes the entire population in a realistic way. This method takes into account the correlation between body dimensions which are disregarded in GEBOD. (For instance tall persons typically have long legs combined with a comparably short trunk.).

A translator has been developed to convert RAMSIS models into MADYMO models. The resulting model contains joint locations, joint ranges of motion, segment masses and centres of gravity and a triangulated skin connected to various body segments. Inertia properties are derived by integration over segment volume assuming a homogeneous density. The conversion can be performed for any anthropometry specified in RAMSIS and examples of such models are shown in Figure 21.22.

2.4.3   Examples of a human body model

Two examples will be shown here, the first one being a multi-body model and the second a full finite element model.

The multi-body model is the 50th percentile male model from Figure 21.22. This model has been validated in a number of crash conditions including frontal, rearward and lateral volunteer tests and several types of cadaver tests

[21.35]. The simulation shown in Figure 21.23 is a 15 g frontal sled tests with human volunteers conducted by the Naval Biodynamics Laboratory in New Orleans. For the neck in this model a global 7-segment model is available with lumped properties as well as detailed neck model with separate representations for the facet joints, intervertebral discs, ligaments and (active) muscle response. Figure 21.24 shows the response of this neck model in comparison with human volunteer response. In the model active muscle response was taken into account. The performance of this neck model appeared to be much more realistic than the neck behaviour of the current Hybrid III crash dummy [21.36]. Also for some of the other body parts, like for the ankle/foot and the head/brain, detailed multi-body and/or finite elements representations are available in this 50th percentile male model. Figure 21.25 illustrates a finite element model of the brains developed at the Eindhoven Technical University.

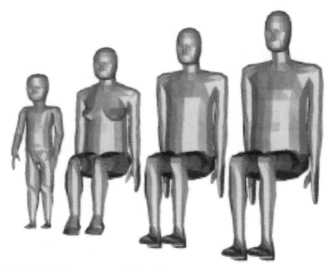

Figure 21.22 MADYMO human models of various body sizes generated from the RAMSIS model, from left to right: 3 year old child, extremely small female, 50th percentile male, extremely large male

The second example is shown in Figure 21.26. It is a finite element model representing a 50th percentile male developed by Lizee et. al. in the RADIOSS programme package [21.37]. Detailed representations of the neck, shoulder, thorax and pelvis have been included and the resulting model has been validated in more than 30 test configurations. The model has more than 10.000 elements. Head and arms and legs were represented as rigid bodies.

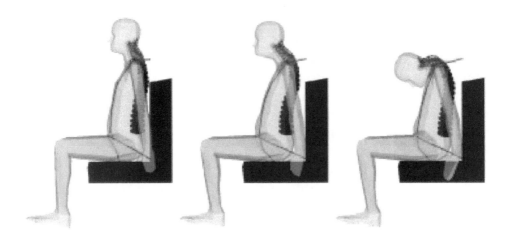

Figure 21.23 Simulation of 50th male human body model in 15g volunteer test [21.35]

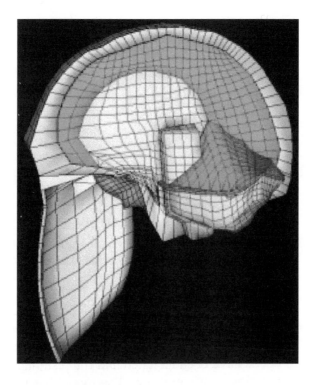

Figure 21.24 MADYMO finite element brain model developed at the Eindhoven Technical University, The Netherlands

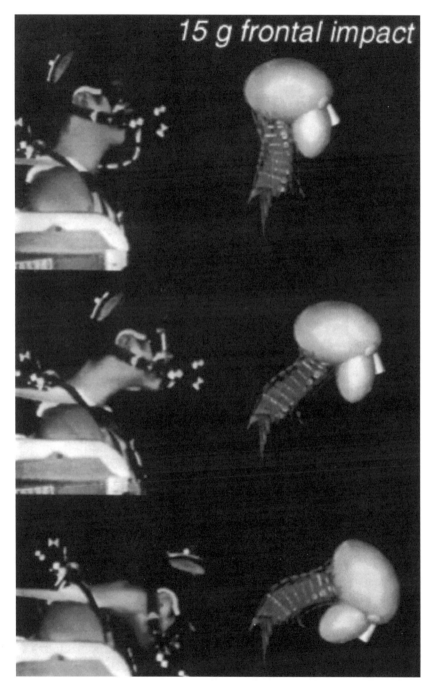

Figure 21.25 Comparison of human neck model response and volunteer response [21.36]

Figure 21.26 Finite element human body model developed by Lizee et. al.[21.37]

21.5   Concluding remarks

The earliest numerical models of the full human body have been based on multibody techniques. More recently also finite element techniques have been used for this purpose. A mayor advantage of the multibody approach is its capability of simulating, in an efficient way, spatial motions of mechanical systems with complex kinematical connections like they are present in the human body and in parts of the vehicle structure. The advantage of the finite element method is the capability of describing (local) structural deformations and stresses in a realistic way. But the creation of a finite element model is a time consuming job and the availability of realistic material data is limited, particular in case of biological tissue response. Furthermore relatively large computer times are required to perform a finite element crash simulation, making the method less attractive for complex optimisation studies involving many design parameters.

A general advantage of computer crash simulations over crash tests with mechanical human substitutes (crash dummies) is that the safety performance of design concepts and the effect of changes in the design can be studied efficiently, sometimes even without a prototype to be build (virtual testing). An important condition for the usage of such models is that well-validated databases of the human body are available. Continuous efforts are needed to further improve the quality of existing human body models in order to allow their usage in even a wider range of applications. Standards for validation procedures and performance criteria are needed in order to further enhance and extent the applicability of crash simulations. In the past some attempts have taken place to develop such standards. Table 21.1 shows a model Validation Index developed in the early eighties by the "Analytical Human Simulation Task Force" of the SAE Human Biomechanics and Simulation Subcommittee (HBSS) [21.38]. The index with levels from 0-8 was agreed upon by the Committee but a number if issues were not resolved like the number and type of tests to be conducted.

Table 21.1 Validation Index proposed by SAE sub committee [21.38]

Class	Characteristics
Level 0	No agreement between predicitions of model and "reference event"
Level 1	Qualitative agreement:
	a) Trends of predicted parameters same
	b) Kinematics correspond qualitatively
	c) Contacts between the occupant and vehicle interior are the same in general
Level 2	HIC and similar indicators predicted by simulation are within 20% of those obtained in reference event
Level 3	Peak values of important occupant responses limited to a relative error of 20% (20% on vector magnitude, 11.31 deg on vector direction)
Level 4	Same as level 3 except 5%
Level 5	Timing of peaks of important vector responses limited to 5% relative difference
Level 6	All peaks and valleys in the duration of time-dependent predictions must match the reference event within 10%
Level 7	Same as level 6 except 5%
Level 8	1% relative error, point-by-point, over the durations of the reference and predicted events

Models of the human body can be subdivided into models of crash dummies and models of the real human body. Many models of crash dummies have been developed in the eighties/nineties and extensive series of validation studies have been conducted with rather impressive results. Also in the field of real human body models in the nineties rather promising results have been achieved. An important advantage of real human body models is that they allow the study of the effect of body size, posture influence as well as muscular activity. A major step forward will have been made once mathematical models reach a stage where they offer a more realistic representation of the human body than current crash test dummies do. Some recent achievements in this field indicate that this stage has been reached closely now.

A unique advantage of a design strategy based on real human body crash models over a design strategy based on crash tests with dummies is the possibility to benefit without delay, in principle, from new scientific knowledge on injury mechanisms and injury criteria obtained through biomechanical research. In case of a crash test dummy based design strategy usually a long period elapses before new findings actually can be implemented in crash dummy hardware. For example the current most frequently used crash dummy, the Hybrid III dummy, is based to a large extent on biomechanical knowledge of more than twenty years old. New scientific findings have so far seldom resulted in improvements in the dummy design particularly since safety regulations which specify this dummy as a regulatory test device tend to freeze the specifications in the regulation for a long period.

Apart from design studies and the analysis of biomechanical tests, an increased usage of computer models also can be observed in the area of accident reconstruction and litigation. Application of computer models in this field should be handled with much care, due to the limited level of development of real human body models for different body sizes, the usually large number of unknown accident parameters and the lack of experimental data available for validation for the case under consideration. Development of a code of practice with guidelines for usage of models in this field is highly recommended.

Several areas can be identified in the field of human body crash simulations where further developments should take place. As far as crash dummies is concerned, in particular realistic models for the foam type structures (skin and damping material) are required. Area's of future developments in the field of real human body models include further improvements in the description of the non-linear dynamic behaviour of muscles (incl. neuro-muscular control), the modelling of complex human joints and the study of constitutive equations and parameters for biological materials (e.g. brain, skin).

This Chapter has concentrated in particular on the usage of multi-body technology for human body modelling. Further refinements of the models using finite element techniques constitutes an increase in model complexity, however, with the advantage that detailed stress and strain analysis can be performed. For detailed studies of injury mechanisms in specific body part this is a necessary and feasible approach. Boundary conditions for such segment models may be obtained from experiments or results obtained through more global models. The usage of finite element techniques coupled with multibody techniques will allow the user to benefit from the capabilities of both approaches and will offer the flexibility of merging more global multi-body models with, whenever needed, detailed representations for certain parts in the model.

REFERENCES

[21.1]   Lobdell T.E., Impact response of the human thorax, In *Human Impact Response: Measurement And Simulation*, Plenum Press, New York, 201-245, 1973.

[21.2]   McHenry, R.R., Analysis of the dynamics of automobile passenger restraint systems, *Proceedings of the 7th Stapp Car Crash Conference*, 207-249, 1963.

[21.3]   *EU compatibility project*, Final report for publication, European Commission, Brussel, 2000.

[21.4]   Shugar, T.A., *A finite element head injury model*, Final report Vol 1 Contract DOT HS 289-3-550-IA, NHTSA , Washington DC, July 1977.

[21.5]   Bruijs, W.E.M., *Subcycling in Transient Finite Element Analysis*. Thesis, Department of Mechanical Engineering, Eindhoven University of Technology, Eindhoven, The Netherlands, 1990.

[21.6]   Robbins, D.H., Bowman, B.M. and Bennett, R.O., The MVMA two-dimensional crash victim simulation, *Proceedings of the 18th Stapp Car Crash Conference*, 657-678, 1974.

[21.7]   Robbins, D.H., Bennett, R.O. and Bowman, B.M., User-oriented mathematical crash victim simulator, *Proceedings of the 16th Stapp Car Crash Conference*, 128-148, 1972.

[21.8]   Huston, R.L., Hessel, R. and Passerello, C., A three-dimensional vehicle-man model of collision and high acceleration studies, SAE paper No. 740725, Society of Automotive Engineers Inc., (1974).

[21.9]   Fleck, J.T., Butler, F.E. and Vogel, S.L., An improved three-dimensional computer simulation of motor vehicle crash victims, *Final Technical Report No. ZQ-5180-L-1*, Calspan Corp,. (4 Vols.), 1974.

[21.10] King, A.I. and Chou, C.C., Mathematical modelling, simulation and experimental testing of biomechanical system crash response, *J. Biomechanics*, **9**, 301-317, 1976.

[21.11] Wismans, J. and Obergefell, L., Data bases and analytical modelling, In *AGARD Advisory Report 330 Anthropomorphic Dummies for Crash and Escape System Testing*, Chapter 8 , AGARD/AMP/WG21, 1996.

[21.12] *MADYMO Theory Manual, Version 5.4*, TNO Automotive, Delft, The Netherlands, 1999.

[21.13] Prasad, P., An overview of major occupant simulation models, In *Mathematical Simulation Of Occupant And Vehicle Kinematics*, SAE Publication P-146, SAE paper No. 840855, 1984.

[21.14] Prasad, P., Comparative evaluation of the MVMA2D and the MADYMO2D occupant simulation models with MADYMO-test comparisons, In *10th International Technical Conference on Experimental Safety Vehicles*, Oxford, 1985.

[21.15] Prasad, P. and Chou, C.C., A review of mathematical occupant simulation models, In *Crashworthiness and occupant protection in transportation systems, Proceedings AMD-Vol. 106, BED-Vol. 13 of the Winter Annual Meeting of ASME*, 1989.

[21.16] Wittenburg, J., *Dynamics of Systems of Rigid Bodies*, B.G. Teubner, Stuttgart, 1977.

[21.17] Prasad, P. and Padgaonkar, A.J., Static-to-dynamic amplification factors for use in lumped mass vehicle crash models, SAE paper No. 810475, 1981.

[21.18] Kaleps, I. and Whitestone, J., Hybrid III Geometrical and Inertial Properties, SAE paper No 880638, 1988.

[21.19] Prasad P., Comparative evaluation of the dynamic response of the Hybrid II and Hybrid III dummies, SAE paper No. 902318, In *Proceedings of the 34th Stapp Conference*, 1990.

[21.20] Obergefell, L., Kaleps, I. and Steele, S., Part 572 and Hybrid III dummy comparisons in sled test simulations, SAE paper No. 880639, SAE PT-44,1988.

[21.21] J. Wismans and Hermans, J.H.A., MADYMO 3D Simulations of Hybrid III Dummy Sled Tests, SAE paper No. 880645, SAE PT-44, 1988.

[21.22] Khatua, T., L. Chang and Pizialli: "ATB simulation of the Hybrid III dummy in sled tests", SAE paper No. 880646, SAE PT-44, SAE Int. Congress and Exposition, Detroit, Society of Automotive Engineers Inc., 1988.

[21.23] Huang Y., A.I. King, and CavanaughJ.M., A MADYMO model of near-side human occupants in side impacts, *J. of Biomechanical Engineering*, **116**, 228-235, 1994.

[21.24] Huang Y., A.I. King, and Cavanaugh, J.M., Finite element modelling of gross motion of human cadavers in side impact, SAE paper No 942207, 1994.

[21.25] Irwin, A.L. *Analysis and CAL3D Model of the Shoulder and Thorax Response of Seven Cadavers Subjected to Lateral Impacts*, Ph.D. Thesis Wayne State University, 1994.

[21.26] Ma D., Obergefell A., and Rizer, A., Development of human articulating joint model parameters for crash dynamics simulations, SAE paper No 952726, 1995.

[21.27] Jakobsson L., Norin H., Jernstrom C., et al. Analysis of different head and neck responses in rear-end car collisions using a new humanlike mathematical model, In *Proceedings of the 1994 IRCOBI Conference*, 109-125, 1994.

[21.28] Kroonenberg, A. van den, Thunnissen, J., and Wismans, J., A human model for low severity rear-impacts, In *Proceedings of the 1997 IRCOBI Conference*, 1997.

[21.29] Prasad, P., and King, A.I., An experimentally validated dynamic model of the spine, *J. Appl. Mech.*, 546-550, 1974.

[21.30] Ishakawa, H., Kajzer, J., and Schroeder, G., Computer simulation of impact response of the human body in car-pedestrian accidents, In *Proceedings of the 37th STAPP Car Crash Conference*, SAE paper No 933129, 1993.

[21.31] Yang, J.K, and Lovsund, P, Development and validation of a human body mathematical model for simulation of car-pedestrian impacts, In *Proceedings of the 1997 IRCOBI Conference*, 1997.

[21.32] Wismans, J., Maltha, J.W., Melvin, J.W. and Stalnaker, R.L., Child restraint evaluation by experimental and mathematical simulation, In *Proceedings of the 23rd Stapp Car Crash Conference*, SAE paper No. 791017, 1979.

[21.33] Baughman, L.D., *Development of an Interactive Computer Program to Produce Body Description Data*, University of Dayton Research Institute, Ohio, USA, Report nr. AFAMRL-TR-83-058, NTIS doc. no. AD-A 133 720, 1983.

[21.34] Geuß H., *Entwicklung eines anthropometrischen Mebsystems für das CAD-Menschmodel Ramsis*, PhD thesis, München University, 1994.

[21.35] R., Happee, M., Hoofman, A.J., van den Kroonenberg, P. Morsink and Wismans, J., A mathematical human body model for frontal and rearward seated automotive impact loading, In *Proceedings of the 42nd Stapp Car Crash Conference*, 1998.

[21.36] Wismans, J., van den Kroonenberg, A.J., Hoofman, M.L.C. and van der Horst, M.J., Neck performance of human subjects in frontal impact direction, RTO Specialist' Meeting *Models for aircrew safety assessment: uses, limitations and requirements*, Dayton, USA, 26-28 October 1998.

[21.37] Lizee, E., Robin, S., Bertholon, N., Le Coz, J.Y., Besnault, B. and Lavaste, F., Development of a 3D Finite Element Model of the Human Body, In *Proceedings of the 42nd Stapp Car Crash Conference*, Tempe, USA, November 2-4, 1998.

[21.38] P. Prasad, Occupant simulation models: experiment and practice, *Crashworthiness of Transportation Systems: Structural Impact and Occupant Protection*, J.A.C. Ambrosio et. al. (eds), Kluwer Academic Publishers, 209-219, 1997.

22. NECK PERFORMANCE OF HUMAN SUBSTITUTES IN FRONTAL IMPACT DIRECTION

In the past several laboratories have conducted human subject tests in order to derive biofidelity performance requirements for crash dummies and computer models. Both human volunteer and human cadaver tests have been conducted. Particularly noteworthy are the human volunteer tests conducted at the Naval Biodynamics Laboratory (NBDL) in New Orleans. In an extensive test program a large number of human subjects were exposed to impacts in frontal, lateral and oblique directions. Detailed analyses of these tests have been conducted and presented in various publications. Based on these results, a set of biofidelity performance requirements was developed. These requirements include trajectories and rotations of the head as well as acceleration requirements and data on the neck loads.

The objective of this paper is to compare the performance of various human neck models with the observed response in the volunteer tests. Concerning mechanical models, the neck of the Hybrid III dummy, which is the dummy currently specified in motor vehicle safety regulations, as well as the neck of the new THOR dummy will be evaluated. It will be shown that the neck of the THOR dummy offers more biofidelity than the Hybrid III dummy neck. Regarding mathematical neck models, a neck model developed in the MADYMO crash simulation program will be evaluated. It will be shown that the mathematical model which includes a representation of vertebrae, ligaments and active muscle response is able to reproduce the observed human subject response more accurately than the available mechanical models.

22.1   Introduction

Since 1967, human volunteers have been exposed to impact acceleration experiments in which the kinematic responses of the head-neck system have been measured. Most noteworthy are the experiments conducted at Wayne State University [22.1-3] and at the Naval Biodynamics Laboratory (NBDL) [22.4-7] in New Orleans. Figure 22.1 illustrates the head-neck motions as observed in 15g frontal impact with human volunteers conducted at NBDL.

The results of this type of experiments have been analysed by a number of investigators and were used to define biofidelic performance requirements for crash dummy development. Well-known are the analyses of Mertz and Patrick [22.3] who defined neck performance requirements on the basis of the relationship between the moment of force about the occipital condyles due to the forces acting on the head and the angular position of the head relative to the torso. The mechanical neck of the Hybrid III dummy, which is the dummy specified in the current motor vehicle safety regulations, has been based on these requirements.

Figure 22.1   Human volunteer head-neck response in a 15.6 g frontal impact at NBDL

However, such a requirement is not a sufficient condition to ensure a humanlike response. In fact an infinitely stiff neck with a flexible joint at the top as a head-junction could theoretically fulfil the conditions defined by Mertz and Patrick. Both Mertz et al. [22.8] and Melvin et al.[22.9] discussed the need for additional displacement requirements, such as the trajectory of the head centre of gravity relative to the torso, but no such requirements were formulated by these investigators.

Wismans et al. [22.10] developed neck performance requirements in which the head trajectories were included. These requirements, which were formulated for frontal, lateral and oblique impacts, were based on the analysis of a large number of human volunteer experiments conducted between 1981 and 1985 at NBDL. In these analyses rotations of T1 were neglected. In 1987, the analyses of the frontal high severity experiments were extended with postmortem human subject (PMHS) experiments conducted at the University of Heidelberg [22.11]. The impact severity ranged from 15 - 23 g and the test set-up was identical to the set-up at NBDL. The head centre of gravity trajectories were of the same order of magnitude as in the volunteer tests but the head rotations were found to be larger in the PMHS experiments. The differences between the volunteers and PMHS response were largely attributed to the absence of muscle activity in the PMHS tests. In the PMHS tests T1 rotations of more than 20 degrees in frontal direction were observed. Comparison with the T1 measurements in the volunteer tests at NBDL revealed some inaccuracies in the measured volunteer T1 kinematic response, due to non-rigid mounting of the T1 instrumentation to the T1 vertebra.

More recently Thunnissen et. al. [22.12] performed a new analysis of frontal human head-neck volunteer tests. A correction method was developed to account for the observed errors in the human volunteer T1 response. As a result for frontal impacts a similar set of response data could be developed but now expressed relative to a rotating T1 co-ordinate system, thereby incorporating the rotations of T1. As a consequence the head excursion became smaller than in the earlier analyses.

The objective of this chapter is to compare the performance of various human neck models with the observed response in the volunteer tests. Regarding mechanical models, the neck of the Hybrid III dummy, which is the dummy currently specified in motor vehicle safety regulations, as well as the neck of the new THOR dummy will be evaluated. Concerning mathematical neck models, a model with passive and active muscle response formulated in the MADYMO crash simulation program will be evaluated.

22.2   Response requirements

The set of biofidelity response requirements to be used here for assessment of the performance of mechanical and mathematical neck models is a subset of data presented in the analysis performed by Thunnissen et al. [22.12]. This subset consists of (see Figure 22.2):

1 the occipital condyle (OC) trajectory relative to T1
2 head rotation (flexion) as a function of time relative to T1
3 the neck angle as function of the head angle
4 the mid-sagittal head rotational acceleration as a function of time
5 the mid-sagittal moment of force with respect to the occipital condyles joint
6 the resultant head centre of gravity linear acceleration as a function of time

The neck angle is defined here as the angle between the T1 z-axes and a line connecting the origin of the T1 frame and the occipital condyles. From the neck angle vs. head angle relation, the so-called "head lag" can be observed during the initial part of the motion: the head translates and shows negligible rotation while the neck is already deforming.

The study of Thunnissen et al. [22.12] was based on 9 sled tests with 5 different subjects. The sled velocity change was 17 m/s and the peak sled acceleration approximately 15 g. The NBDL volunteers were strapped tightly into their seat, allowing just the head, neck and a small part of the upper thoracic spine to deform during the impact. The corridors shown in Figure 22.2 represent the mean values minus and plus the standard deviation.

22.3   Mechanical dummy neck performance

Two mechanical neck models were evaluated: the neck of the 50th percentile Hybrid III dummy and the neck of the THOR dummy (fig 22.3). The THOR dummy is a new frontal crash test dummy which has been developed in the US by GESAC under contract by NHTSA (US National Highway Traffic Safety Administration).

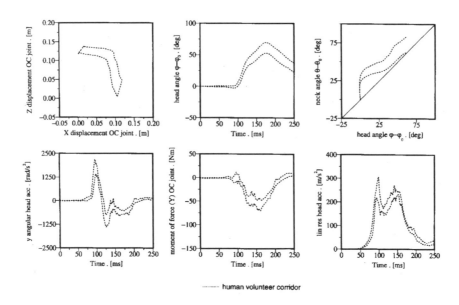

Figure 22.2   Response corridors for frontal impacts based on Thunnissen et. Al. [22.12]

Rather than testing the complete Hybrid III and THOR dummy, only the head-neck assemblies were evaluated . These are placed directly on a HyGe sled. Thunnissen et. al. [22.12] concluded that the linear T1 acceleration in X direction is the most relevant T1 motion parameter for the head-neck response during frontal impact at this high g level. Therefore, the two dummy head-neck systems are accelerated with a sled pulse equivalent to the average X-acceleration measured at the first thoracic vertebra of the NBDL volunteers (Figure 22.4).

Figure 22.5 shows the results for both neck systems together with the frontal neck response corridors. More details of these tests are given by Hoofman et al. [22.13]. The most striking difference between the two dummy necks concern the occipital condyle trajectories: the trajectory of the THOR dummy approximates the response corridor quite well, only the forward displacement is somewhat too large. The excursion of the occipital condyles joint of the Hybrid-III neck is too small compared to the response corridors, particularly in Z-direction. The head rotation for the THOR dummy appears to be slightly too large. The shape of the head lag curve (neck angle as function of head angle) for the THOR neck is similar to the head lag performance requirement corridor, but still not within the corridor. The Hybrid-III head-neck system shows no head lag. The head angular acceleration of the THOR dummy appears to be quite realistic (the Hybrid-III head was not equipped with a nine accelerometer array). The torque at the occipital condyles joint of the Hybrid-III is too small compared to the requirement (no torque was available for the THOR occipital condyles). The resultant linear head centre of gravity accelerations of THOR and Hybrid-III are similar: both acceleration signals are close to the performance requirement corridor.

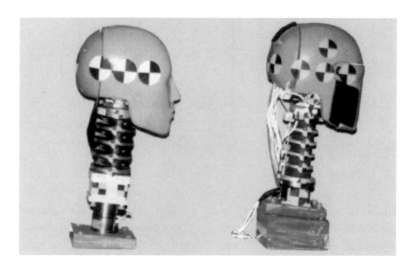

Figure 22.3   Hybrid-III Head-Neck System (left) and THOR Head-Neck System (right)

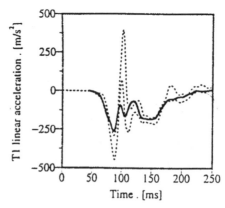

Figure 22.4   Sled test acceleration in the dummy neck tests based on T1 accelerations in the NBDL volunteer tests

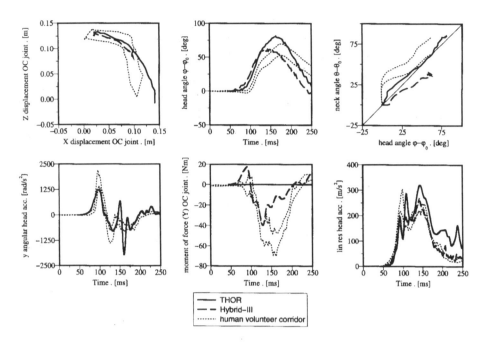

Figure 22.5 Results for the Hybrid III and THOR neck in 15 g frontal tests together with a corridor derived from human volunteer tests

22.4   Mathematical neck model performance

De Jager [22.14] developed a detailed 3D mathematical model of the neck in the crash simulation program MADYMO. The model included the vertebrae as rigid bodies, the facet joints and ligaments. The most important muscles in the neck were represented by simple cord elements connecting the muscle attachment points. Active muscle behaviour was simulated. The model was compared with the frontal and lateral volunteer tests conducted at NBDL.

It was concluded that modelling the neck muscles as simple cords was not satisfactory. For this reason the neck model is enhanced by using a curved muscle model (Figure 22.6). Input for the model was the T1 acceleration measured in volunteer tests. Results were presented by van der Horst [22.15]. The model was validated for frontal volunteer tests at different severity levels (3g, 6g, 8g, 10g, 12g and 15g). The study showed that muscle contraction has a large influence on the head-neck response.

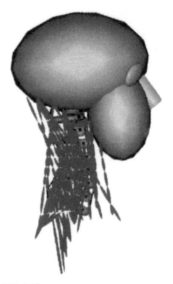

Figure 22.6 3D neck model [22.15]

Figure 22.7 shows the model predictions together with the response corridors from the volunteer tests for the 15 g frontal pulse. Results for 2 different muscle activation levels are included, i.e. for passive muscle response (no muscle activation) and for 100% activated muscles. In case of passive muscle response the neck appears to be too flexible: trajectories and head rotations are much too large. In case of 100% activated muscles, the head-neck response appears to lie largely within the specified corridors for trajectories and head rotations. Also the head lag appears to be predicted accurately by the model with 100% activation level for the muscles. In case of passive muscle response the initial part of the head lag response in not affected. Furthermore it can be observed that the angular and resultant head accelerations are hardly affected by the muscle response. The torque-time histories show difference between the passive and active muscle response, and the active model is closer to the volunteer response.

In the low severity simulations conducted by van der Horst et al. [22.15] it was observed that a lower activation level of the muscles and a larger reflex delay had to be introduced in the model in order to achieve realistic predictions for the neck response.

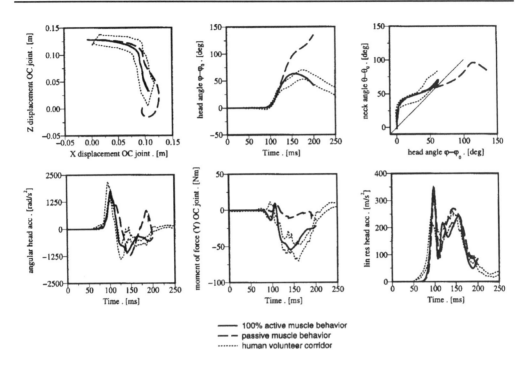

Legend:
——— 100% active muscle behavior
— — passive muscle behavior
·········· human volunteer corridor

Figure 22.7   Results from the 3D MADYMO neck model of van der Horst et al. [22.15] for passive and 100% active muscle behaviour in 15 g frontal tests together with a corridor derived from human volunteer tests

22.5   Discussion and conclusions

A human-like mechanical or mathematical neck model must not only fulfil requirements on the external anthropometry and mass, but it should also describe the dynamic human head and neck response in a realistic way. In this study a set of response data is applied, based on analyses of human volunteer tests performed by Thunnissen et al. [22.12]. The set consists of occipital condyle trajectories relative to T1, head flexion time histories, the so-called "head-lag" which specifies the relation between the head and neck flexion, the occipital condyle torque time history and angular and resultant linear head acceleration time histories. These response data are more extended than the current Mertz response requirements for the Hybrid III, i.e. head rotation versus moment at the occipital condyles.

In interpreting these response data it should be realised that the tested subjects were young males, recruited from the U.S. navy population (active duty) and screened for good health, no history of cervical spine pain or injury and that they pre-tensed their muscles in order to reduce the effects of the impact. For a general population a less stiff response is expected. Other limitations to be considered include the effect of the volunteer head instrumentation mount, the effect of the initial head-neck position of the volunteers, and the limited impact severity in the volunteer tests compared to real-world crashes causing injuries. See for a discussion on these limitations Thunnissen et al. [22.12].

The behavior of the Hybrid III and THOR head-neck system has been compared to these response requirements. The comparison shows that the Hybrid III neck is too stiff to respond in a human-like manner during frontal impact. This seems to be a major limitation of the Hybrid III neck since in case of head impact the head might impact the vehicle structure in a different location than when a more biofidelic neck is used. The new THOR neck shows a more realistic response in particular for the head trajectories. But it should be noted that the design of the THOR neck is rather different compared to the Hybrid III neck. For example, the THOR neck is longer than the Hybrid III neck: the T1 location of the THOR neck is located about 65 mm above the neck base [22.13]. The effect of this is discussed in Ref. [22.13]. Due to this longer neck it can be questioned whether the selected test method where the measured T1 acceleration in the volunteer tests is applied to the neck base, is valid for the THOR neck. This should be further investigated.

The 3D computer neck model has been formulated in the MADYMO crash simulation program. This program combines both the multibody and the finite element techniques. For the model presented only multibody techniques were employed. In order to achieve accurate model predictions simulation of active muscle response were of critical importance. A model with 100% activated muscles appeared to lie almost completely within the specified response corridors, which appeared to be more humanlike than the THOR neck.

A final interesting observation is that all models, including the models that show only limited biofidelity for the relative head trajectory, closely approximate the acceleration response corridors. Therefore, acceleration based response requirements seem not to be a good discriminator for judging the biofidelity of neck motions.

REFERENCES

[22.1]   Patrick, L.M. and Chou, C.C., *Response Of The Human Neck In Flexion, Extension And Lateral Flexion*, Report no. VRI-7.3, Wayne State University, Detroit, 1976.

[22.2]   Cheng, R., Mital, N.K., Levine, R.S. and King, A.I., Biodynamics of the living human spine during -Gx impact acceleration, In *Proceedings of the 23th Stapp Car Crash Conference*, SAE paper No 791027, 1979.

[22.3]   Mertz, H.J. and Patrick, L.M., Strength and response of the human neck, In *Proceedings of the 15th Stapp Car Crash Conference*, SAE paper No 710855, 1971.

[22.4]   Ewing, C.L. and Thomas, D.J., Torque versus angular displacement response of human head to -Gx impact acceleration, In *Proceedings of the 17th Stapp Car Crash Conference*, SAE paper No 730976, 1973.

[22.5]   Ewing, C.L., Thomas, D.J., Lustick, L., Becker, E., Willems, G., and Muzzy, W.H. III, The effect of the initial position of the head and neck on the dynamics response of the human head and neck to -Gx impact acceleration, In *Proceedings of the 19th Stapp Car Crash Conference*, SAE paper No. 751157, 1975.

[22.6]   Ewing, C.L., Thomas, D.J., Lustick, L., Muzzy, W.H., Willems, G. and Majewski, P.L., The effect of duration, rate of onset, and peak sled acceleration on the dynamic response of the human head and neck." In *Proceedings of the 20th Stapp Car Crash Conference*, SAE paper No. 760800, 1976.

[22.7]   Ewing, C.L., Thomas, D.J. and Lustick, L.S., Multi-axis dynamic response of the human head and neck to impact acceleration, In *Proceedings Aerospace Medical Panel's Specialist Meeting*. AGARD No. 153, 1978.

[22.8]   Mertz, H.J., Neathery, R.F. and Culver, C.C., Performance requirements and characteristics of mechanical necks, *Human Impact Response*, W.F. King and H.J. Mertz (Eds.), Plenum Press, NY, 1973.

[22.9]   Melvin, J.W., McElhaney, J.H. and Roberts, V.L., Evaluation of dummy neck performance, *Human Impact Response*, W.F. King and H.J. Mertz (Eds.), Plenum Press, NY, 1973.

[22.10] Wismans, J.S.H.M., van Oorschot, H. and Woltring, H.J., Omni-Directional human head-neck response, In *Proceedings of the 30th Stapp Car Crash Conference*, San Diego, 313-332, 1986.

[22.11] Wismans, J., Philippens, M., Oorschot, E., Kallieris, D. and Mattern, R., Comparison of human volunteer and cadaver head-neck response in frontal flexion, In *Proceedings of the 31th Stapp Car Crash Conference*, SAE paper No 872194, 1987.

[22.12] Thunnissen, J., Wismans, JS.H.M., Ewing, C.L. and Thomas, D.J., Human volunteer head-neck response in frontal flexion: a new analysis, In *Proceedings of the 39th Stapp Car Crash Conference*, SAE paper No 592721, 1995.

[22.13] Hoofman, M, van Ratingen, M, and Wismans, J.S.H.M., Evaluation of the dynamic and kinematic performance of the THOR dummy : neck performance, In *Proceedings of the 1998 International IRCOBI Conference on the Biomechanics of Impact*, Göteborg, Sweden, September 16-18, 1998

[22.14] Jager, de M., Sauren, A., Thunnissen, J. and Wismans, J.S.H.M., A three-dimensional head-neck model: Validation for frontal and lateral impacts, In: *Proceedings of the 38th Stapp Car Crash Conference*, SAE paper No 942211, 1994.

[22.15] Horst, van der M.J., Thunnissen, J.G.M., Happee, R. Haaster, van R.M.H.P., Wismans, J.S.H.M., The influence of muscle activity on head-neck response during impact, In *Proceedings of the 41th Stapp Car Crash Conference*, SAE paper No 973346, 1997.

Part VI

INJURY BIOMECHANICS

D. Viano
General Motors Corporation, Warren, MI, USA

23. HEAD INJURY BIOMECHANICS

Impact to the head causes translational and rotational acceleration of the skull that sets up differential motion with the brain. This causes strain in neural tissues and at points of tethering. With a severe enough impact, the relative brain motion can also tear bridging veins at the cortex, lacerate the base of the brain and contuse brain tissue adjacent to internal membranes and tethering points. Brain function can be damaged by diffuse axonal injury, related to high strain and strain-rate deformation of neural tissue. The Head Injury Criterion (HIC) assesses injury risks from translational acceleration. It is proportional to $\int a^{2.5} dt$, where a is the resultant head acceleration. HIC = 1000 represents a 16% risk of serious brain injury. Limits are also proposed for peak 3 ms translational acceleration, rotational velocity and rotational acceleration of the head. Various mechanical and mathematical tools are used as human surrogates to simulate impact events and assess injury risks.

23.1 Injury mechanisms

Relative motion of the brain surface with respect to the rough inner surface of the skull base can result in surface contusions on the inferior surfaces of the frontal and temporal lobes. Figure 23.1 shows that motion at the smooth surface of the cortex can tear bridging veins between the brain and the dura mater, the principal membrane protecting the brain beneath the skull. The irregular geometry and surface of intracranial bones and membranes contributes to deformation of brain tissue during severe head impact that can result in injury.

Rotational acceleration of the head can cause a diffuse injury to the white matter of the brain, as evidenced by retraction balls along axons of injured nerves [23.1]. This injury was described as diffuse axonal injury (DAI) in the white matter of autopsied human brains [23.2]. Researchers [23.3] have been able to cause DAI in the brain of a ferret by the application of direct impact to the brain. Others [23.4] have produced it solely with rotational acceleration of the head, so both forms of head acceleration can cause the injury. DAI is the most important factor in severe head injury, as it is irreversible and leads to incapacitation and dementia [23.5]. Unfortunately, DAI cannot be detected by staining techniques at autopsy unless the patient survives the injury for several hours.

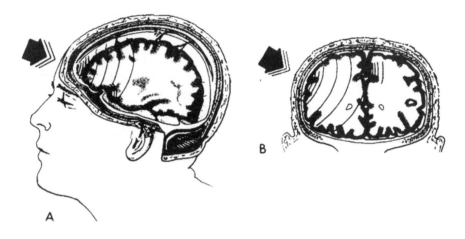

Figure 23.1 Response of the brain within the skull to frontal and lateral head impact.

Among the other theories of brain injury due to blunt impact, are changes in intracranial pressure and the development of shear strains in the brain. Positive pressure increases are found in the brain behind the site of impact on the skull. Rapid acceleration of the head, in-bending of the skull and the propagation of a compressive pressure wave are proposed as mechanisms for the generation of intracranial compression, which causes local contusion of the brain tissue. At the contrecoup site, there is an opposite response in the form of a negative pressure pulse, which may also causes bruising. It is not clear whether the injury is due to the negative pressure (tensile loading) or to a cavitation phenomenon (compression loading). The pressure differential across the brain results in a pressure gradient, which can give rise to shear strains within the deep structures of the brain. A review of head injury biomechanics is available [23.6].

23.2 Mechanical response of the head

Many cadaver studies on blunt head impact have been carried out over the past 50 years. The head was impacted by rigid and padded surfaces and by impactors of varying shapes to simulate flat surfaces and knobs encountered in the automotive environment. In general, the impact responses were described in terms of head acceleration or impact force. Both of these responses depend on a variety of factors, including the inertial properties of the head and surface impacted. The inertial properties of the head will be provided for impact against a flat rigid surface. It should be noted that while response data against surfaces with a variety of shapes and stiffness are of interest, the only generally applicable and reproducible data are for impacts on flat rigid surfaces.

23.2.2. Inertial properties of the head

There are several sources of data on the inertial properties of the human head [23.7, 8]. All data was analyzed by Hubbard and McLeod [23.9] who found that 31 heads had dimensions close to the average male. The average head mass is 4.54 kg with a specific gravity of 1.097. Information is also available on the average mass moment of inertia of the head [23.10, 23.11]. They are $I_{xx} = 22.0$, $I_{yy} = 24.2$, and $I_{zz} = 15.9 \times 10^{-3}$ kgm^2.

23.2.3. Cranial impact response

Impact response of the head was determined in drop tests against a flat rigid surface using embalmed cadavers [23.8, 23.12]. For skull fracture, peak force was 6 kN and acceleration 200-250 g for a 400 mm drop height. The responses increased linearly to 10 kN and 250-300 g for a 1000 mm drop. For non-fracture, peak force was 4 kN and acceleration 125-200 g for 200 mm drops. The data were for frontal, lateral and occipital impact directions although there was a large scatter in the peak values. For this reason, the data were pooled and individual data points were not shown.

The difficulty with acceleration measurements in head impact is twofold. The head is not a rigid body and accelerometers cannot be mounted at the cg of the head. The cg acceleration can be computed if head angular acceleration is measured but the variation in skull stiffness cannot be corrected for easily. That is why measurement of head angular acceleration is recommended. Several methods for measuring this parameter have been proposed. At present, the most reliable method uses an array of 9 linear accelerometers arranged in a 3-2-2-2 cluster [23.13].

For injury producing head impacts, the motion of the brain inside the skull has not been studied exhaustively. There is evidence that relative motion of the brain with respect to the skull occurs, particularly during angular acceleration of the head. However, this motion does not fully explain injuries seen in the center of the brain and in the brain stem. More research is needed to explore the mechanical response of the brain to both linear and angular acceleration and to relate this response to observed injuries, such as, diffuse axonal injury.

23.3 Injury tolerances of the head

The most commonly measured parameter during head impact is acceleration. It is therefore natural to express human tolerance in terms of head acceleration. The first known tolerance criterion is the Wayne State Tolerance Curve [23.14]. The curve was modified with the addition of animal and volunteer data to the original cadaver responses [23.15]. The modified curve is shown in Figure 23.2.

The head can withstand higher accelerations for shorter durations and any exposure above the curve is injurious. When this curve is plotted on logarithmic paper, it becomes a straight line with a slope of -2.5. This slope was used as an exponent by Gadd [23.16] in his proposed severity index, now known as the Gadd Severity Index and is determined by GSI = $\int a^{2.5}\, dt$, where a is the instantaneous head acceleration. If the integrated value exceeds 1000, a severe injury is expected [23.17].

A modified form of the GSI, known as the Head Injury Criterion (HIC), which identifies the most damaging part of the acceleration pulse by finding the maximum value of the same function [23.18]. A severe but not life-threatening injury is expected if HIC reaches or exceeds 1000. Subsequently, a probabilistic method of assessing head injury [23.19]. For a HIC = 1000 for a maximum duration of 15 ms, approximately 16% of the population would sustain a severe to fatal injury. This criterion is widely used in automotive safety testing and the tolerance level has been reduced to 700 in automotive testing. HIC is also used in the evaluation of protective equipment for the head, such as football and bicycle helmets.

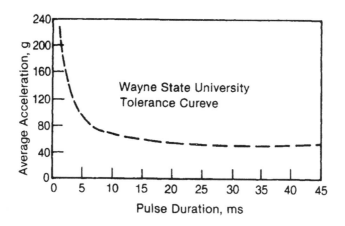

Figure 23.2 Blunt impact tolerance as a function of acceleration magnitude and duration.

Ommaya [23.20] addressed the injurious potential of angular acceleration causing cerebral contusion of the brain surface and rupture of the parasagittal bridging veins between the brain and the dura mater. A limit for angular acceleration is 4500 rad/s^2, based on a mathematical model [23.21]. This limit has not received universal acceptance, and volunteer data shows tolerance to up to 17,000 rad/s^2 may be possible for short durations. Many other criteria have been proposed but HIC is the current measure for Federal Motor Vehicle Safety Standard (FMVSS) 208, 214 and 210.

23.4 Human surrogates

23.4.1. Experimental surrogates

The most effective experimental surrogate for impact biomechanics research is the unembalmed cadaver. This is also true for the head and neck, despite the lack of muscle tone because the duration of impact is usually too short for the muscles to respond adequately. It is true, however, that muscle pre-tensioning in the neck may have to be added under certain circumstances, since it influences head and neck kinematics. Similarly, for the brain, the cadaver brain cannot develop DAI and the mechanical properties of brain change rapidly after death. If the pathophysiology of central nervous system is to be studied, the ideal surrogate is an animal brain. Currently, the rat is frequently used as the animal of choice and there is some work in progress using the mini-pig.

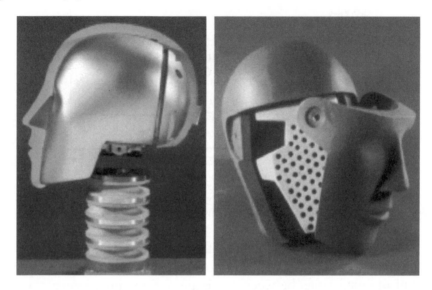

Figure 23.3 Hybrid III head and neck, and frangible face.

23.4.2. Injury assessment tools

The response and tolerance data acquired from cadaver studies have been used to design humanlike surrogates, known as anthropomorphic test devices (ATD). These surrogates are required to have biofidelity, the ability to simulate the essential characteristics of the human response. They also need to provide physical measurements that are representative of human injury, and are designed to be repeatable and reproducible. The current frontal impact dummy is the Hybrid III, which is humanlike in many of its responses, including that of the head and neck. The head consists of an aluminum headform covered by an appropriately designed vinyl skin to yield humanlike acceleration responses for frontal and lateral impacts against a flat rigid surface.   Two-dimensional physical models of the brain have been developed using a silicon gel in which inscribed grid lines would deform under angular acceleration [23.22, 23.23]. No injury criterion is associated with the gel models.

23.4.3. Computer models

Models of head impact first appeared over 50 years ago [23.24]. Extensive reviews of such models are available [23.25]. The use of the finite element method (FEM) to simulate the various components of the head appears to be the most effective and popular means of modeling brain response. Despite the large number of nodes and elements used, these models are still not detailed enough to predict displacement and distortion of the brain and the location of DAI following a given impact. The research is also hampered by the limited amount of animal DAI data currently available to validate model responses.

REFERENCES

[23.1]   Gennarelli, T.A., Head injuries in man and experimental animals: Clinical aspects, *Acta neurochirurgica Suppl.* **32**, 1-13, 1983.

[23.2]   Strich, S.J., Shearing of nerve fibers as a cause of brain damage due to head injury, *The Lancet*, **2**,443-448, 1961.

[23.3]   Lighthall, J.W., Goshgarian, H.G. and Pinderski, C.R., Characterization of axonal injury produced by controlled cortical impact, *J. Neurotrauma*, **7**(2), 65-76, 1990.

[23.4]   Gutierrez, E., Huang, Y., Haglid, K., Feng, B., Hansson, H.A., Hamberger A. and Viano, D.C., A new model for diffuse brain injury by rotational acceleration: I model, gross appearance and astrocytosis, *Journal of Neurotrauma*, (in print), 2000.

[23.5]   Adams, J.H., Doyle, D., Graham, D.I., Lawrence, A.E. and McLellan, D.R., Gliding contusions in nonmissile head injury in humans, *Arch. Pathol. Lab. Med.*, **110**, 485-488, 1986.

[23.6]   King, A.I. and Viano, D.C., Mechanics of head and neck, Chapter 25, in *The Biomedical Engineering Handbook*, J. D. Bronzino Ed., CRC Press, Inc. and IEEE Press, Boca Raton, FL, 357-368, 1995.

[23.7]   Walker, L.B.Jr., Harris, E.H. and Pontius, U.R., Mass, volume, center
         of mass, and mass moment of inertia of head and neck of human body,
         Proceedings of the 17th Stapp Car Crash Conference, 525-537, 1973.

[23.8]   Hodgson, V.R. and Thomas, L.M., Comparison of head acceleration
         injury indices in cadaver skull fracture, *Proceedings of the 15th Stapp
         Car Crash Conference*, 190-206, 1971.

[23.9]   Hubbard, R.P. and McLeod, D.G., Definition and development of a
         crash dummy head,   *Proceedings of the 18th Stapp Car Crash
         Conference*, 599-628, 1974.

[23.10] Reynolds, H.M., Clauser, C.E., McConville, J., Chandler, R. and
         Young, J.W., Mass distribution properties of the male cadaver, SAE
         Paper No. 750424, Society of Automotive Engineers, Warrendale, PA,
         1975.

[23.11] Beier, G., Schuller, E., Schuck, M., Ewing, C., Becker, E. and Thomas,
         D., Center of gravity and moments of inertia of human head,
         *Proceedings of the 5th International Conference on the Biokinetics of
         Impacts*, 218-228, 1980.

[23.12] Hodgson, V.R. and Thomas, L.M., *Head impact response*, Vehicle
         Research Inst., Soc. Automotive Engrs., Warrendale, PA, 1975.

[23.13] Padgaonkar, A.J., Krieger, K.W. and King, A.I., Measurement of
         angular acceleration of a rigid body using linear accelerometers, *J.
         Appl. Mech.*, **42**, 552-556, 1975.

[23.14] Lissner, H.R., Lebow, M. and Evans, F.G., Experimental studies on the
         relation between acceleration and intracranial pressure changes in man,
         Surg. Gynecol. Obstet., **111**, 329-338, 1960.

[23.15] Patrick, L.M., Kroell, C.K. and Mertz, H.J., Forces on the human
         body in simulated crashes, *Proceedings of the 9th Stapp Car Crash
         Conference*, SAE, 237-260, Society of Automotive Engineers,
         Warrendale, PA, 1965.

[23.16] Gadd, C.W., Criteria for injury potential, In *Impact Acceleration Stress Symposium*, Nat. Res. Council Publication No. 977, Nat. Acad. Sci., Washington, DC, 141-144, 1961.

[23.17] Patrick, L.M., Lissner, H.R. and Gurdjian, E.S., Survival by design - Head protection, *Proceedings of the 7th Stapp Car Crash Conference*, 483-499, 1965.

[23.18] Versace, J., A review of the severity index, *Proceedings of the 15th Stapp Car Crash Conference*, 771-796, 1970.

[23.19] Prasad, P. and Mertz, H.J., The position of the United States delegation to the ISO working group 6 on the use of HIC in the automotive environment, SAE Paper No. 851246. Soc. of Automotive Engrs., Warrendale, PA, 1985.

[23.20] Ommaya, A.K., Biomechanics of head injury, In *The Biomechanics of Trauma*, A.M. Nahum and J.W. Melvin Eds, Appleton-Century-Crofts, Norwalk, 1984.

[23.21] Lowenhielm, P., Mathematical simulation of gliding contusions, *Journal of Biomechanics,* **8**, 351-356, 1975.

[23.22] Margulies, S.S., Thibault, L.E. and Gennarelli, T.A., Physical model simulation of brain injury in the primate, *Journal of Biomechanics*, **23**, 823-836, 1990.

[23.23] Ivarsson, J., Viano, D.C., Lovsund, P. and Aldman, B., Strain relief from the cerebral ventricles during rotational acceleration of the head - an experimental study with physical models, *Journal of Biomechanics*, **33**, 181-189, 2000.

[23.24] Holbourn, A.H.S., Mechanics of head injury, *The Lancet*, **2**, 438-441, 1943.

[23.25] King, A.I. and Chou, C.C., Mathematical modeling, simulation and experimental testing of biomechanical systems crash response, *Journal of Biomechanics*, **9**, 301-317, 1976.

24. SPINAL INJURY
BIOMECHANICS

Injuries of the cervical and thoracolumbar spine have the potential for long-term disabling consequence, including para- and quadriplegia, paralysis and paresis. Protection of the spinal cord is a critical area of safety performance. Mechanisms of spinal injury are described for frontal and lateral loading of the body, which involves bending, shear and tensile loads on the vertebrae. For the neck, the current injury criteria is the Nij, which combines the normalized neck tension and bending into a single criterion, and reflects the combination of loads typically acting on the cervical spine during occupant restraint. While the Hybrid III dummy neck is widely used for the evaluation of injury risks in frontal and lateral crashes, the dummy has a rigid thoracic spine and a flexible lumbar joint. The spine does not articulate like in the human. For low-speed rear crashes, the BioRID dummy has been developed to assess injury risks to the spine. In this case, the NIC injury criterion is used to reflect the differential motion at the top and bottom of the neck, and to assess the risk of whiplash injury.

24.1 Cervical spine

24.1.1. Cervical spine injury mechanisms

Injuries to the upper cervical spine, particularly at the atlanto-occipital joint, are considered to be more serious and life threatening than those at a lower level. The atlanto-occipital joint can be dislocated either by an axial torsion load or a shear force applied in the anteroposterior direction, or vice versa. A large compression force can cause the arches of C_1 to fracture, breaking it up into two to four sections. The odontoid process of C_2 is also a vulnerable area.

Hyperflexion of the neck is a common cause of odontoid fractures, and a large percentage of these injuries are related to automotive crashes of largely unrestrained occupants [24.1]. Fractures through the pars interarticularis of C_2, commonly known as "hangman's" fractures in automotive collisions, are the result of a combined axial compression and extension (rearward bending) of the cervical spine. Impact of the forehead and face of unrestrained occupants with the windshield can result in this injury, which has been discussed in relation to hanging [24.2]. A British judiciary committee estimated that the energy required to cause a hangman's fracture was 1,708 Nm (1,260 ft-lb).

In automotive crashes, load on the neck is generally due to head contact forces and a combination of an axial or shear load with bending. Bending loads are almost always present, and the degree of axial or shear force depends on the location and direction of the contact force. For impacts near the crown of the head, compressive forces predominate. If the impact is principally in the transverse plane, there is less compression and more shear. Bending can occur in any direction because impacts can come from any angle around the head. The following injury modes are considered the most prominent: tension-flexion, tension-extension, compression-flexion and compression-extension in the midsagittal plane, and lateral bending.

24.1.1.1. Tension-Flexion Injuries

Forces resulting from inertial loading of the head-neck can result in flexion of the cervical spine while it is being subjected to a tensile force. In experiments with restrained subjects in forward deceleration [24.3], atlanto-occipital separation and C_1-C_2 separation occurred in subhuman primates at 120 g. Similar injuries in human cadavers were found at 34 to 38 g with a pre-inflated driver airbag to restrain the thorax but not the head, which was allowed to rotate over the airbag [24.4].

24.1.1.2. Tension-Extension Injuries

The most common injury due to combined tension and extension of the cervical spine is the "whiplash" syndrome. However, a large majority of such injuries involve the soft tissues of the neck and the pain is believed to reside in the joint capsules of the articular facets of the cervical vertebrae [24.5]. In severe cases, teardrop fractures of the anterior-superior aspect of the vertebral body can occur. Alternately, separation of the anterior aspect of the disc from the vertebral endplate can occur. More severe injuries occur when the chin impacts the instrument panel or when the forehead impacts the windshield. In both cases, the head rotates rearward and applies a tensile and bending load on the neck. In the case of windshield impact by the forehead, hangman's fracture of C_2 can occur. It is caused by spinal extension combined with compression on the lamina of C_2, causing the pars to fracture.

24.1.1.3. Compression-Flexion Injuries

When force is applied to the posterior-superior quadrant of the head or when a crown impact occurs while the head is in flexion, the neck is subjected to a combined load of axial compression and forward bending. Anterior wedge fractures of vertebral bodies are commonly seen, but with increased load, burst fractures and fracture dislocations of the facets can result. The latter two conditions are unstable and tend to disrupt or injure the spinal cord; the extent of the injury depends on the penetration of the vertebral body or its fragments into the spinal canal. Burst fractures of lower cervical vertebrae have been reproduced in cadavers by a crown impact to a flexed cervical spine [24.6, 24.7]. Fracture dislocations of the cervical spine can occur early in the impact event (within the first 10 ms) and subsequent motion of the head or bending of the cervical spine cannot be used as a reliable indicator of the mechanism of injury [24.8].

24.1.1.4. Compression-Extension Injuries

Frontal impacts to the head with the neck in extension will cause compression-extension injuries. These involve the fracture of one or more spinous processes and possibly, symmetrical lesions of the pedicles, facets and laminae. If there is a fracture-dislocation, the inferior facet of the upper vertebra is displaced posterior and upward and appears to be more horizontal than normal on x-ray.

24.1.1.5 Injuries Involving Lateral Bending

If the applied force or inertial load on the head has a significant component out of the midsagittal plane, the neck will be subjected to lateral or oblique bending along with axial and shear loads. The injuries characteristic of this type of bending are lateral wedge fractures of the vertebral body and fractures to the posterior elements on one side of the vertebral column.

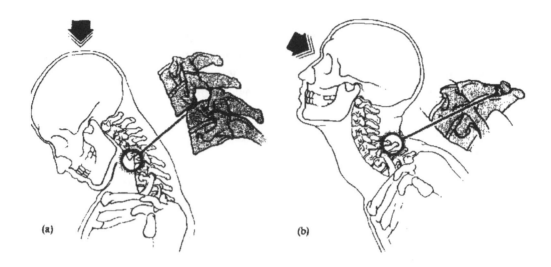

Figure 24.1 Downward impact on the head can flex (forward bending) or extend (rearward
bending) the neck with the potential for fracture-dislocation of the vertebrae and
damage to the spinal cord.

Whenever there is lateral or oblique bending, there is the possibility of
twisting the neck. The associated torsional loads may be responsible for
unilateral facet dislocations or unilateral locked facets [24.9]. However, pure
torsional loads on the neck are rarely encountered in automotive accidents. In a
purely lateral impact, the head rotated axially about the cervical axis while it
translated laterally and vertically, and rotated about an antero-posterior axis
[24.10]. These responses were obtained from lateral impact tests performed by
the Naval Biodynamics Laboratory on human subjects who were fully restrained
at and below the shoulders.

24.1.2. Mechanical responses of the neck

The mechanical response of the cervical spine is usually quantified by the
rotation of the head relative to the torso and the bending moment at the occipital
condyles [24.11-24.14]. Figure 2 shows response corridors for flexion and
extension. A definition of the impact environments used to evaluate dummy
necks for these corridors can be found in SAE J1460 [24.15]. The basis for the
curves is volunteer data; extension of the corridors to dummy tests in the injury
producing range is uncertain.

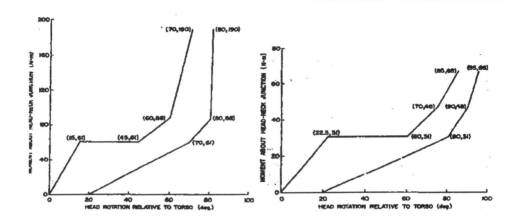

Figure 24.2 Loading corridors for neck flexion (left) and extension (right).

Static and dynamic lateral response data is shown in Figure 24.3 [24.16]. A limited amount of volunteer data was analyzed for lateral and sagittal flexion [24.17]. The rotations were represented in 3D by a rigid link of fixed length pivoted at T1 at the bottom and within the head at the top. In terms of torque at the occipital condyles and head rotation, the results fell within the earlier corridors for forward and lateral flexion.

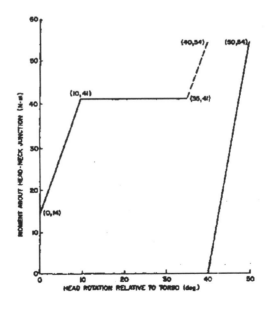

Figure 24.3 Lateral flexion response corridors.

24.1.3 Neck injury tolerances

A new neck injury criterion has been specified for frontal crash testing in FMVSS 208. Figure 24.3 shows the limits on maximum neck tension and compression, and a calculation of a combined tension and bending response, Nij. This alters the earlier criteria, which are shown. These responses are for risks of serious neck injury. For lower severity injuries of the whiplash type, there is no accepted injury criteria, although head extension angle, and neck moment, tension and shear are commonly used to assess the relative risks for injury.

24.1.4. Human surrogates

24.1.4.1. Experimental Surrogates

The most effective experimental surrogate for impact biomechanics research is the unembalmed cadaver. This is also true for the head and neck, despite the lack of muscle tone because the duration of impact is usually too short for the muscles to respond adequately. It is true, however, that muscle pre-tensioning in the neck may have to be added under certain circumstances, since it influences head and neck kinematics. If the pathophysiology of central nervous system is to be studied, the ideal surrogate is an animal brain. Currently, the rat is frequently used as the animal of choice and there is some work in progress using the mini-pig.

24.1.4.2. Injury Assessment Tools

The response and tolerance data acquired from cadaver studies have been used to design humanlike surrogates, known as anthropomorphic test devices (ATD). These surrogates are required to have biofidelity, the ability to simulate the essential characteristics of the human response. They also need to provide physical measurements that are representative of human injury, and are designed to be repeatable and reproducible. The current frontal impact dummy is the Hybrid III, which is humanlike in many of its responses, including that of the head and neck. The head consists of an aluminum headform covered by an appropriately designed vinyl skin to yield humanlike acceleration responses for frontal and lateral impacts against a flat rigid surface. The dummy neck was designed to yield responses in flexion and extension, which fit within the corridors shown in Figure 24.2 and 24.3. The principal function of the dummy neck is to place the head in the approximate position of a human head in the same impact involving a human occupant. More recently, the BioRID dummy has been developed to assess injury risks in low-speed rear end crashes. It simulates the individual vertebrae of the spine, and uses tension straps to simulate muscle tension seen in volunteer tests. For this dummy, the NIC measures the differential acceleration at the top and bottom of the neck to assess whiplash risks.

24.1.4.3. Computer Models

Many neck and spinal models have been developed over the past four decades. Kleinberger [24.18] provides a review of models. However, the method of choice for modeling the response of the neck is the finite element method, principally because of the complex geometry of the vertebral components and the interaction of several different materials. A fully validated model for impact response is still not available.

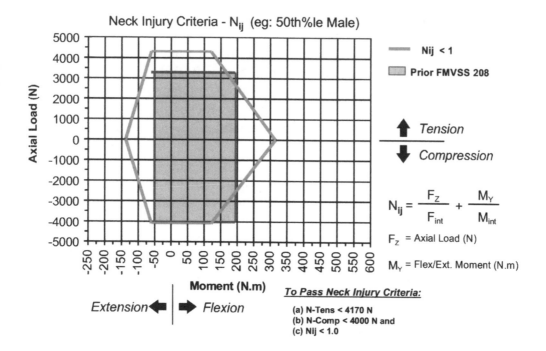

Figure 24.4 Nij neck injury criteria based on the combined tension and bending moment.

24.2. Thoracolumbar spine

24.2.1. Thoracolumbar spine injury mechanisms

Although injuries to the bony portion of the thoracolumbar spine are rare in automotive crashes, paraplegia can result if the spinal cord is involved and complaints of low back pain are a common claim. Impact spinal biomechanics attempts to explain the various injury mechanisms and provides data on human response and tolerance. Surrogates in the form of a dummy spine or a computer model are developed based on the results of these biomechanical studies.

Historically, the study of thoracolumbar spinal injuries were motivated by the pilot ejection problem. The vertebral body of the lower thoracic and upper lumbar spine tends to sustain anterior wedge fractures when military pilots eject from disabled jet aircraft. The vertical acceleration necessary to effect a successful ejection is about 20 g. The mechanism of wedge fractures was due to a combined compressive load and a forward flexion moment [24.19]. The load through the spine is transmitted from one level to another via the intervertebral disc and the articular facets [24.20]. In hyperextension, the inferior tips of the superior facets bottom out onto the lamina of the vertebra below and create a load path which relieves the vertebral body of some of the compressive load it carries due either to vertical acceleration or muscular compression [24.21].

Since research on the impact biomechanics of the spine was originally motivated by the pilot ejection problem, most of the work is related to the effects of caudocephalad acceleration. It has led to a better understanding of the effects of compressive loading on the spine, mechanism of load transmission, and injury. In fact, it has contributed to the search for causes of low back pain by identifying the facets as load bearing elements of the spine. Automotive related injuries to the thoracolumbar spine are rarely encountered. However, injuries involving fracture-dislocation of an intervertebral joint can result in paraplegia. Two types of such injuries were identified: Chance fractures in improperly lap-belted rear seat occupants and spinal fractures due to the shoulder belt. In terms of soft tissue injuries, there is no causal relationship between a single impact to the spine and subsequently diagnosed disc ruptures, if the impact does not result in spinal fractures.

Of the several major categories of thoracolumbar injuries, the mechanism of injury usually involves an applied force accompanied by a bending or twisting moment. This combined loading causes not only anterior wedge fractures but also dislocation and fracture-dislocation of the vertebrae, rotational injuries, Chance fractures and hyperextension injuries. The only injury in which a moment may not be involved is burst fracture of the vertebral body, which could be reproduced in the laboratory by dropping a weight onto a spinal specimen [24.22]. However, the study did not determine if the fracture occurred only if spine was slightly flexed and the facets were non-load bearing.

In the automotive setting, the Chance fracture is a belt-induced injury, which can occur to rear seat occupants who are usually restrained by a lap belt over the abdomen or slouching in the seat [24.23]. Because of the flat angle of the rear belt relative to the horizontal plane, the lap belt tends to ride up above the iliac crest in a horizontal crash. The abdominal organs are compressed by the belt, which then bears against the spine as the torso flexes. This causes the supraspinous and interspinous ligaments to rupture and the vertebral body to split horizontally, starting along its posterior aspect. The spinal cord is stretched and paraplegia can result.

Another automotive-related belt induced injury that may be on the increase is anterior wedge fracture of the thoracolumbar lumbar vertebral body, similar to that seen in pilots who eject from disabled aircraft. In this case, the shoulder harness imposes a large load across the torso of an occupant in a severe frontal crash, causing the curved thoracic spine to straighten out. As a result, the compressive force is generated in the thoracolumbar spine to push the head and neck upward and the rest of the torso downward [24.24]. This load was measured on the seat pan in tests on three-point belted cadavers and volunteers but was not observed when an anthropomorphic test device (dummy) was used.

In terms of soft tissue injuries, the complaint of low back pain is often associated with a diagnosis of disc rupture. The incident provoking such symptoms can range from a minor rear end impact to a very severe frontal crash. However, predominant findings in the literature indicate that disc rupture is a slow degenerative process and that an extremely violent single loading event is needed to cause the nucleus pulposus to extrude from the disc, in association with bony fracture. The disc does not herniate like a balloon and minor events, which result in pain, may not be due a defect in the disc. There are many sources of back pain, such as in the capsules of facet joints, and a causal relationship between an impact and a rupture usually does not exist [24.25].

24.2.2. Mechanical responses

Much information exists on the static mechanical response of a functional spinal unit (two vertebrae and a disc) but very much less is available regarding the response of the entire thoracic or lumbar spine. In terms of dynamic response in an impact, much of the work on the intact spine was done in relation to the pilot ejection problem. Thus, the early data were concerned with spinal response to a caudocephalad or $+G_Z$ acceleration. The first known whole-body cadaver experiments were performed on a vertical accelerator housed in an elevator shaft of the School of Medicine at Wayne State University. The response data originally took the form of vertebral body strain as a function of time [24.26]. Subsequently, spinal force-time histories were produced [20]. However, because the development of a surrogate spine was not a primary aim, response data were obtained primarily on the mechanisms of load transmission and injury. A limited amount of moment and rotation response data was acquired from volunteer subjects in a frontal crash or $-G_X$ environment.

24.2.3. Human Tolerances

Tolerance data for the spine are available from several sources for a variety of loading conditions. These conditions range from static loading of individual vertebrae to dynamic loading of the whole body in impact tests using human cadavers. While data on the strength of the intact spine under dynamic loading are of particular importance in impact biomechanics, data sources for strength of individual vertebrae and functional spinal units are provided. Yamada [24.27] provides a summary of results [24.28] for tensile, compressive and torsional strength characteristics of cervical, thoracic and lumbar vertebrae and the intervertebral discs. The data include variation in strength along the column and its dependence on age. Quasi-static compressive loads were applied to individual bodies, portions of the excised spine with ligaments intact and torsos of complete cadavers [24.29]. Isolated thoracic and lumbar vertebral bodies were loaded in compression between parallel plates in the superior-inferior direction. Excised spinal segments and intact cadavers were also compressed to yield fracture loads, which were lower than the loads borne by the individual bodies. The level of shear force and bending moment that may have been present was not measured.

As early as 1959, human tolerance limits to impact acceleration from several directions [24.30]. For caudocephalad acceleration, the available data from humans and animals indicates a 20 g limit for ejection seats, which is used by the U.S. Army Air Force. The limit assumed a trapezoidal acceleration pulse acting on a subject fully restrained by a military restraint system. Limits for frontal crash, rear end crash and downward acceleration were also proposed. For impact durations of less than 100 ms, a 40 g limit was proposed for well-restrained seated individuals.

24.2.4. Human Surrogates

The thoracolumbar spine in dummies is not humanlike. The spine is a rigid steel box section and the lumbar spine is a relatively rigid curved rubber cylinder, which has a steel cable running through its center. An attempt was made to include a joint in the middle of the thoracic column to simulate the kyphotic curve [24.31]. This represents an improvement but the entire column is too rigid to represent the human spine or injury of an individual vertebrae.

An alternative is to use mathematical or computer models of the spine to predict forces and moments under various loading conditions. There are many mathematical models of the spine but most of them deal with quasi-static loading. Models were developed to simulate ejection seat impacts [24.32]. A two-dimensional discrete parameter model was developed and validated against cadaver data [24.33]. It was this model which first predicted spine loads due to the shoulder belt in a frontal crash. Finite element models of the thoracolumbar spine simulating impact events do not seem to have appeared in the literature as yet, primarily because injuries are rare in frontal crashes and the pilot ejection problem is considered by the military as a mature technology, not requiring further research.

REFERENCES

[24.1]   Pierce, D.A. and Barr, J.S., Fractures and dislocations at the base of the skull and upper spine, In *The Cervical Spine*. R.W. Baily Ed., 196-206, Lippincott, Philadelphia, PA, 1983.

[24.2]   Garfin, S.R. and Rothman, R.H., Traumatic spondylolisthesis of the axis (Hangman's fracture), *The Cervical Spine*, R.W. Baily Ed., 223-232, Lippincott, Philadelphia, PA, 1983.

[24.3]   Thomas, D.J. and Jessop, M.E., Experimental head and neck injury, In *Impact Injury of the Head and Spine*, C.L. Ewing et al. Eds., Charles Thomas, Springfield, IL, 177-217, 1983.

[24.4]   Cheng, R., Yang, K.H., Levine, R.S, King, A.I. and Morgan, R., Injuries to the cervical spine caused by a distributed frontal load to the chest, *Proceedings of the 26th Stapp Car Crash Conference*, 1-40, 1982.

[24.5]   Lord, S., Barnsley, L. and Bogduk, N., Cervical zygapophyseal joint pain in whiplash, In *Cervical Flexion, Extension, Whiplash Injuries*, R.W. Teasell and A.P. Shapiro Eds., 355-372. Hanley & Belfus, Inc., Philadelphia, PA, 1993.

[24.6]   Pintar, F.A., Yoganandan, N., Sances, A. Jr, Reinartz, J., Harris, G.M. and Larson, S.J., Kinematic and anatomical analysis of the human cervical spinal column under axial loading, *Proceedings of the 33rd Stapp Car Crash Conference*, 191-214, 1989.

[24.7]   Pintar, F.A., Sances, A. Jr, Yoganandan, N., Reinartz, J., Maiman, D., Suh, J.K. and Unger, G., Biodynamics of the total human cadaveric spine, *Proceedings of the 34th Stapp Car Crash Conference*, 55-72, 1990.

[24.8]   Nightingale, R.W., McElhaney, J.H., Best, T.M., Richardson, W.J. and Myers, B.S., *Proc. 39th Meeting Orthopedic Res. Soc.*, 233, 1993.

[24.9]   Moffat, E.A., Siegel, A.W. and Huelke, D.F., The biomechanics of automotive cervical fractures, *Proc. 22nd Conf. of Am. Assoc. for Automotive Med.*, 151-168, 1978.

[24.10] Wismans, J. and Spenny, D.H., Performance requirements for mechanical necks in lateral flexion, *Proceedings of the 27th Stapp Car Crash Conference*, 137-148, 1983.

[24.11] Mertz, H.J. and Patrick, L.M., Investigation of the kinematics and kinetics of whiplash, *Proceedings of the 11th Stapp Car Crash Conference*, 267-317, 1967.

[24.12] Mertz, H.J. and Patrick, L.M., Strength and response of the human neck, *Proceedings of the 15th Stapp Car Crash Conference*, 207-255, 1971.

[24.13] Schneider, L.W., Foust, D.R., Bowman, B.M., et al. Biomechanical properties of the human neck in lateral flexion, *Proceedings of the 19th Stapp Car Crash Conference*, 455-486, 1975.

[24.14] Mertz, H.J., Neathery, R.F. and Culver, C.C., Performance requirements and characteristics of mechanical necks, In *Human Impact Response. Measurement and Simulation*, W.F. King and H.J. Mertz Eds., Plenum Press, New York, NY, 263-288, 1973.

[24.15] Patrick, L.M. and Chou, C., Response of the human neck in flexion, extension, and lateral flexion, *Vehicle Res. Inst. Report No. VRI-7-3*, Soc. of Automotive Engrs., Warrendale, PA, 1976.

[24.16] Society of Automotive Engineers Human Mechanical Response Task Force, "*Human mechanical response characteristics*," SAE J1460. Society of Automotive Engrs., Warrendale, PA, 1985.

[24.17] Wismans, J.and Spenny, D.H., Head-neck response in frontal flexion, *Proceedings of the 28th Stapp Car Crash Conference*, 161-171, 1984.

[24.18] Kleinberger, M., Application of finite element techniques to the study of cervical spine mechanics, *Proceedings of the 37th Stapp Car Crash Conference*, 261-272, 1993.

[24.19] Ewing, C.L., King, A.I. and Prasad, P., Structural consideration of the human vertebral column under +Gz impact acceleration, *J. of Aircraft*, **9**, 84-90, 1972.

[24.20] Prasad, P., King, A.I., An experimentally validated dynamic model of the spine, *J. Appl. Mech.*, **41**, 546-550, 1974.

[24.21] El-Bohy, A.A., Yang, K.H. and King, A.I., Experimental verification of facet load transmission by direct measurement of facet/lamina contact pressure, *Journal of Biomechanichs*, **22**, 931-941, 1989.

[24.22] Oxland, T.R., *Burst Fractures Of The Human Thoracolumbar Spine: A Biomechanical Investigation*, Ph.D. Dissertation, Yale University, 1992.

[24.23] Chance, G.O., Note on a type of flexion fracture of the spine, *Br. J. Radiol.*, **21**, 452-453, 1948.

[24.24] Begeman, P.C., King, A.I. and Prasad, P. Spinal loads resulting from - Gx acceleration, *Proceedings of the 17th Stapp Conference*, SAE Paper No. 730977, 343-360, 1973.

[24.25] King, A.I., Injury to the thoraco-lumbar spine and pelvis, In *Accidental Injury: Biomechanics and Prevention*, A. Nahum and J. Melvin Eds. , Springer-Verlag, New York, 429-459, 1993.

[24.26] Evans, F.G., Lissner, H.R. and Patrick, L.M., Acceleration-induced strains in the intact vertebral column, *J. Appl. Physiol*, **17**, 405-409, 1962.

[24.27] Yamada, H., *Strength of Biological Materials*, F.G. Evans Ed., Williams and Wilkins, Baltimore, 75-80, 1970.

[24.28] Sonoda, T., Studies on the strength for compression, tension and torsion of the human vertebral column, *J. Kyoto Prefectural U. of Med., Med. Soc.*, **71**, 659-702, 1962.

[24.29] Myklebust, J., Sances, A., Maiman, D., et al. Experimental spinal trauma studies in the human and monkey cadaver, *Proceedings of the 27th Stapp Conference*, 149-161, 1983.

[24.30] Eiband, A.M., *Human Tolerance to Rapidly Applied Acceleration. A Survey of the Literature*, National Aeronautics and Space Administration, Washington DC, NASA Memo No. 5-19-59E, 1959.

[24.31] Schneider, L.W., Haffner, M.P., Eppinger, R.H., et al., Development of an advanced ATD thorax system for improved injury assessment in frontal crash environments, *Proceedings of the 36th Stapp Conference*, SAE Paper No. 922520, 129-155, 1992.

[24.32] Latham, F., A study in body ballistics: Seat ejection, *Proc. Royal Soc. (B)*, **147**, 121-139, 1957.

[24.33] Prasad, P., King, A.I. and Ewing, C.L., The role of articular facets during +Gz acceleration, *J. Appl. Mech.*, **41**, 321-326, 1974.

25. CHEST AND ABDOMEN INJURY BIOMECHANICS

The biomechanics of chest and abdomen injury is related to the amount and rate of deformation that occurs in an impact. Since the human body is viscoelastic, the reaction force developed to resist deformation increases with the speed of loading. This allows survival in high-speed impacts at force levels that would statically crush the body. The risk of high-speed impact injury is described by the viscous response (VC), which combines the amount and rate of deformation into a single injury criterion. For blunt chest and abdomen impact, the maximum viscous response for serious injury is VC = 1.0 m/s. For the same level of risk, the maximum compression is 32-40%. Historically, crash testing with dummies has also set a limit on the maximum chest acceleration at 60 g for no more than 3 ms. For abdominal impacts, the risk of submarining injury is related to compression, which can be evaluated with a frangible insert. Crush of the insert is correlated to injury risks.

25.1. Injury Mechanisms

Chest and abdomen injury is related to energy from an impacting object as well as its shape, stiffness, point of body contact and orientation. The torso has a viscoelastic behavior; load developed by the chest and abdomen increases with the speed of impact as internal structures resist deformation [25.1, 25.2]. The biomechanical response of the body has three components: (1) inertial resistance by acceleration of body masses, (2) elastic resistance by compression of stiff structures and tissues, and (3) viscous resistance by rate-dependent properties of the body and its tissues. For low impact speeds, the elastic stiffness is critical to protection from crush injuries; whereas, for high rates of body deformation, the inertial and viscous properties determine the force developed to limit deformation.

In any impact situation, the inertial resistance of body masses and the elastic and viscous properties of soft tissues combine to develop dynamic load, resist deformation, and prevent injury. Impact force produces acceleration and deformation of the body, and each relates to injury when the soft tissues are deformed beyond their recoverable limit. In most situations, the viscoelastic properties of the body protect vitals organs by absorbing energy and developing high forces which resistance body deformation during impact.

Deformation beyond a recoverable limit is the general injury mechanism in blunt chest and abdominal impact. This mechanism relates to compression or strain, defined as the change in dimension over the original thickness of the body, tissue or organ. For example, the typical anteroposterior thickness of the chest is about 22.2 cm (8.74"), so 20% compression involves 4.44 cm (1.75") deflection of the chest. Even in this condition, the primary types of strain that produce tissue damage are tensile and shear strain, which can fracture ribs and lacerate, rupture and avulse vessels. A third type is compressive strain which produces crush injury.

When compression of the torso exceeds the ribcage tolerance, fractures occur and internal organs and vessels can be contused or ruptured. In some chest impacts, however, internal injury occurs without skeletal damage. This can happen particularly during high-speed loading. It is due to the viscous or rate-sensitive nature of human tissue responses. Thus, biomechanical response and injury differs for low and high-speed impact.

When organs or vessels are loaded slowly, the input energy is absorbed gradually through deformation, which is resisted by elastic properties and pressure build-up in tissue. When loaded rapidly, reaction force is proportional to the speed of tissue deformation as the viscous properties of the body resist deformation and provides a natural protection from impact. However, there is also a considerable inertial component to the reaction force. In this case, the body develops high internal pressure and injuries can occur before the ribs deflect much. The ability of an organ or other biological system to absorb impact energy without compression failure is called the viscous tolerance. Internal organs and vessels can also be torn from attachment points during torso impact or high-level acceleration causing rapid motion of the body.

If an artery is stretched beyond its tensile strength, the tissue will tear. Organs and vessels can be stretched in different ways, which result in different types of injury. Motion of the heart during chest compression stretches the aorta along its axis from points of tethering in the body. This elongation generally leads to a transverse laceration when the strain limit is exceeded.

The abdomen is more vulnerable to injury than the chest, because there is little bony structure below the ribcage to protect internal organs in front and lateral impact. Blunt impact of the upper abdomen can compress and injure solid organs, such as the liver and kidneys, before significant whole-body motion or acceleration occurs. In the liver, compression increases the intrahepatic pressure and generates tensile or shear strains. If the tissue is sufficiently deformed, laceration of the major hepatic vessels can result in hemoperitoneum.

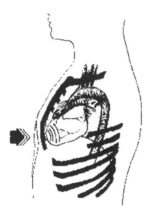

Figure 25.1 Blunt chest impact deforms the ribs and internal organs with potential injury by stretching the aorta.

25.2 Injury tolerances of the chest and abdomen

25.2.1. Acceleration injury

Stapp [25.3] conducted rocket-sled experiments that demonstrated the effectiveness of belt-restraint systems in achieving high tolerance to long-duration, whole-body acceleration. This improved the protection of military personnel exposed to rapid but sustained acceleration. The experiments demonstrated that the tolerance to whole-body acceleration increased as the exposure duration decreased (Figure 25.2), which links human tolerance and acceleration for exposures of 2-1000 ms in duration. The tolerance data is based on average sled acceleration rather than the acceleration of the volunteer subject, which would be higher due to compliance of the restraint system used in the tests. Even with this limitation, the data provide useful early guidelines for the development of crash restraint systems for military and civilian personnel. Analysis of the data also indicated that rate of onset affected acceleration tolerance, since high peaks could be tolerated if reached over a greater period of time.

More recent tests of side impact injury have led to other acceleration formulas for chest injury tolerance [25.4]. Based on rigid, sidewall cadaver tests TTI, thoracic trauma index, was developed. It is the average rib and spine acceleration of the chest, and has a tolerance limit of 85-90g in vehicle crash tests. Somewhat better injury assessment has been achieved using average spinal acceleration (ASA), which is obtained by integrating the thoracic spinal acceleration and determining the average slope. This term relates to the rate of momentum transfer to the body during side impact, and a value of 30g is proposed. In most cases, the torso can withstand 60-80g whole-body acceleration by a well-distributed load.

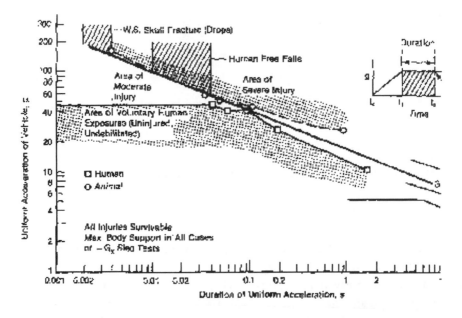

Figure 25.2 Blunt impact tolerance as a function of acceleration magnitude and duration.

25.2.2 Force injury

The basis for whole-body tolerance is Newton's second law of motion: acceleration of a rigid mass is proportional to the force acting on it, or the well-known $F = ma$. Although the human body is not a rigid mass, a well distributed restraint system allows the torso to respond as though it were fairly rigid if loads are applied through the shoulder and pelvis. The greater the acceleration, the greater the force acting on the body and the greater the risk of injury. For a high-speed frontal crash, a restrained occupant can experience 60g acceleration (60 times the force of gravity). For a body mass of 76 kg, the inertial load is 44.7kN (10,000 lb) and is tolerable if distributed over strong skeletal elements.

The ability to withstand high acceleration for short durations implies that tolerance is related to momentum transfer, because an equivalent change in velocity can be achieved by increasing the acceleration and decreasing its duration, as $\Delta V = a\Delta t$. The implication for occupant protection systems is that the risk of injury can be decreased if the crash deceleration is extended over a greater period of time. For occupant restraint in 25 ms (0.025s), a velocity change of 53 kmph (14.7 m/s) occurs with 60g whole-body acceleration. This duration of deceleration can be achieved by the use of crushable vehicle structures and occupant restraints.

Tests by Patrick et al. [25.5] demonstrated that blunt chest loading of 3.3 kN (740 lb) could be tolerated with minimal risk of serious injury. This is a pressure of 187 kPa. Subsequent experiments demonstrated that tolerance was as high as 8.0 kN (1800 lb) if the load was distributed over the shoulders and chest by a properly designed steering wheel and column. More recent side impact tests show that the torso can tolerate similar forces as in frontal impacts and that shoulder loading is an important load-path. However, the loads are a conservative threshold of injury.

25.2.3 Compression injury

Tolerance of the chest and abdomen must consider body deformation. Force acting on the body generates two simultaneous responses: (1) compression of the compliant structures of the torso, and (2) acceleration of body masses. The previously neglected mechanism of injury was compression, which causes the sternum to displace toward the spine as ribs bend and possibly fracture.

The importance of chest deformation was confirmed by Kroell [25.6] in a series of blunt thoracic impacts of unembalmed cadavers. Both peak spinal acceleration and impact force were poorer injury predictors than the maximum compression of the chest, as measured by the percent change in the anteroposterior thickness of the body. A relationship between injury risk and compression involves the concept of energy stored by elastic deformation of the body. Stored energy (E_s) by a spring representing the ribcage and soft tissues is related to the displacement integral of force: $E_s = \int F dx$. Force in a spring is proportional to deformation: $F = kx$, where k is a spring constant. Stored energy is $E_s = k \int x dx = 0.5kx^2$. Over a reasonable range, stored energy is proportional to deformation or compression, so $E_s \approx C$.

Tests with human volunteers showed that compression up to 20% during moderately long duration loading produced no detectable injury and was fully reversible. Cadaver impacts at levels of compression greater than 20% showed an increase in rib fractures and internal organ injury as the compression increased up to 40%. The original tolerance for chest deflection was set at 8.8 cm (3.5") for moderate but recoverable injury. This represents 39% compression. However, at this level of compression, multiple rib fractures and a range of serious injury can occur so a more conservative tolerance of 32% is used to avert the possibility of flail chest. This reduces the risk of direct loading on the heart, lungs and internal organs by a loss of the protective function of the ribcage.

25.2.4. Viscous injury

The velocity of body deformation is determined by the rate of loading and is an important factor in high-speed, non-penetrating injury. For example, when a fluid-filled organ is compressed slowly much of the applied energy can be absorbed through tissue deformation without damage. When loaded rapidly, however, the organ cannot deform fast enough and rupture may occur without significant change in shape, even though the load on the organ has increased substantially over the level occurring in the slow loading condition.

Research on soft-tissue injury has made it increasingly evident that the body is not merely an elastic structure, but rather is viscoelastic for impacts causing body deformation velocities greater than 3 m/s. For higher speeds of deformation, such as occupant loading by the door in a side impact or for an unrestrained occupant or pedestrian, maximum compression does not adequately address viscous injury risks.

Viano and Lau [25.7, 25.8] proposed a viscous injury mechanism for soft biological tissues. The viscous response (VC) is defined as the product of velocity of deformation (V) and compression (C), which is a time-varying function in an impact. The parameter has physical meaning to absorbed energy (E_a) by a viscous dashpot under impact loading. Energy is related to the displacement integral of force: $E_a = \int Fdx$, and force in a dashpot is proportional to the velocity of deformation: $F = cV$, where c is a dashpot parameter. Absorbed energy is: $E_a = c\int Vdx$, or a time integral by substitution: $E_a = c\int V^2 dt$. The integrand is composed of two responses, so: $E_a = c(\int d(Vx) - \int axdt)$, where a is acceleration across the dashpot. The first term is the viscous response and the second an inertial term related to the deceleration of fluid set in motion. Absorbed energy is given by: $E_a = c(Vx - \int axdt)$, or $E_a \approx VC$. The viscous response is proportional to absorbed energy during the rapid phase of impact loading prior to peak compression, and correctly determines when injury occurs [25.9].

Figure 25.3 summarizes torso injury criteria associated with impact deformation. For low speeds of deformation, the limiting factor is the risk of crush injury from high compression of the body (C). This occurs at about 35-40% depending on the contact area and orientation of loading. For deformation speeds above 3 m/s, a similar level of injury risk is determined by the peak viscous response (VC). In a particular situation, there is a potential for injury related to compression or viscous responses; either or both mechanism can occur during impact. At extreme rates of loading, such as in a blast wave exposure, injury occurs with less than 10-15% compression by high-energy transfer to viscous elements of the body.

25.3. Biomechanical Responses

The reaction force developed by the chest varies with the velocity of impact, so biomechanics is best characterized by the force-deflection response of the torso [25.6]. The dynamic compliance is related to viscous, inertial and elastic properties of the body.

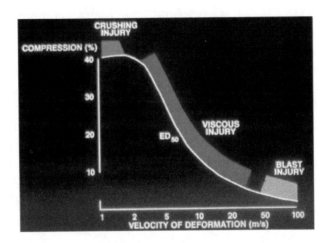

Figure 25.3 Biomechanics for chest and abdominal injury.

There is an initial rise in force, which is related to inertial responses as the sternal mass is rapidly accelerated to the impact speed. This is followed by a plateau in force, which is related to the viscous response and is rate-dependent, and a superimposed stiffness component related to chest compression. By analyzing frontal biomechanics, the chest response can be modeled as an initial stiffness $k = 0.26 + 0.60(V-1.3)$ and a plateau force $F = 1.0 + 0.75(V-3.7)$, where k is in kN/cm, F is in kN and the velocity of impact V is in m/s. The force F reasonably approximates the plateau level for lateral chest and abdominal impact, but the initial stiffness is lower at $F = 0.12(V-1.2)$ for side loading.

25.4. Injury Risk Assessment

Over years of study, tolerance levels have been established for most responses to assess injury from chest and abdomen impact. Reviews by Cavanaugh [25.4] and Rouhana [25.10] provide current tolerance information. For the chest, the compression tolerance is 32% and Viscous tolerance is 1.0 m/s in sternal impact. These are single level thresholds, which are commonly used to evaluate safety systems. The implication is that for biomechanical responses below tolerance, there is no injury, and for responses above tolerance, there is injury. An additional factor is biomechanical response scaling for individuals of different size and weight. The commonly accepted procedure involves equal stress and velocity, which enabled Mertz et al. [25.11] to predict injury tolerances and biomechanical responses for different size dummies.

Injury risk assessment is frequently used. It evaluates the probability of injury as a function of biomechanical response. A Logist function relates injury probability p to a biomechanical response x by $p(x) = [1 + \exp(\alpha - \beta x)]^{-1}$ where α and β are parameters derived from statistical analysis of biomechanical data. A sigmoidal function is typical of human tolerance because it represents the distribution in weak through strong subjects in a population exposed to impact.

25.5. Anthropometric Test Devices

Anthropomorphic test devices, or dummies, are mechanical analogs of the human body, which are routinely used to evaluate the effectiveness of restraint systems, protective clothing, safety devices and automotive designs in preventing injury. Dummies are designed to simulate the size, shape, mass, stiffness and energy absorption of the human body during impact. The most sophisticated frontal dummy is the Hybrid III. The family of Hybrid III dummies includes the 50th percentile and 95th percentile adult male, 5th percentile adult female, 3 and 6 year old child, and a range of infants for child safety seat testing. There is also a range of side impact dummies representing small to mid-size adults.

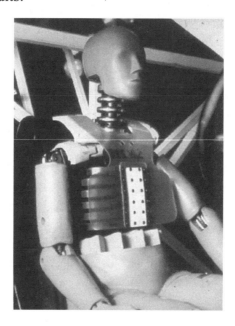

Figure 25.4 Hybrid III chest structure with frangible abdomen for belt submarining assessment [25.12].

Each dummy mimics the trajectory, acceleration and impact deformation experienced by a human of that size during crash deceleration or impact. The body of biomechanical information has been used to define the humanlike response and tolerance of each member of the dummy family to impact, and the mechanical dummies realistically approximate the human response for a range of impact severities from non-injury through serious/fatal injury. The dummies are durable and repeatable in response, and are well-accepted in the industry and by governments worldwide. A recent summary by Mertz [25.13] provides background information and technical details on the available frontal and side impact dummies.

REFERENCES

[25.1]   Viano, D.C., King, A.I., et al., Injury biomechanics research: an essential element in the prevention of trauma, *Journal of Biomechanics*, **22**, 403-417, 1989.

[25.2]   Viano, D.C. and King, A.I., Biomechanics of chest and abdomen impact, Chapter 26, In *The Biomedical Engineering Handbook*, J. D. Bronzino, Ed., CRC Press, Inc. and IEEE Press, Boca Raton, FL, 369-380, 1995.

[25.3]   Stapp, J.P., Voluntary human tolerance levels, In *Impact Injury and Crash Protection*, E.S. Gurdjian, W.A. Lange, L.M. Patrick and L.M. Thomas, Eds, Charles C Thomas, Springfield, IL, 308-349, 1970.

[25.4]   J.M. Cavanaugh, The biomechanics of thoracic trauma, In *Accidental Injury: Biomechanics and Prevention*, A.M. Nahum and J.W. Melvin, Eds., Springer-Verlag, New York, 362-391, 1993.

[25.5]   Patrick, L.M., Mertz, H.J. and Kroell, C.K., Cadaver knee, chest, and head impact loads, *Proceedings of the 11th Stapp Car Crash Conference*, SAE Paper No. 670913, Society of Automotive Engineers, Warrendale, PA, 168-182, 1967.

[25.6]   Kroell, C.K., Schneider, D.C. and Nahum, A.M., Impact tolerance and response to the human thorax II, *Proceedings of the 18th Stapp Car Crash Conference*, SAE Paper No. 741187, Society of Automotive Engineers, Warrendale, PA, 383-457, 1974.

[25.7]   Viano, D.C. and Lau, I.V., A viscous tolerance criterion for soft tissue injury assessment, *Journal of Biomechanics*, **21**, 387-399, 1988.

[25.8]   Lau, I.V. and Viano, D.C., The viscous criterion-bases and application of an injury severity index for soft tissue, *Proceedings of the 30th Stapp Car Crash Conference*, SAE Paper No. 861882, Society of Automotive Engineers, Warrendale, PA, 123-142, 1986.

[25.9]   Lau, I.V. and Viano, D.C., How and when blunt injury occurs: implications to frontal and side impact protection, *Proceedings of the 32nd Stapp Car Crash Conference*, SAE Paper No. 881714, Society of Automotive Engineers, Warrendale, PA, 81-100, 1988.

[25.10] Rouhana, S.W., Biomechanics of abdominal trauma, In *Accidental Injury: Biomechanics and Prevention*, A.M. Nahum and J.W. Melvin , Eds, Springer-Verlag, New York, 391-428, 1993.

[25.11] Mertz, H.J., Irwin, A., et al. Size, weight and biomechanical impact response requirements for adult size small female and large male dummies, SAE Paper No. 890756, Society of Automotive Engineers, Warrendale, PA, 1989.

[25.12] Rouhana, S.W., Viano, D.C., Jedrzejczak, E,A. and McCleary, J.D., Assessing submarining and abdominal injury risk in the hybrid III family of dummies, *Proceedings of the 33rd Stapp Car Crash Conference (P-227)*, SAE Technical Paper No. 892440, Society of Automotive Engineers, Warrendale, PA, 257-279, 1989.

[25.13] Mertz, H.J., Anthropomorphic test devices, In *Accidental Injury: Biomechanics and Prevention*, A.M. Nahum and J.W. Melvin , Eds., Springer-Verlag, New York, 66-84, 1993.

26. REAR CRASH SAFETY

In high-speed rear crashes, the seatback needs to be sufficiently strong to manage energy transfer while maintaining the occupant on the seat. Integration of occupant load over seatback displacement or moment over angle change gives the energy transfer capability of a seat. Seatback rotation correlates with occupant kinetic energy transfer, which is determined by the rear crash delta V. The energy transfer is about 2000 J in a 32 kmph rear delta V crash. In low-to-high speed crashes, the head restraint and upper seatback need to reduce relative motion between the head and neck, thus controlling kinematics to prevent "whiplash". A head restraint height above the head center of gravity and close to the back-of-head provides favorable neck responses. However, the trajectory of the head restraint is forward and downward in a rear crash. This promotes neck extension. A self-aligning head restraint gives a more horizontal direction of head restraint motion and a more upright head and neck orientation. The head restraint moves forward and upward by occupant load on the seatback. This closes the gap behind the head.

26.1   Introduction

Pioneering safety research was conducted on seat designs, which control seatback rotation and support the head in rear crashes [26.1-3]. Subsequently, there have been many studies on the epidemiology of rear crash injuries and the required engineering performance of seats and head restraints.

However, severe rear crashes involve vehicle delta Vs of 17 to 22 mph as occurring in FMVSS 301-type rear barrier impacts. The kinetic energy transfer to the occupant in these crashes is well beyond the capability of seats that exceed even twice the moment specified in FMVSS 207. Strother, James and Warner [26.4-6] provide an engineering analysis and rationale for seatback strength requirements and the need for a yielding seatback. This provides energy absorbing restraint as the seatback yields rearward. Seatback moments near the FMVSS 207 criterion were argued as necessary to balance the needs for "whiplash" injury prevention and occupant retention. These studies furthered the idea that there is an underlying design conflict between occupant retention in a severe, but infrequent, rear crash by a "stiff" seat and the need for a "compliant" seatback to prevent "whiplash" in the more frequent, minor severity rear impacts.

A thorough analysis of FARS and NASS field accident data indicates that nearly 5% (1700 cases) of all fatal passenger-car crashes are rear end impacts, which involve 3.5% (843) of all fatalities [26.7]. When all crashes are considered, rear impacts involve 12% (1,433,000) of exposed occupants but 23% (613,000) of all those injured. Based on selected state accident data, NHTSA GES-1993 estimated that rear impacts cause 23% (352,000) of minor-moderate injury and 9% (22,000) of serious-fatal injury in a total of 1,360,000 (18%) USA car crashes annually. The statistics are only slightly different for light trucks, vans and utility vehicles, where rear impacts cause 21% (78,000) of minor-moderate injury and 7% (5,000) of serious-fatal injury in a total of 522,000 (21%) USA crashes annually.

"Whiplash" injury claims are common after motor-vehicle collisions. While some cases lead to prolonged symptoms and disability, the vast majority has a favorable outcome. The Insurance Corporation of British Columbia found that 41.3% of "whiplash" claims were closed without payment within a year. However, the remainder of the claims accounted for 54.9% of injury claim losses. Nygren [26.8] found chronic pain and long-term disability in 10% of patients who initially complained of neck pain after rear crashes.

The introduction of higher head restraints in the late 1960s reduced the incidence of neck injury. Integral head restraints designed to FMVSS 202 were found to reduce the overall risk of injury in rear end crashes by 17% [26.9]. Adjustable head restraints reduced injury risks by 10%; and, the injury rate correlated with the height adjustment of the head restraint. The higher the height, the lower the incidence of injury with a 9% reduction in injury.

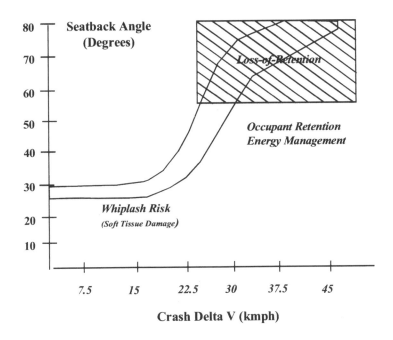

Figure 26.1 Seatback rotation in relation to rear crash severity involving either "whiplash" injury at low speeds and "loss of retention" at high speeds.

While "whiplash" claims are a significant component of motor-vehicle insurance costs, there is a widespread suspicion that exaggerated and fraudulent claims are a substantial factor masking real injury patterns. The inability to objectively diagnose the underlying pathology of "whiplash" injury further complicates the issue, since the claim of neck pain is often sufficient evidence to initiate treatment and rehabilitation.

26.2   High-speed crashes: the quasistatic seat test (QST)

Federal Motor Vehicle Safety Standard (FMVSS) 207 provides a test of seatback strength by loading the top cross-member of the seatback rearward. It involves a concentrated load usually orthogonal to the torso line, and performance is measured by multiplying the supported load by the distance to the seating reference point or occupant H-point. FMVSS 207 is a subsystem test of the seat but it does not involve an occupant-loading interface, so many details of occupant interactions with the seat in a real-world crash are not comprehended. The occupant exerts a horizontal load and moment low on the seatback, whereas the FMVSS 207 loads high on the seatback so there is a moment about the recliner but not a substantial horizontal load.

A Quasistatic Seat Test (US Patent 5,379,646, January 10,1995) was developed to assess occupant interactions with the seat in rear impacts. The kinematic sequence in the test mimics the dynamics of torso loading and ramping in a rear end crash. The responses include the moment about the H-point (M_h):

$$M_h = F_x*(z_h + z)$$
(26.1)

where F_x is the horizontal load of the dummy on the seat, z_h is the initial offset from the centerline of the ram to the H-point of the dummy.

26.2.1. Energy Management

Occupant load and displacement determine the energy transfer to the seat (E):

$$E = \int F_x * dx$$
(26.2)

A rear end crash involves energy transfer to the occupant by the seatback in proportion to crash severity (delta V):

$$V = [2E/m]^{0.5}$$
(26.3)

where m is the effective mass of the occupant, which is 70% of body mass. For purposes of comparing seats, a 60° seatback angle from vertical is used as a limiting rotation angle from vertical. For a 50th percentile male, a seat with 1000 J capability performs for rear crashes up to 22 kmph (6.2 m/s); whereas, a 2000 J seat provides performance up to 31.4 kmph (8.7 m/s), a 40% increase in rear crash delta V. This involves a peak occupant load of approximately 7.5 kN through a moment arm of 22 cm from the seatback pivot. This gives a 1700 Nm H-point moment.

Federal Motor Vehicle Safety Standard 301 specifies a stationary vehicle impact in the rear by a rigid, moving barrier at 50 kmph (13.5 m/s) to evaluate fuel system performance. The barrier weighs 5,000 lbs (2270 kg). The typical vehicle experiences a 31 kmph (8.7 m/s) change in velocity of the struck vehicle with a range of 27- 36 kmph (7.6 - 9.9 m/s).

26.2.2 Safety belts use

Safety belts are 49% effective in preventing fatalities in rear end crashes. The lap belt contributes about half of the effectiveness by holding the occupant on the seat and preventing ejection. However, due to typical forward orientation of the lap belt, the pelvis displaces rearward building up energy before the belt tightens. Pretensioning the lap belt in rear end crashes reduces the rearward and upward pelvic displacement. This provides earlier coupling of the occupant to the vehicle.

26.3   Low speed rear crashes: whiplash prevention

26.3.1 Head restraint positions from observed driving

A study was completed of seating positions and head restraint placement during normal driving [26.10]. Since 73% of the vehicles were fitted with adjustable head restraints, the overall driving situation has relatively low head restraint heights in relation to seated drivers. In fact, only 10% of the drivers had the most favorable condition of a small gap (less than 10 cm) and a high head restraint (top of head restraint above the ear).

26.3.2 Hyge sled tests of head restraint height and gap

Low-speed sled tests were conducted to simulate the relative risk differences for the situations observed during normal driving. Five different head restraint heights were used from below the chin to above the head center of gravity. There was approximately 2.5" (62.5 mm) difference in head restraint height between each level for the five conditions. Three gaps were used, including 5 cm, 15 cm and 25 cm.

The biomechanical data in Figure 26.2 support the position that the height of the head restraint should be at or above the head center of gravity. A small gap is preferred to improve early head restraint loading during vehicle acceleration or ride-up in a rear end crash. The particular design of the head restraint and upper seatback are key factors in the eventual head and neck interaction during rear impact. In particular, the energy absorption properties of the foam and trim design should minimize rebound so the elastic properties do not increase the head delta V and biomechanical responses.

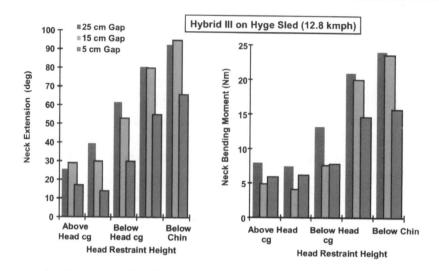

Figure 26.2 Low-speed rear crashes with various head restraint heights and gaps.

The relative risk differences varied by a factor of six with a three-fold difference for head restraint height and a two-fold difference for gap. The fraction of drivers in each group was next multiplied by the associated relative risk and summed over the distribution. This provided a relative risk of 3.4 for the current driving condition relative to the most favorable condition. A similar calculation was made for the distribution assuming each of the adjustable head restraints had been placed in the full up position and that this change shifted those drivers to the next higher head restraint height category. This calculation led to a relative risk of 2.4. Proper placement of head restraints would reduce the risks of neck injury by 42%.

The tests confirmed that control of neck biomechanical responses, such as extension angle, neck moment and tension require a head restraint positioned to support the head at or above the head center of gravity. Head restraint performance at low speeds is quite important. Occupant kinematics in rear end crashes tend to involve horizontal accelerations. This implies that the load distribution, which supports the body, needs to be considered in occupant kinematic control since the primary direction of restraining force on the head/neck is horizontal. However, the trajectory of the head restraint is downward, since the seat rotates rearward as it displaces forward in a crash. For favorable kinematics during head restraint contact and to reduce neck extension, the top of head restraint should be high enough so reasonable kinematics are achieved even as the head restraint trajectory is downward.

26.3.3 Self-aligning head restraint (SAHR)

A recurring observation from the rear impact testing is that the pelvis and lower torso displace into the seatback early in the crash, before sufficient loads develop on the shoulders and upper back to initiate neck extension. These kinematics led to the concept of a self-aligning head restraint (US Patent No. 5,378,043, January 3, 1995).

Figure 26.3 shows the basic principle of self-aligning, which involves the early displacement of the torso into the upper seatback region. The rearward displacement of the torso is used to load a platen in the seatback at the chest level. The platen is attached to the head restraint posts, which pass through the seatback frame. By using a mechanism to support the platen, a see-saw action can be achieved. Rearward displacement of the platen by occupant movement rotates the front surface of the head restraint forward into earlier head contact and upward to achieve a higher position with respect to the seated occupant

By early rotation of the head restraint forward, the head and neck benefit from a lower relative velocity of impact on the head restraint, thus lowering the forces, moments, and extension of the neck. In addition, the self-aligning action provides an improved direction of loading on the head. Even in low-speed crashes, the occupant loads the seatback causing a rearward rotation of the seatback. This causes a downward trajectory of the head restraint with respect to the head. In contrast, self-aligning motion tends to compensate the rearward seatback rotation. This results in a relatively horizontal, if not upward, direction of head restraint loading on the head. This results in a slight flexion and a more vertical orientation of the neck.

One important application of the self-aligning head restraint is for occupants who have a high relative head position with respect to the head restraint. This is usually tall occupants, individuals with long necks, or people who sit or drive with a high head position. For these individuals, the head restraint may not be able to be adjusted high enough during normal driving to fully extend over the head cg without obstructing the rear view of shorter

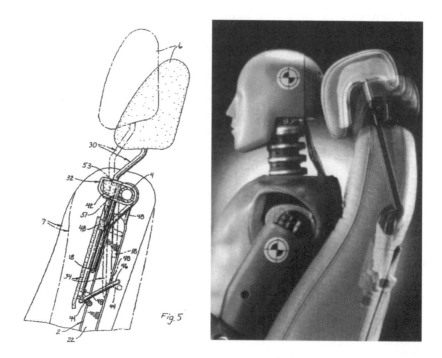

Figure 26.3 Self-Aligning Head restraint concept showing the upward and forward movement of the head restraint by occupant loading of the seatback.

drivers. Self-aligning provides an increase in head restraint height as it rotates forward [26.11]. This would allow a low position for an unobstructed view during normal driving and an automatic adjustment up during a crash to benefit the taller occupants.

REFERENCES

[26.1]   Severy, D.M., Mathewson, J.H. and Bechtol, C.O., Controlled automotive rear-end collisions, an investigation of related engineering and medical phenomina, *Canadian Services Medical Journal*, **VII**, 727-759, 1955.

[26.2]   Severy, D.M., Brink, H.M. and Baird, J.D., Preliminary findings of head support designs, *Proceedings of the 11th Stapp Car Crash Conference,* SAE Paper N. 670921, Society of Automotive Engineers, Warrendale PA, 337-405, 1967.

[26.3]   Severy, D.M., Brink, H.M. and Baird, J.D., Backrest and head restraint design for rear-end collision protection, SAE Paper N. 680079, Society of Automotive Engineers, Warrendale PA, 1968.

[26.4]   James, M., Strother, C., Warner, C., et al. Occupant protection in rear-end collisions: I safety priorities and seat belt effectiveness, *Proceedings of the 35th Stapp Car Crash Conference*, SAE Paper N. 912913, Society of Automotive Engineers, Warrendale, PA, 1991.

[26.5]   Strother C. and James, M., Evaluation of seat back strength and seat belt effectiveness in rear impacts, *Proceedings of the 31st Stapp Car Crash Conference*, SAE Paper N. 872214, Society of Automotive Engineers, Warrendale, PA, 1987.

[26.6]   Warner, C., Strother, C., et al., Occupant protection in rear-end collisions ii: the role of seat back deformation in injury reduction, *Proceedings of the 35th Stapp Car Crash Conference*, SAE Paper N. 912914, Society of Automotive Engineers, Warrendale, PA, 1991.

[26.7]   Malliaris, A.C., *Current Issues of Occupant Protection in Car Rear Impacts*, Data Link, Inc., NHTSA Docket 89-20-No1-021, February, 1990.

[26.8]   Nygren A, Gustafsson H, Tingvall C "Effects of different types of head restraints in rear-end collisions," 10th ESV Conference, p.85-90, US DOT/NHTSA, 1985.

[26.9]   Kahane, C.J., *An Evaluation of Head Restraints - Federal Motor Vehicle Safety Standard 202*, USDOT/NHTSA Technical Report DOT HS 806 108, 1982.

[26.10] Viano, D.C. and Gargan, M.F., Seating position and headrest location during normal driving: implications to neck injury risks in rearend crashes, *Accident Analysis & Prevention*, **28**(6), 665-674, 1996.

[26.11] Wiklund, K. and Larsson, H., Saab active head restraint (SAHR) – seat design to reduce the risk of neck injuries in rear impacts, SAE Paper N. 980297, Society of Automotive Engineers, Warrendale, PA, 1997.

27. OCCUPANT RESTRAINTS

The fundamentals of occupant protection involve vehicle crash worthiness where are strong occupant compartment resists intrusion and crushable front and rear structures deform and absorb energy in a crash. This combination provides a controlled vehicle deceleration and survival space in the occupant compartment. The use of lap-shoulder belts and airbags provides ride-down of the vehicle crush, containment on the seat, and load distributing forces on the pelvis, should and upper body to decelerate the occupant. This has proven an effective means of restraining the occupant and reducing the risk of serious injury and death in a crash. Lap-shoulder belts are 42% effective in preventing death. This includes the highest effectiveness of 77% in rollovers and lowest of 27% in near-side impacts. The addition of the frontal airbag raises the level to 47%. The addition of belt system enhancements for comfort and performance and introduction of side airbags and curtains for rollover protection further improve the safety in automotive crashes.

27.1 Introduction

Airbags and safety belts are complements for occupant protection in a crash, as the safety community understands that there is no single solution to occupant protection. Rather, a system of technologies is needed to provide maximum safety in the wide variety of real-world crashes [27.1]. While safety belts effectively prevent fatal injuries, they must be worn to protect vehicle occupants. Safety belts provide the greatest relative effectiveness in rollover crashes where airbags provide little protection. However, the greater load distribution provided by airbags in frontal crashes enhances the protection of safety belts and provides substantial protection for unbelted occupants.

The current view that safety belts and airbags work together for occupant protection also embraces the concepts of passive and active protection. In many cases, the most effective public health approaches involve passive protection, because they don't rely on the voluntary action of the public. However, maximizing fatality prevention in the wide range of crash types and injury mechanisms involved in motor vehicle crashes requires a combination of approaches. A complement of built-in vehicle safety, active safety belt use, and driving behavior improvement is widely viewed as a strategy to advance road safety.

Occupant restraints enhance the safety of currently available vehicle technologies such as energy absorbing front and rear structures, a built-in safety cage around the occupant compartment, and friendly interior components. However, even with all of these safety measures, the severity of many crashes is so great that serious injury and fatality still occur, even as many thousands of lives are being saved by belt use and supplemental airbags.

27.2 Occupant Protection

27.2.1 Vehicle crashworthiness

From 1930-1950, the automotive industry emphasized structural integrity of the passenger compartment to contain the occupant in frontal and rollover crashes. By the 1960's, the concept of energy management through crushable front- and rear-end structures was added. This combined approach attempted to preserve the occupant's space, or "room to live," while the vehicle's crushing structures absorbed crash energy. This also lengthens the stopping time and distance of the passenger compartment, and reduces impact accelerations acting on the occupant. Further improvements are achieved by isolating front-end structures from the passenger compartment to minimize intrusion or deformation around the occupant.

27.2.2 Friendly interiors

For an unrestrained occupant, the controlled deceleration of the vehicle in a frontal crash is followed by impact of the occupant against the vehicle interior. Removing hard knobs and sharp edges that occupants tended to hit, just as rigid door ornaments have been removed for the benefit of pedestrians, eliminated many injuries. Further protection was achieved by using energy-absorbing interior structures and load-distributing surfaces that minimize the occupant's impact acceleration while spreading the forces over a broader portion of the body's strongest parts. The concept of impact energy absorption has two aspects. First, the change in occupant velocity must be extended over as long a time as possible. Having the occupant hit something that deforms in the direction of impact increases the body's stopping distance. Second, it is important that yielding structures not spring back at the occupant, but rather deform permanently or recover slowly, reducing the rebound velocity.

Energy-absorbing steering system. This is a steering column that crushes at a prescribed load, which is not great enough to cause significant rib fracture. Deforming the steering system under occupant impact increases the driver's stopping distance, decreases thoracic deceleration, and absorbs impact energy. The system includes a deformable element and steering wheel with improved load distribution and stiffness. When the load of the driver on the steering wheel exceeds the compressive force of the energy-absorbing element, the column slips out of the shear capsule, compresses, and absorbs energy.

High-penetration-resistant windshields. Early safety glass could cause significant facial laceration because it was fairly brittle and would break and be penetrated by the head in severe crashes. Improved occupant protection was achieved if the head was kept from passing through the glass during impact while at the same time ensuring that the head is safety decelerated to protect against concussion injury.

Figure 27.1 Sequence of restraint action in a frontal crash.

27.3 Restraint Systems

Although interior safety achieved tremendous gains in occupant crash protection in the 1960s, the crush distance available in even the friendliest interior is only a fraction of that needed to achieve safe occupant decelerations in high-speed crashes. The use of safety belts and inflatable restraints further improved crash safety by better restraining the occupant. Figure 27.1 shows the action of safety belt and airbag restraint in a frontal crash.

Seat belt restraints. A snug-fitting lap-shoulder belt ties the occupant directly to the passenger compartment and allows the occupant to "ride-down" the crash as the vehicle's front-end crushes. This coupling and ride-down decelerates the occupant more gradually than possible with energy-absorbing interiors. Belts are also designed to distribute restraining loads over strong skeletal structures of the shoulder and pelvis.

Inflatable restraints. Airbags were originally developed to overcome the primary weakness of belt systems: to be effective the occupant must fasten the belts in advance of the crash. Using a pyrotechnic device to generate nitrogen gas, a bag can be rapidly inflated during the early phase of vehicle frontal crush without action by the occupant. The bag then "fills" some of the space between the occupant and the interior, which couples the occupant to the passenger compartment and achieves the safety benefits of ride-down and load distribution. This coupling is only temporary, however, because the bag must be vented and deflate, so that it will not act as a spring. However, the rapid deployment speed of an airbag can present a risk to occupants, including children, who may be close to the bag during inflation. Its design requires a trade-off between a long inflation time to reduce the risk of inflation injury and a rapid inflation to quickly fill the space between the occupant and the interior.

Because airbags neither remain inflated nor provide lateral restraint, seat belts are needed to adequately control occupant kinematics over the range of crash types, including rollovers and side impacts. The current safety thinking, therefore, is to use inflatable restraints as a supplement to seat belts. The lap-shoulder belts provide the primary coupling to the vehicle and control kinematics, while the airbag provides the additional protection of load distribution and crash energy absorption in the more severe frontal crashes. The combination of safety technologies work with a crashworthy vehicle structure and friendly interiors to enhance occupant protection.

27.4 Restraint biomechanics and performance

27.4.1 Driver airbag

The driver airbag takes advantage of rapid filling of a concealed bag to provide a cushion in front of an occupant in a crash [27.2]. This provides a large area to gradually decelerate the driver. Although compressed gas in a cylinder was one of the early concepts, the eventual first production systems took advantage of a relatively small mass (70-100 g) of sodium azide. The chemical rapidly inflates a 80-100 1 driver airbag by ignition and conversion to harmless nitrogen gas when sensors detect a severe frontal crash. The airbag is stored inconspicuously in the interior and deploys only during a severe frontal crash. Sensors are located in the passenger and engine compartments to detect rapid decelerations of the vehicle during a crash and an electronic system is used to monitor and initiate deployment of the airbags.

27.4.2 Lap-belts

An early use of the lap-belt was to complement crash protection of the chest by a driver airbag. In the absence of pelvic restraint, loads would be applied through the knees and seat pan to restrain the occupant and control the upright posture of the driver. The lap-belt complemented early airbag designs for passenger protection by a padded dashboard and also found their way into rear seating positions to supplement padded seat backs. Although much of the development work on restraints has involved frontal barrier and sled testing, the importance of the lap-belt was recognized for preventing ejection, which was known by crash investigation as the leading cause of fatality in the 1950's.

27.4.3 Lap-shoulder belts

In the 1960s, the combination of a lap and diagonal shoulder belt gained rapid acceptance, because it provides occupant restraint by routing safety belt loads over the bony structures of the pelvis and shoulder. This takes advantage of the relatively high tolerance to impact force for these regions of the skeleton and avoids concentrating load on the more complaint abdominal and thoracic regions.

The fundamentals of a high quality belt restraint system involve occupant kinematic controls [27.3-5], which maintain the lap belt low on the pelvis through adequate seat cushion support. This minimizes pelvic rotation and reduces the tendency for the lap-belt to slide off the ileum and directly load the abdomen. Forward rotation of the upper torso to slightly greater than 90° upright posture directs a major portion of the upper torso restraint into the shoulder.

Injury patterns associated with safety belt use are understood. The potential for improper use has also continued, particularly placement of the shoulder harness under the arm and wearing of the lap-belt high on the abdomen with poor seating posture. In rollover crashes, safety belt use virtually eliminates the risk of paralyzing cervical injury and ejection.

27.5 Restraint effectiveness

The 1980s saw the introduction of statistical methods for epidemiological analyses of field accident data. Evans [27.6] developed the double-pair comparison method to isolate the effectiveness of belt use from other confounding factors in automotive crashes. The work has been extended to investigate how rear lap belt use, airbags, alcohol use, occupant age, seating position, and direction of the crash affect fatality risks. The fatality prevention effectiveness of front seat lap-shoulder belts was found to be $(41 \pm 4)\%$ in preventing fatality for front-seat occupants, $(42 \pm 4)\%$ for drivers and $(39 \pm 4)\%$ for right-front passengers [27.7,27.8].

Safety belts provide at least two essential components of occupant protection [9]. One is protection against ejection, which is primarily due to the lap portion of the belt system, and the other is mitigation of interior impact, which is largely due to upper body restraint by the shoulder harness. Figure 27.2 shows the effectiveness of driver restraints for different crash types and directions. In frontal crashes, belts are 43% effective in preventing fatality, 9% of the effectiveness comes is by the prevention of ejection from the lap-belt. The remainder comes from the shoulder belt reducing upper body impacts.

The relative safety contribution by eliminating ejection with the lap belt depends on the type of crash. It is highest at 63% in rollovers where ejection risks are the greatest. Overall, safety belts are 77% in rollovers. Lap-shoulder belts are least effective in preventing fatalities in near-side impacts for the driver. Nonetheless, they are at 27%.

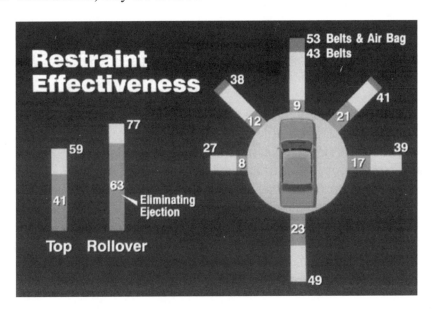

Figure 27.2 Effectiveness of occupant restraints in preventing fatalities.

Zador and Ciccone [27.10] evaluated crash data on airbag equipped vehicles and have provided the early estimates of fatality and injury prevention. As a supplement to safety belts, the driver airbag and belt combination provides an overall 47% effectiveness in preventing fatalities. While this represents a 5% incremental improvement over the 42% effectiveness of belts alone, the supplemental airbag reduces the risk of fatality by 9%. The 47% level coincides with earlier projections of safety benefit. Fatalities of unbelted drivers are 18% lower with airbag restraint. This level is higher than earlier estimates, but within the expected confidence interval. In addition, injury claims data indicate substantially lower hospital admissions and seriousness of injuries. The airbag is providing a 24% lower hospitalization rate and a 25%-29% lower incidence of serious injuries for the current mix of safety belt use in field accidents.

Figure 27.3 summarizes the overall effectiveness of occupant restraints and includes estimates for other restraint combinations in preventing occupant fatalities [27.8, 27.9]. The results show that lap-shoulder belts are the most effective restraint in protecting occupants in severe crashes. Airbags and other interior components do not restrain the occupant in a seating position and have minimal effectiveness in reducing ejection. This and potentially other factors result in a lower overall effectiveness for airbags only. Their principal contribution to crash protection is by excellent interior impact mitigation and load distribution, which provides about half the overall benefit of lap-shoulder belt use.

The fact that a lap-shoulder belt and airbag is only 47% effective in preventing fatalities underscores that absolute protection is not achievable by occupant restraints, and that injury and fatality will continue to occur to belt wearers even with an overall net safety gain as active restraint usage increases and airbags have become widely available [27.11]. The limit reflects the severity of many fatal crashes which may involve extreme vehicle damage and forces on the passenger compartment, unusual crash configurations and causes of death, unique situations associated with particular seating positions, crash dynamics and human tolerance limits.

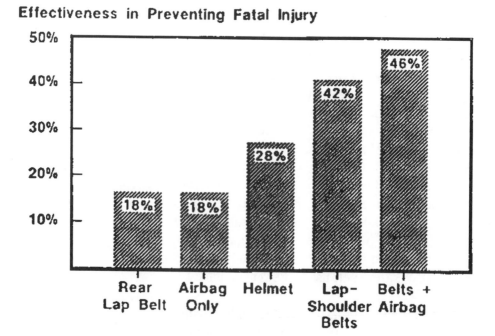

Figure 27.3 Overall effectiveness of occupant restraints.

Efforts continue to increase the overall effectiveness of occupant restraints above the 47% level. This has included the use of features to increase the comfort and restraint performance of the lap-shoulder belts by pre-tensioning, load-limiting, and height adjusting the shoulder belt attachment to the B-pillar. Side airbags and curtains are now available to pad impacts in side crashes and restrain the occupants from full or partial ejection in rollovers.

REFERENCES

[27.1]   Viano, D.C., Cause and control of automotive trauma, *Bulletin of the New York Academy of Medicine*, Second Series, **64**(5), 376-421, 1988.

[27.2]   Struble, D., Airbag technology: what it is and how it came to be, SAE Paper N. 980648, Society of Automotive Engineers, Warrendale, PA, 1998.

[27.3]   Adomeit, D. and Heger, A.l., Motion sequence criteria and design proposals for restraint devices in order to avoid unfavorable biomechanic conditions and submarining, *Proceedings of the 19th Stapp Car Crash Conference*, SAE Paper N. 751146, Society of Automotive Engineers, Warrendale, PA., 139-166, 1975.

[27.4]   Adomeit, D., Evaluation methods for the biomechanical quality of restraint systems during frontal impact, *Proceedings of the 21st Stapp Car Crash Conference*, SAE Paper N. 770936, Society of Automotive Engineers, Warrendale, PA, 911-932, 1977.

[27.5]   Adomeit, D., Seat design – a significant factor for safety belt effectiveness, *Proceedings of the 23rd Stapp Car Crash Conference*, SAE Paper N. 791004, Society of Automotive Engineers, Warrendale, PA, 39-68, 1979.

[27.6]   Evans, L., Double pair comparison - a new method to determine how occupant characteristics affect fatality risk in traffic crashes, *Accident Analysis and Prevention*, **18**, 217-227, 1986.

[27.7]   Evans, L., The effectiveness of safety belts in preventing fatalities, *Accident Analysis and Prevention*, **18**, 229-241, 1986.

[27.8]   Evans, L., *Traffic Safety and the Driver*, Van Nostrad Reinhold, 1991.

[27.9]   Viano, D.C., Effectiveness of safety belts and airbags in preventing fatal injury, In *Crash Safety Technologies for the 90's*, SAE Paper N. 910901, Society of Automotive Engineers, Warrendale PA, 1991

[27.10] Zador, P. and Ciccone, M., *Driver Fatalities in Frontal Impacts: Comparisons Between Cars with Airbags and Manual Belts*, Insurance Institute for Highway Safety, Arlington VA, October 1991.

[27.11] Viano, D.C., Limits and challenges of crash protection, *Accident Analysis and Prevention*, **20**(6), 421-429, 1988.

28. CRASH COMPATIBILITY

by
D.BUZEMAN-JEWKES[1], R.W.THOMSON[1], and D.C. VIANO

Undesirable - or incompatible - crash interactions between one vehicle and its collision partner can be due to differences in structure, mass, stiffness and geometry. These differences can increase deformation and acceleration levels in one or both of the vehicles and thereby increase occupant injury risks. Real world statistics indicate that the ratio of occupant fatalities between two vehicle types can range from 1.6:1 (cars and light trucks in frontal collisions) to 23:1 (cars struck in the side by large vans). The mix of vehicle types and potential impact conditions present a challenge when designing crash protection strategies for motor vehicles and the transportation network.

Mass is a predominant incompatibility factor. However, incompatibilities due to stiff front-structures and high bumper-heights in some vehicles may reflect greater safety consequences than mass alone. Historically, crash-testing legislation has focused on occupant protection of the tested vehicle. Now, deformable and load-cell barrier tests are being considered to improve local stiffness and geometric compatibility, leading to greater protection of the collision partner.

28.1 Introduction

Injuries to vehicle occupants are caused by two principle mechanisms during a crash. One injury source is due to the *acceleration* of the vehicle structure that is transferred to the occupants. Another injury source is the occupant impact on the vehicle interior. This impact is due to the relative motion of the occupant within the vehicle and may be accentuated by deformation or *intrusion* of the passenger compartment.

[1] Crash Safety Division; Department of Machine & Vehicle Design; Chalmers University of Technology; SE-412 96 Goteborg, Sweden

For many years, occupant compartment acceleration has been evaluated in full frontal crash tests (ECE 12, FMVSS 208 and NCAP). As a result, acceleration management has improved considerably for crashes of 30 to 35 mph. While vehicle structures have improved to control vehicle deceleration and increase restraint performance, injuries associated with compartment intrusion have become a major focus of recent research. Intrusion is more severe in asymmetric vehicle loading and in side impacts, where structural integrity is a key factor for occupant safety [28.1-3]. The need to understand and control intrusion related injuries has led to the development of frontal offset tests, such as Euro-NCAP, as well as other crash test configurations. However, the various crash test configurations have not yet been able to fully address the *incompatibilities* that may exist in colliding vehicles. Physical differences between colliding vehicles almost always result in one vehicle experiencing higher accelerations and/or intrusions that increase the risk of occupant injury. Improvements to vehicle compatibility will increase overall traffic safety levels.

Over many years, occupant compartment acceleration has been evaluated in full frontal tests (ECE 12, FMVSS 208 and NCAP), and acceleration management has improved considerably for crashes of 30 to 35 mph. Occupant injuries are caused by restraining loads related to vehicle acceleration, and by impact on the vehicle interior or compartment intrusion. While countermeasures to control vehicle deceleration have improved vehicle structures and restraint performance, injuries associated with compartment intrusion and incompatibilities in crashes have become a major focus with the development of frontal offset tests, such as in the EuroNCAP and other crash tests. Intrusion is more severe in asymmetric loading and side impacts, where structural integrity is a key factor for occupant safety [28.1-3].

28.2. Incompatibility definitions

To understand how incompatibilities arise, it is important to understand how vehicle structures are designed for collision protection. Figure 28.1 shows some of the important structural components in a modern vehicle. Both longitudinal beams are designed to crumple in a controlled manner and absorb kinetic energy during the collision. Lateral connections help stabilize and transfer loads between each side of the vehicle. This load transfer is important for collisions where only one longitudinal beam is loaded directly. Loads from the longitudinal beams are transferred into the passenger compartment through the firewall into the A-pillars, tunnel and sill. These structures should not deform, maintaining an intact passenger capsule during the collision. Important components not seen in Figure 28.1 are the engine and wheels. Both of these components can contact the firewall with sufficient deformation of the front structure. It is these contacts that usually introduce frontal compartment intrusion and the increased risk of occupant injury.

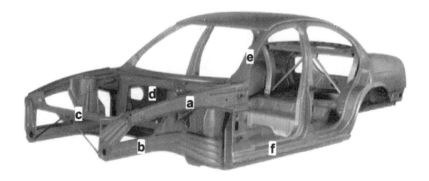

Figure 28.1 Structural components in frontal crash protection: a – upper longitudinal, b – lower longitudinal, c – lateral connection, d – firewall, e – A-pillar, f - sill

In frontal impact situations, the goal is to have the structures in Figure 28.1 interact with the corresponding structures on the collision partner. However, for other impact conditions (oblique, side, and rear impacts) this is not possible. As seen in Figure 28.1, there are no corresponding structures in the side of the vehicle, other than the door (not shown in Figure 28.1). Similarly in rear impacts, the structures present may not interact with the front of another vehicle. Thus, compatibility issues between collision partners are not simply solved by making the front of every vehicle identical. Every surface of the vehicle must be considered as a collision object with differing crash characteristics and load paths.

To investigate crash compatibility, it is useful to identify unique compatibility concepts. One approach is to consider the four definitions identified in Figure 28.2: structural, mass, stiffness and geometrical incompatibility. These descriptions of compatibility are not independent from each other. However, they provide the possibility to systematically study vehicle collisions and describe how the structures interact during the crash.

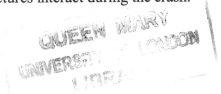

28.2.1 Structural incompatibility

Structural incompatibility relates to the mismatch between the structural capacity of the vehicle and the crash configuration. It is more pronounced in asymmetrical loading, like frontal offset and pole impacts, where the front structure may not distribute the load over the full vehicle front-end [28.1]. This usually causes larger deformation (D) and intrusion in crashes with smaller contact areas. The deformation, D2, in the offset crash is thus larger than D1 in full frontal impact of equal energy.

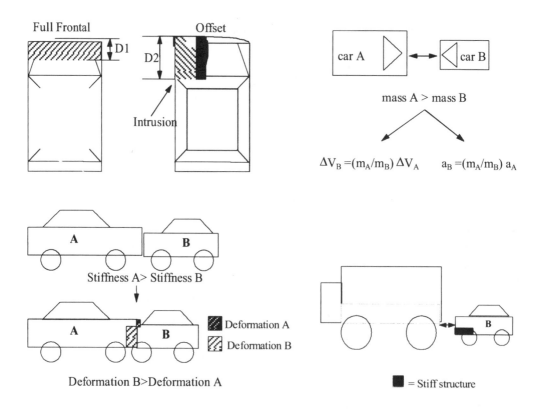

Figure 28.2 (top left) structural incompatibility in offset and full frontal crashes, (top right) mass incompatibility causing higher delta Vs and accelerations in the lighter vehicle, (bottom left) stiffness incompatibility causing more deformation in the less stiff structure, and (bottom right) geometrical incompatibility with a mismatch in the interaction of vehicle structures.

28.2.2 Mass incompatibility

Mass incompatibility is due to the difference in mass of two colliding vehicles of masses, m_A and m_B. The conservation of momentum equation shows that the accelerations of colliding vehicles are inversely related to their mass. Mass incompatibility causes higher acceleration of the lighter vehicle, a_B, (see Eq. (28.11)).

28.2.3 Stiffness incompatibility

Stiffness incompatibility is the difference in (local) stiffness between colliding vehicles. The weaker vehicle deforms most (see Eq. (28.8), resulting in a higher risk of compartment intrusion.

28.2.4 Geometrical incompatibility

Geometrical incompatibility involves the misalignment of stiff, energy-absorbing structures of colliding objects (see Eq. (28.9), where K depends on the vehicle impact location), which often involve higher compartment intrusion.

28.3 Mathematical Description of Vehicle Crashes

Vehicle acceleration and compartment intrusion can be estimated in a collision using the laws of physics. First, conservation of momentum and energy relate the amount of absorbed energy of two colliding vehicles to their mass ratio and closing speed at impact. Consider the case shown in Figure 28.3. The pre-crash momentum of the two vehicles equals that post-crash (Eq. 28.1), while the kinetic energy difference between pre and post crash is absorbed by vehicle damage or crush ($E_{abs,1}$ and $E_{abs,2}$), and by the tire-work (W_{tires}) with the road (Eq. 28.2):

$$M_1 V_{1,pre} + M_2 V_{2,pre} = M_1 V_{1,post} + M_2 V_{2,post} \qquad (28.1)$$

$$\tfrac{1}{2} M_1 V_{1,pre}^2 + \tfrac{1}{2} M_2 V_{2,pre}^2 = \tfrac{1}{2} M_1 V_{1,post}^2 + \tfrac{1}{2} M_2 V_{2,post}^2$$
$$+ E_{abs,1} + E_{abs,2} + W_{tires,1} + W_{tires,2} \qquad (28.2)$$

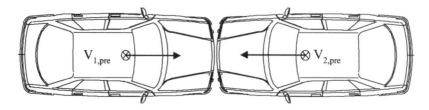

Figure 28.3   Centroidal collision of two vehicles

If we consider that the momentum vectors pass through the vehicles' centroids, there will be no rotation of the vehicles after the impact. The post-impact speeds will be equal for both vehicles if the restitution is negligible. The energy absorbed by both vehicles can then be written as a function of the closing speed, $|V_{cl}| = |V_{1,pre}| + |V_{2,pre}|$ by substituting (28.1) in (28.2). Other assumptions in this analysis are the square of the post impact speed is small compared to V_{cl} and the influence of the tire forces (W_{tires}) can be neglected.

$$E_{abs,1} + E_{abs,2} = \tfrac{1}{2}\frac{M_1 M_2}{M_1 + M_2} V_{cl}^2 \tag{28.3}$$

Second, the velocity change, or collision severity sustained by the occupants, is proportional to the closing speed and mass ratio:

$$\Delta V_1 = \frac{M_2}{M_1 + M_2} V_{cl} \tag{28.4}$$

$$\Delta V_2 = \frac{M_1}{M_1 + M_2} V_{cl} \tag{28.5}$$

Third, the energy loss, E_{abs}, is absorbed by vehicle crush.   The relationship between absorbed energy and vehicle crush (Eq. 28.7) is often based on a linear force deflection (F-X) characteristic of the vehicle's structure:

$$F_1 = K_1 X_1 \tag{28.6a}$$
$$F_2 = K_2 X_2 \tag{28.6b}$$

$$E_{abs} = E_{abs,1} + E_{abs,2} = \tfrac{1}{2}K_1 X_1^2 + \tfrac{1}{2}K_2 X_2^2 \tag{28.7}$$

The stiffness, K, varies with the vehicle and the impact location on the vehicle. The effective stiffness can be considered proportional to the impact area on the vehicle. Newton's third law makes it possible to express the damage of vehicle 1 in terms of vehicle 2's damage (Eq. 28.8), such that vehicle crush is a function of impact speed and vehicle masses:

$$X_1 = \frac{K_2}{K_1} X_2 \tag{28.8}$$

$$\tfrac{1}{2}(\frac{K_2^2}{K_1} + K_2)X_2^2 = \tfrac{1}{2}\frac{M_1 M_2}{M_1 + M_2}V_{cl}^2 \tag{28.9}$$

Compartment intrusion is related to the amount of crush depth, X, and occurs when the structural crush zone is exceeded (or the compartment strength becomes less than the structural crush force). Equation (28.9) provides information, which can be used to study relationships among compartment intrusion, collision speed, vehicle masses and vehicle stiffnesses.

Finally, the relationship between vehicle crush and vehicle acceleration is explained by Newton's second law, using the linear force-deflection characteristic (Eq. 28.6). This law states that the sum of the forces working on an object of mass M_i, causes acceleration a_i:

$$\sum F = M_i a_i, \quad with \sum F = K_i X_i \quad i{=}1, 2 \tag{28.10}$$

$$\Leftrightarrow \quad K_i X_i = M_i a_i \tag{28.11}$$

Equation (28.9) can be used to calculate the force and acceleration in Eq. (28.11).

The exercise above shows that vehicle acceleration and intrusion depend on the crash speed, impact configuration and vehicle properties relative to structure, mass, stiffness and geometry of the object struck, since stiffness varies with each vehicle, impact location and area. These properties affect the occupant's protection as well as the collision partner's in a vehicle-to-vehicle collision. A mismatch or incompatibility of vehicle masses causes a higher acceleration in the lighter car. Furthermore, incompatible structures, stiffness or geometry may cause compartment intrusion and related injuries in the car with the weaker structure.

28.4 Real world accident data

An important component of vehicle safety research is the review of collision statistics and associated injuries. Using the information in different databases (e.g. police reports, hospital records, etc), an understanding of the injuries sustained in car crashes can be developed. In the following sections, the different compatibility definitions are discussed in relation to the data collected for motor vehicle collisions.

28.4.1 Structural Incompatiblity

Real world accidents have been studied to analyze injury occurrence and risk in frontal offset collisions. Injury probability reflects the relative proportion of injuries in a specific crash to other crash conditions, while the risk indicates the proportion of injured per number exposed individuals in a certain crash. Injuries have been related to occupant compartment intrusion in asymmetric or offset loading with high (1/3 to 2/3) overlap for upper body injuries [28.3]. However, distributed, >2/3 overlap [28.4] and < 1/3 overlap crashes [28.2, 28.4-6] also cause risks in frontal crashes. Crashes with low acceleration and high intrusion cause higher risk than other combinations for a velocity change (delta-V) between 36 and 65 km/h [28.7]. However, the results indicate that the estimated risk may be influenced by the choice of data set and statistical method.

Buzeman et al. [28.8] conducted an epidemiology study to determine the effect of overlap on occupant injury in frontal crashes of comparable severity. Drivers and passengers were combined into groups seated near the impact (near group) or opposite from the contact location (far group) to reduce biasing effects from sex and age. Crashes with an equivalent barrier speed (speed at which the vehicle damage reflects a rigid barrier impact) greater than 32 km/h (20 mph) were chosen to reduce the potential biasing effect of impact severity. Frontal collisions with 2/3 overlap caused the majority of front occupant injury, followed by distributed collisions.

Figure 28.4 shows that the risk for drivers and front-passengers is higher in near 1/3 overlap collisions than in any other frontal crash. Low overlap crashes involve a relatively low acceleration and high compartment intrusion, since intrusion increases as overlap decreases [28.2, 28.3, 28.5-7]. The high injury risk in <1/3 overlap indicates that the injury mechanism may be more related to intrusion than to acceleration. In terms of overlap, a 40% offset test seems a satisfactory compromise to address high risk in 1/3 overlap and the high injury probability in 2/3 and full overlap collisions [28.8]. A 40% overlap frontal test has been adopted in Europe (EuroNCAP) and has been performed by the Insurance Institute for Highway Safety (IIHS) for car-safety ratings since 1995.

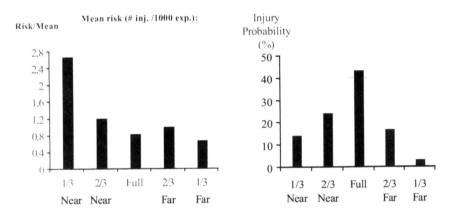

Figure 28.4 Relative injury risk and injury probability for near and far groups with overlap crashes of equivalent barrier speed >32 km/h (20 mph).

28.4.2 Mass Compatibility

Statistical analyses of real-world crashes indicates that vehicle-to-vehicle crashes occur about twice as often as single car crashes. Occupant injuries and fatalities increase for decreasing car mass, while partner injuries and fatalities decrease [28.3, 28.9-10]. Heavier vehicles, like sport utility vehicles, experience lower delta-V's and the occupants have a lower injury risk than other passenger cars [28.11]. Figure 28.5 shows the relative fatality ratio for car occupants involved in fatal collisions with small trucks through large vans. The ratio varies from 1.6:1 to 6.0:1 for car occupants versus the collision partner in head-on crashes depending on the mass, size and structure of the collision partner.

In crashes with similar mass vehicles, lighter vehicles offer less protection than heavy vehicles. This indicates that the mass effect on injury may include other factors like vehicle size, stiffness and inherent protection [28.1, 28.9]. Figure 28.6 shows the injury distribution for passenger vehicle

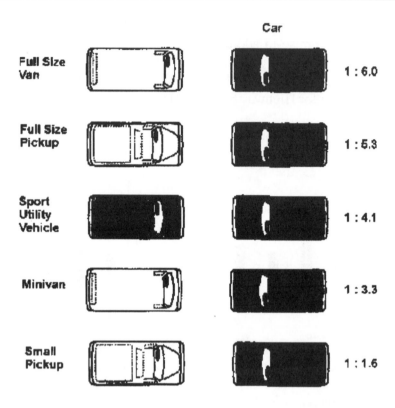

Figure 28.5 Relative fatality risk in collisions between a passenger car and other vehicles (from [28.11]).

occupants. In this database of 40,000 exposed vehicle passengers, most front occupants were injured in collisions with other passenger cars [28.8]. However, if risk in terms of injury per 1000 exposed individuals is considered, the worst collision partner is a heavier vehicle. These results demonstrate the importance of mass incompatibilities.

28.4.3 Stiffness Compatibility

Occupant injury seems to be lower for stiffer vehicles, especially for small 'city' cars. On the other hand, stiffer vehicles are more aggressive to other vehicles, causing higher partner injury [28.3]. Hobbs [28.1] observed the potential effects of incompatible local stiffness, which may lead to compartment intrusion and related injuries.

28.4.4 Geometrical Compatibility

More compatible geometries are beneficial for traffic safety, regardless of crash configuration and struck object. Proposals have been made to restrict front heights to 350-550 mm above road level in low deformation zones, and to 700 mm in higher crush zones. When the car is struck in the side, the fatality ratio varies from 11:1 to 23:1 when the striking vehicle varies from a small truck to a large van. These vehicles have structures that load the struck vehicle above the sill (see Figure 28.1), the strongest structure in the vehicle side. This results in intrusion levels near the occupant's head and chest. For comparison, side impacts between passenger cars result in the relative fatality risk of 6:1 for the occupants in the struck car versus the striking car of similar mass.

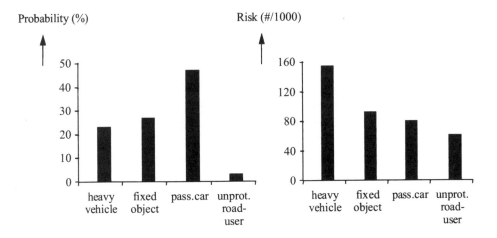

Figure 28.6   Injury probability and risk for collisions with various objects (from [28.8]).

28.5. Crash Testing

Full-scale crash tests have been performed and analyzed to examine the relationship between occupant injury and compatibility parameters. Car-to-car tests allow the relative risk between vehicles to be assessed (partner aggressivity) while barrier tests allow the self-protection performance of the vehicle to be studied.

28.5.1 Structural Compatibility

Front offset tests have investigated how dummy responses vary with increasing overlap amount. Higher dummy injury measures occur with increased overlap from 50% to 70%, as well as for overlaps from 25% to 70%, when focusing on upper body injuries. However, most researchers observe a maximum in dummy responses for an overlap between 25-100% [28.2, 28.4, 28.7]. Results often relate to which body segment is studied. Dummy responses are higher in 40-50% overlaps than in 25-30% or 100% overlap frontal crashes, when lower extremity and pelvic injuries are considered. Buzeman-Jewkes et al. [28.12] found higher dash intrusion in 50% overlap testing than in 100% overlap testing, even though the speed in the former tests was more than 15% lower.

28.5.2 Mass Compatibility

A draw back with the current crash tests legislated by government regulations is that all vehicles have the same test speed. Because the vehicle masses vary considerable, smaller vehicles are not subjected to the same impact energy as a larger vehicle. Frontal moving barrier impact tests have been proposed as a means to evaluate vehicle safety in car-to-car impacts and to assess mass-incompatibility. Another recommendation is to conduct rigid barrier tests at higher speeds for lighter vehicles to address mass incompatibility in real-world car-to-car crashes. Load-cell barrier tests can be used to compare the stiffness and mass aggressivity of vehicles at similar impact speeds [28.12]. The effect of vehicle mass has been studied in side impact tests. A heavier striking mass results in higher chest and pelvic dummy responses, although the effect of striking mass was relatively small compared to that of striking front stiffness.

28.5.3 Stiffness Compatibility

Frontal rigid barrier tests (ECE R12, FMVSS 208, NCAP) focus on occupant protection, but are unable to assess stiffness incompatibility between different car sizes. Campbell [28.13] measured front stiffness changes with increasing crush, and others have measured stiffness in high-speed tests and repeated low-speed crash-tests [28.14]. The use of a deformable barrier enables assessment of local stiffness effects of the vehicle front on occupant risk in car-to-car crashes, and more realistically reproduce car-to-car collisions

[28.1, 28.3-7]. However, some researchers have cautioned that the deformable barrier may increase stiffness incompatibility, since the deformable barrier has limited energy absorption when used in tests with heavier vehicles. The heavier vehicles may experience a more severe condition when impacting the rigid barrier after bottoming out of the deformable barrier.

Buzeman-Jewkes et al. [28.12] conducted repeated crash-tests to determine local stiffness at low-to-high crash speeds, bumper-force, vehicle acceleration and compartment intrusion. Compartment intrusion stiffness was measured using displacement transducers. They found stiffness localized around the energy absorbing structures or power-train rails and the engine as deformation increased. The engine increased the bumper-force and both vehicle acceleration and intrusion. Repeated crash tests with a load-cell barrier are an important tool in assessing local stiffness compatibility.

28.5.4 Geometrical Incompatibility

Geometrical effects have been tested in side-impacts where the bumper level of the striking vehicle was varied. The results indicate that an increased bumper level of the striking car caused higher responses for the nearside impacted occupant dummy. The geometrical effects exceed those of increased striking vehicle mass.

28.6. Mathematical Modeling

The continued development of computer simulations has contributed significantly to the understanding of vehicle compatibility during crashes. Mathematical models allow individual components of the vehicle to be investigated, reporting information not possible in crash tests. In addition, changes to vehicle descriptions can be studied in a theoretical model without devoting considerable resources into constructing prototypes for testing.

The investigation of vehicle safety for the traffic network can also be investigated using computer models. Different researchers have used statistical models to estimate system wide collision rates. The collision severities and injury risks can then be investigated through "what if" scenarios for different vehicle and fleet characteristics.

28.6.1 Structural Compatibility

Frontal crashes with varying overlap have been simulated using finite element and lumped-mass vehicle models. The results have been used as input for crash victim simulation programs like Madymo and the Articulated Total Body program. Buzeman-Jewkes et al. [28.15] developed and validated an integrated vehicle-occupant model in Madymo to predict vehicle acceleration, local deformation of the front structure and compartment intrusion, Figure 28.7. Vehicle acceleration reduced slightly for frontal crashes with greater offset (lower overlap), while compartment intrusion significantly increased. Lower extremity risks were higher in collisions with low overlap and high intrusion, while upper body injuries were more strongly related to high overlap collisions and thus with acceleration. The results indicate that higher moderate and severe injury risk occurred with greater overlap amount.

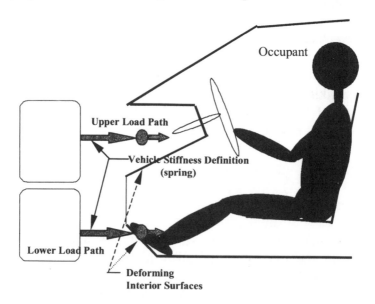

Figure 28.7   Integrated occupant-vehicle model of Buzemen-Jewkes

28.6.2 Mass Incompatibility

Several researchers have investigated the influence of vehicle mass on safety levels in a transportation network [28.5, 28.9, 28.10, 28.16]. An important feature developed in these analyses has been the definition of relationships between occupant injury risk and vehicle mass. For example, Evans [28.9] investigated mass effects by calculating the delta-V for a collision based on average impact speeds and the masses of the colliding cars (Eqs 28.4 and 28.5). In this analysis, the system injury risk decreases as vehicle mass decreased in car-to-car crashes.

Several studies [28.16, 28.10] have shown that the injuries vary more between models of the same mass class, than between various mass classes. This indicates that downsizing consequences for traffic safety may not be as significant as other factors, like inherent vehicle protection, driver behavior and impact speed. Using mathematical models, it has been estimated that vehicle mass had a relative low effect (3% to 5%) on overall injury or fatality rates in comparison to inherent vehicle protection parameters and impact speed distribution (27% to 40%).

28.6.3 Stiffness Compatibility

Crash victim simulation programs such as Madymo are frequently used to study dummy responses in frontal (offset) and side impacts. Vehicle structures can be characterized as lumped-mass models for general analyses or modeled using the finite element method (FEM) for more detailed analysis. The integrated model of Buzeman-Jewkes was used to analyze the effects of vehicle stiffness. The original deformation and intrusion stiffness characteristics were obtained from repeated crash-tests [28.12]. By increasing the vehicle stiffness in the model, increases in the bumper contact forces and vehicle accelerations were observed. If the stiffness was increased only for one vehicle, the intrusion in the stiffened car was reduced and that of the partner car was increased. Through these studies with the integrated model, vehicle stiffness was found to be the most important compatibility factor.

28.6.4 Geometrical Compatibility

Using the integrated vehicle-occupant model, geometrical changes to the vehicle structure were conducted. If the bumper levels of the two vehicles did not align, the calculated injury risks increased. The higher injury risks were due higher dash intrusion in both cars even though the crash acceleration levels were unaffected. An increased bumper height resulted in significantly higher risk, and had a stronger effect than mass, but a less strong effect than stiffness.

28.6.5 Interactions of Mass, Stiffness and Geometry

A significant feature of mathematical models is the ability to conduct sensitivity analyses for many variables. The integrated vehicle model was used to determine the relative importance of interactions between mass, stiffness and geometric compatibility in frontal collisions with 33%, 67% and 100% overlap to a fixed flat object or other passenger car. From this series of simulations, any combination of increases in mass, stiffness and bumper level caused a higher risk. The interaction between stiffness and mass implies that lighter cars should have higher stiffnesses than heavier cars to improve overall safety. This result is in agreement with the recommendation to stiffen light cars for compatibility improvements [28.17]. Bumper-level and vehicle mass had the strongest interaction, which reflects the incompatibility of heavier off-road vehicles [28.18-19].

28.7 Compatibility Between Vehicles and Other Traffic Elements

With the previous descriptions of compatibility between different motor vehicles, it is important to realize there are other potential collision partners in the traffic system. One significant incompatibility that is quickly identified in the collision statistics is that between motor vehicles and unprotected road users. The presence of pedestrians and cyclists in traffic situations is a particularly difficult problem for the vehicle designer. The human frame is much more fragile than that found in the vehicle. Collision outcomes for pedestrians and cyclists are often catastrophic, while vehicle occupants experience minor or no injuries at all (see Figure 28.6).

Even with the considerable differences between vehicles and humans, it is possible to develop more compatible protection strategies starting with the compatibility definitions developed for vehicle-vehicle collisions. Although there will always be a mass difference working against the human, investigations into stiffness, geometry, and structural compatibility can be conducted [28.21]. Tools and resources outlined in the earlier sections can be used in these activities. New pedestrian safety testing procedures, such as those introduced into Euro-NCAP, will increase research into this area and promote safer vehicle surfaces for pedestrians and cyclists.

The roadside infrastructure is also an area that can be investigated using the previously discussed compatibility concepts. Potential collision objects like guardrails, posts, and crash cushions can be considered as a collision partner with vastly different characteristics. Poles and trees will present challenges to the structural compatibility of vehicles. Their narrow contact areas can increase the vehicle intrusions and decelerations if the vehicle structures do not suitably support and dissipate crash loading. Guardrail and crash cushion designs are presented with challenges in terms of mass and stiffness compatibility [28.22]. These structures face all potential colliding vehicle types and must respond favorably in all cases. As in vehicle-vehicle compatibility, an array of research activities covering crash analysis, mathematical modeling, and crash testing is being used to develop refined protection strategies.

28.8 Conclusions

Design of safer vehicles and transportation systems is a multifaceted problem requiring an array of research tools. Compatibility in the crash performance of vehicles with other collision partners is one activity that will lead to increased traffic safety levels for all road users.

The problem of compatibility is a combination of different elements representing the physics of car crashes. When analyzing a crash, variables representing the mass, structure, stiffness, and geometry of the event should be analyzed systematically to maximize the understanding of the problem. Unfortunately, there are many inter-relations between these variables. However, the application of different research approaches can help identify the significance of these different parameters and their interaction effects. The combination of field crash analysis, mathematical modeling, and crash testing provides is the most effective approach to understand and solve the problems presented in crash compatibility.

REFERENCES

[28.1]   Hobbs, C.A. and Williams, D.A., The development of the frontal offset deformable barrier test, *Proceedings of the 14th Int. Techn. Conf on ESV*, Paper No. 94-S8-O-06, Munich, Germany, 1994.

[28.2]   Thomas, P., Real world collisions and appropriate barrier tests, *Proceedings of the 14th Int. Techn. Conf. on ESV*, Paper No. 94-S8-W-19, Munich, Germany, 1450-1460, 1994.

[28.3]   Tarriere, C., Morvan, Y., Steyer, C. and Bellot, D., Accident research and experimental data useful for an understanding of the influence of car structural incompatibility on the risk of accident injuries, *Proceedings of the 14th Int. Techn. Conf. on ESV*, Munich, Germany, 593-609, 1994.

[28.4]   O'Neill, B., Lund, A.K., Zuby, D.S. and Preuss, C.A., Offset frontal impacts-a comparison of real-world crashes with laboratory tests, *Proceedings of the 14th Int. Techn. Conf. on ESV*, Paper No. 94-S4-O-19; Munich, Germany, 649-670, 1994.

[28.5]   Tarriere, C., Car safety and ratings: comparative study of facts and practices -proposals, *Proceedings of IRCOBI Conf. on Biomechanics*; Lyon, France, 165-179, 1994.

[28.6]   Kullgren, A., Ydenius, A. and Tingvall, C., Frontal impacts with small partial overlap: real life data from crash recorders, *Proceedings of the. 16th Int. Techn. Conf. on ESV*, Paper No. 98-S1-O-13, Windsor, Canada, 1998.

[28.7]   Thomas, C., Foret-Bruno, J.Y., Brutel, G., LeCoz, J.Y., Got, C, and Patel, A., Front passenger protection: what specific requirements in frontal impact?, *Proceedings of the IRCOBI Conf. on Biomechanics*, Lyon, France, 205-216, 1994.

[28.8]   Buzeman, D.G., Viano, D.C. and Lovsund, P., Injury probability and risk in frontal crashes: the effect of sorting techniques on priorities for offset testing, *Accident Analysis and Prevention*, **30**(5), 583-595, 1998.

[28.9] Evans, L., *Traffic Safety and the Driver*, Van Nostrad Reinhold, 1991.

[28.10] Broughton, J., The theoretical basis for comparing the accident record of car models, *Accident Analysis and Prevention*, **28**, 89-99, 1996.

[28.11] Gabler, H.C. and Hollowell, W.T., The crash compatibility of cars and light trucks, *Journal of Crash Prevention and Injury Control,* **2**(1), 19-31, 2000.

[28.12] Buzeman-Jewkes, D.G., Viano, D.C. and Lovsund, P., Use of repeated crash-tests to determine local longitudinal and shear stiffness of the vehicle front with crush, SAE Paper No. 1999-01-0637, Soc. Of Autom. Engineers, Warrendale, PA, , 1999.

[28.13] Campbell, K.L., Energy basis for collision severity, SAE Paper No. SAE 740565, Soc. Of Autom. Engineers, Warrendale, PA, 1974.

[28.14] Wood, D.P., Safety and the car size effect: a fundamental explanation, *Accident Analysis and Prevention*, **29**, 139-151, 1997.

[28.15] Buzeman-Jewkes, D.G., Viano, D.C. and Lovsund, P., A Multi-body integrated vehicle-occupant model for compatibility studies in frontal crashes, *Journal of Crash Prevention and Injury Control,* **1**(2), 143-154, 1999.

[28.16] Buzeman, D.G., Viano, D.C. and Lovsund, P., Car occupant safety in frontal crashes: a parameter-study of vehicle mass, impact speed and inherent vehicle protection, *Accident Analysis and Prevention*, **30**(6), 713-722, 1998.

[28.17] Zeidler, F., The influence of frontal crash test speeds on the compatibility of passenger cars in real world accidents, *Int. Symp. on RWCIR*, Leicestershire, England, 1997.

[28.18] Hobbs, C.A., Williams, D.A. and Coleman, D.J., Compatibility of cars in frontal and side impact, *Proceedings of the 15th Conference on Enhanced Vehicle Safety*, Paper No. 96-S4-O-05, Melbourne, Australia, 617-625, 1996.

[28.19] MacLaughlin, T.F., Saul, R.A. and Daniel, S.Jr., Causes and measurement of vehicle aggressiveness in frontal collisions, *Proceedings of the 24th Stapp Conference*, Paper No. 801316, 639-700, 1980.

[28.20] Buzeman, D.G., Viano, D.C. and Lovsund, P., Occupant risk, partner risk and fatality rate in frontal crashes: estimated effects of changing vehicle fleet mass in 15 years. *Journal of Crash Prevention and Injury Control*, **2**(1), 1-10, 2000.

[28.21] Yang, J.K., Lovsund, P., Cavallero, C. and Bonnoit J., A human-body 3d mathematical model for simulation of car-pedestrian impacts, *Journal of Crash Protection and Injury Control*, **2**(2), 2000.

[28.22] Reagan, J.A., Vehicle Compatibility With Roadside Safety Hardware, *Public Roads*, **59**(2), Federal Highways Administration, Department of Transportation, 1995.